ALGORITHMS:
Their Complexity and Efficiency
Second Edition

WILEY SERIES IN COMPUTING

Consulting Editor
Professor D. W. Barron
Department of Computer Studies, University of Southampton, UK

BEZIER · Numerical Control—Mathematics and Applications
DAVIES and BARBER · Communication Networks for Computers
BROWN · Macro Processors and Techniques for Portable Software
PAGAN · A Practical Guide to Algol 68
BIRD · Programs and Machines
OLLE · The Codasyl Approach to Data Base Management
DAVIES, BARBER, PRICE, and SOLOMONIDES · Computer Networks and their Protocols
RUS · Data Structures and Operating Systems
BROWN · Writing Interactive Compiliers and Interpreters
HUTT · The Design of a Relational Data Base Management System
O'DONOVAN · GPSS—Simulation Made Simple
LONGBOTTOM · Computer System Reliability
AUMIAUX · The Use of Microprocessors
ATKINSON · Pascal Programming
KUPKA and WILSING · Conversational Languages
SCHMIDT · GPSS–Fortran
PUZMAN/PORIZEK · Communication Control in Computer Networks
SPANIOL · Computer Arithmetic
BARRON · Pascal—The Language and its Implementation
HUNTER · The Design and Construction of Compilers
MAYOH · Problem Solving with ADA
AUMIAUX · Microprocessor Systems
CHEONG and HIRSCHEIM · Local Area Networks
BARRON and BISHOP · Advanced Programming—A Practical Course
DAVIES AND PRICE · Security in Computer Networks
KRONSJÖ · Computational Complexity of Sequential and Parallel Algorithms
AMMERAAL · C For Programmers
KRONSJÖ · Algorithms · Their Complexity and Efficiency · 2nd edition

ALGORITHMS:
Their Complexity and Efficiency

Second Edition

Lydia Kronsjö

The Centre for Computing and Computer Science
The University of Birmingham

A Wiley–Interscience Publication

JOHN WILEY & SONS
Chichester · New York · Brisbane · Toronto · Singapore

Library of Congress Cataloging-in-Publication Data:

Kronsjö, Lydia.
Algorithms: their complexity and efficiency.

(Wiley series in computing)
Bibliography: p.
Includes index.
1. Electronic digital computers—Programming.
2. Algorithms. 3. Numerical analysis—Data processing.
4. Computational complexity. I. Title. II. Series.
QA76.6.K76 1987 511 86–15689
ISBN 0 471 91201 8

British Library Cataloguing in Publication Data:

Kronsjö, Lydia
Algorithms: their complexity and efficiency
—Second Edition—(Wiley series in computing)
1. Algorithms
I. Title
511′.8 QA9.58

ISBN 0 471 91201 8

Printed and bound in Great Britain

Preface to the Second Edition

In this second edition substantial changes and additions have been made to most chapters, in particular chapters on sorting and searching have been fully rewritten. The old-fashioned step-by-step implementation of the algorithms has been replaced by a pseudoform presentation. Many new exercises, some with solutions, have been incorporated in most chapters.

As before, no explicit reference is made to the NP-completeness theory, and the algorithms are not discussed in terms of the P-class or NP-class properties. In our opinion the theme of the NP-completeness is far too important to be mentioned in passing, and at the same time its proper treatment in one volume with all the material already included would make the book too large. The reader is instead advised to refer to a separate text by the author, *Computational Complexity of Sequential and Parallel Algorithms*, John Wiley & Sons, for an introduction to the two fundamental concepts for further improvement of the algorithms' efficiency, the notion of the non-deterministic automata, and the parallel processing approach. In fact, it can be mentioned that the present book and the *Computational Complexity* book form a systematic and a comprehensive coverage of the modern science of algorithms.

LYDIA KRONSJO
Birmingham, April 1986

Preface to the First Edition

This book is concerned with the study of algorithms and evaluation of their performance. Analysis of algorithms is a new area of research, and emerged as a new scientific subject during the sixties and has been quickly established as one of the most active fields of study, becoming an important part of computer science. The reason for this sudden interest in the study of algorithms is not difficult to trace as the fast and successful development of digital computers and their uses in many different areas of human activity has led to the construction of a great variety of computer algorithms. At present it is often the case that several different algorithms exist for the solution of a single problem or a class of problems and these algorithms need to be carefully analysed in order to provide a basis for selecting the best one for the purpose. In many cases, analysis of algorithms also leads to the revelation of completely new algorithms that are even faster than all algorithms known before. On the other hand, the study of algorithms has brought about many no less startling discoveries of certain natural problems for which all algorithms are inefficient.

The book is intended as a text for an intermediate course in computer science or computational mathematics which focuses on the basic principles and concepts of algorithms. It requires general familiarity with computers, preferably some courses on programming and introductory computer science. Each chapter is devoted to one particular class of problems and their algorithms. Chapter 1 introduces the subject, while Chapters 2, 3, 4, and 5 discuss four different classes of problems that are termed as numerical, i.e. the mathematical problems, solution of which is of a numerical nature. For such problems numerical accuracy of the computed results is of particular importance. Chapter 6 discusses the problem of (asymptotically) fast multiplication of two numbers and is based on the fast Fourier transforms and Chapters 7, 8, and 9 discuss sorting and searching, the most common non-numerical problems encountered in computing.

The exercises at the end of each chapter are used to provide examples as well as to complete or generalize some proofs.

All algorithms in the text are given in a form of a sequence of steps, each step describing the actions to be undertaken, in natural English. This seems to be a most neutral way of introducing the algorithms. Actual implementation details, e.g. a specific programming language, various programming tricks etc., are left to an interested reader. The reference list at the end of the book contains published sources for the algorithms and theoretical results discussed in the text.

I would like to thank all the people who have critically read various portions of the manuscript and offered many helpful improvements. In particular, I would like to thank Peter Jarratt, Stuart Hollingdale, Michael Atkinson, Nelson Stevens, and Tom Axford. Sincere thanks go to Ilsie Browne for her excellent typing of the manuscript.

L.I.K.
Birmingham, November 1978

Table of Notations

Written	Denotes
$a \in A$	a is contained in the set A
$\lceil a \rceil$	the least integer greater than or equal to a
$\lfloor b \rfloor$	the greatest integer less than or equal to b
$\ln a$	natural logarithm
$a \approx b$	a approximately equal to b
$\mathbf{A}^*$	conjugate transpose of the matrix $\mathbf{A}$
$\mathbf{a}^*$	conjugate transpose of the vector $\mathbf{a}$
$\mathbf{A}^{\mathrm{T}}$	transpose of the real matrix $\mathbf{A}$
$\mathbf{a}^{\mathrm{T}}$	transpose of the real vector $\mathbf{a}$
$\mu[y]$	mean value of the statistical variable y
$\sigma[y]$	standard deviation of thc statistical variable y
$a \equiv b$	a is equivalent to b
$x \gg y$	x is much larger than y
$ABS(x)$	the absolute value of x
$SQRT(x)$	square root of x

Contents

1

Introduction

This book will be chiefly concerned with an investigation of algorithms in order to evaluate their performance. This comparatively new field of study is known as algorithmic analysis and forms part of the more general discipline of computer science. In practical terms, a goal of algorithmic analysis is 'to obtain sufficient understanding about the relative merits of complicated algorithms to be able to provide useful advice to someone undertaking an actual computation' (Gentleman, 1973). In broader interpretation, however, algorithmic analysis includes the study of all aspects of performance in computational problem solving, from the preliminary formulation, through the programming stages, to the final task of interpreting the results obtained. We would also like to prove lower bounds on the computation time of various classes of algorithms. In order to show that there is no algorithm to perform a given task in less than a certain amount of time, we need a precise definition of what constitutes an algorithm. First, then, what is an algorithm?

1.1 Definition

> A procedure consisting of a finite set of unambiguous rules which specify a finite sequence of operations that provides the solution to a problem, or to a specific class of problems, is called an algorithm.

Several important features of this definition must now be emphasized.

First, each step of an algorithm must be unambiguous and precisely defined. The actions to be carried out must be rigorously specified for each case.

Secondly, an algorithm must always arrive at a problem solution after a finite number of steps. Indeed, the general restriction of finiteness is not sufficient in practice, as the number of steps needed to solve a specific problem, although finite, may be too large for practicable computation. A useful algorithm must require not only a finite number of steps, but a reasonable number.

Thirdly, every meaningful algorithm possesses zero or more inputs and provides one or more outputs. The inputs may be defined as quantities which are given to the algorithm initially, before it is executed, and the outputs as quantities which have a specified relation to the inputs and which are delivered at the completion of its execution.

Fourthly, it is preferable that the algorithm should be applicable to any member of a class of problems rather than only to a single problem. This property of generality, though not a necessity, is certainly a desirable attribute of a useful algorithm.

Finally, we would like to mention that although the concept of an algorithm is a very broad one, in this book we restrict ourselves to algorithms designed to be executed on a computer. Such an algorithm must be embodied in a computer program (or set of programs), and so in the sequel the two terms will be used interchangeably.

1.2 Measures of Efficiency

It is relatively easy to invent algorithms. In practice, however, one wants not only algorithms, one wants good algorithms. Thus, the objective is to invent good algorithms and prove that they are good. The 'goodness' of an algorithm can be appraised by a variety of criteria. One of the most important is the time taken to execute it. There are several aspects of such a time criterion. One might be concerned with the execution time required by different algorithms for solution of a particular problem on a particular computer. However, such an empirical measure is strongly dependent upon both the program and the machine used to implement the algorithm. Thus, a change in a program may not represent a significant change in the underlying algorithm but may, nevertheless, affect the speed of execution. Furthermore, if two programs are compared first on one machine and then another, the comparisons may lead to different conclusions. Thus, while comparison of actual programs running on real computers is an important source of information, the results are inevitably affected by programming skill and machine characteristics.

A useful alternative to such empirical measurements is a mathematical analysis of the intrinsic difficulty of solving a problem computationally. Judiciously used, such an analysis provides an important means of evaluating the cost of algorithm execution.

The performance time of an algorithm is a function of the size of the computational problem to be solved. However, assuming we have a computer program which eventually terminates, solving a particular problem requires only sufficient time and sufficient storage. Of more general interest are algorithms that can be applied to a collection of problems of a certain type. For these algorithms, the time and memory space required by a program will vary with the particular problem being solved. Consider, for example, the following classes of problems, and note the role of the value of the parameter n.

(a) Find the largest in a sequence of n integers.

(b) Solve a set of linear algebraic equations $\mathbf{Ax} = \mathbf{b}$, where $\mathbf{A}$ is an $n \times n$ real matrix and $\mathbf{b}$ is a real vector of length n.
(c) Let W be a one-dimensional array of n distinct integers. Sort the entries of W into descending order.
(d) Evaluate a polynomial $P_n(x) = \sum_{k=0}^{n} a_{n-k}x^k$ at $x = x_0$.

In each of these problems the parameter n provides a measure of the size of the problem in the sense that the time required to solve the problem, or the memory space required by the algorithm, or both, will increase with n.

1.3 Complexity of Algorithms

In order to measure the cost of executing a program, we customarily define a complexity (or cost) function F, where $F(n)$ is a measure of the time required to execute the algorithm on a problem of size n, or a measure of the memory space required for such execution. Accordingly, we speak of the time complexity and the space complexity functions of the algorithm. We can also refer to either kind of function as simply a complexity (or cost) function of the algorithm. In this book our principal concern will be time complexity functions, but we shall distinguish, where appropriate, between the time complexity and the computational complexity of an algorithm. The latter term will be used to refer to estimates of the computational power required to solve the problem and will be measured by the number of arithmetic or logical operations required, e.g. the number of multiplications and additions needed to evaluate a polynomial, the number of comparisons between two entries in a sort algorithm, or the number of function evaluations in an iterative root-finding algorithm. In this context, the computational complexity measure will be considered as a branch of the time complexity measure.

In general, the cost of obtaining a solution increases with the problem size, n. If the value of n is sufficiently small, even an inefficient algorithm will not cost much to run, so the choice of an algorithm for a small problem is not critical (unless the problem is to be solved many times). In most cases, however, as n increases one eventually arrives at a situation when the algorithm can no longer be executed in an acceptable period of time. This point is illustrated in Tables 1.3.1 and 1.3.2. Table 1.3.1 shows the growth of some common complexity functions as n increases, while Table 1.3.2 gives the maximum sizes of problems which can be solved using algorithms with these complexity functions, in a specified length of time.

In practice, of importance is performance of the algorithms for large values of n, that is asymptotic behaviour of the complexity function. Asymptotic complexity is the growth in the limit of the complexity function with the size parameter n. So, the asymptotic (time or space) function ultimately determines the size of the problem that can be solved by the algorithm. In terms of the terminology, we say that the (time) complexity of the algorithm is $O(\log n)$,

Table 1.3.1 A comparison of the growth of some common complexity functions

Time complexity function	Problem size n			
	10	10^2	10^3	10^4
$\log_2 n$	3.3	6.6	10	13.3
n	10	10^2	10^3	10^4
$n \log_2 n$	0.33×10^2	0.7×10^3	10^4	1.3×10^5
n^2	10^2	10^4	10^6	10^8
n^3	10^3	10^6	10^9	10^{12}
2^n	1024	1.3×10^{30}	$> 10^{100}$	$> 10^{100}$
$n!$	3×10^6	$> 10^{100}$	> 10	100

read 'order log n' if the processing by the algorithm of the problem instance of size n takes the time proportional to log n. Generally throughout the text, in relation to, say, the time complexity of an algorithm the following terminology will be used in an equivalent sense:

(a) the time complexity of the algorithm is of order log n; this can also be written as 'O(log n)';
(b) the algorithm is executed in O(log n) time;
(c) the amount of work required by the algorithm is proportional to log n, or is of O(log n).

Where appropriate the term 'unit of time' will be used in the sense equivalent to the term 'one basic operation', as in Chapter 6 for example, and the following additional terminology will be used in the sense equivalent to (a) to (c): the algorithm requires a number of (basic) operations proportional to log n.

Table 1.3.2 Limits on sizes of problems which can be solved using algorithms with some common complexity functions (one operation per microsecond is assumed)

Time complexity function	Maximum problem size for execution in:		
	1 sec	1 min	1 hour
$\log_2 n$	2^{10^6}	$2^{6 \times 10^7}$	$2^{3.6 \times 10^9}$
n	10^6	6×10^7	3.6×10^9
$n \log_2 n$	62746	2.8×10^6	1.3×10^8
n^2	10^3	7746	60 000
n^3	10^2	3.9×10^2	1.5×10^3
2^n	20	25	32
$n!$	9	11	12

1.4 Computational Complexity of Numerical Algorithms

By a numerical algorithm we mean an algorithm that solves a problem where a numerical content is essential. Examples of such problems are many, and we give here two:

(a) the computation of the roots of a non-linear equation or of a set of non-linear equations;
(b) the evalution of a polynomial for a given argument value.

A numerical algorithm produces some numerical output which is the computed solution to the problem. Chapters 2 to 5 of this book discuss the numerical algorithms for solving problems of the polynomial evaluation, the computation of a root of a non-linear equation and of a set of non-linear equations, the solution of a set of linear algebraic equations, and the computation of the Fourier transforms.

In Chapter 6 a fast multiplication of two integers is discussed. This problem can be referred to as a seminumerical problem, because though this problem is about the numbers, the solution to it is treated from the point of view of designing a super-efficient computational scheme rather than emphasizing its numerical content.

Non-numerical algorithms, on the other hand, solve problems, the solution of which, although in some cases expressed by a number (or a set of numbers), is of a non-numerical nature. Two important examples of non-numerical problems are sorting in order a given set of unordered items (note that the items can be expressed in the form of either numerals or strings of letters), and searching for a specific item in a given set of items. Chapters 7 to 9 of the text are concerned with algorithms for solving such problems.

In the analysis of computational complexity of the numerical algorithms it is convenient to distinguish between (a) algebraic and (b) analytic complexity measures, both of which represent branches of the computational complexity. The objectives of the algebraic complexity are manifold. It attempts to estimate the number of arithmetic operations required by a given algorithm. It further attempts to estimate the minimum number of arithmetics that is needed to solve the given problem. It also seeks to discover an algorithm or algorithms which solve the problem in a minimum number of arithmetics. The evaluation of a polynomial at a given point or points and the solution of a set of linear equations by direct methods are examples of the problems which are studied in terms of algebraic complexity. An important feature of the algorithms which are used to solve these problems is the strictly finite number of steps that are needed to achieve the solution. This obviously implies a finite number of arithmetic operations required.

Analytical complexity, on the other hand, addresses the question of how much computation has to be performed to obtain a result with a given degree of accuracy, and focuses on those computational processes which in a certain sense never end. Iterative processes provide an obvious example. Here, the process is interrupted at some point and, if the current value of the result lies within some error bound, that value is accepted; otherwise the computation is continued until a satisfactory result is obtained. In this case one estimates the number of arithmetic operations per iteration step, and a 'best' algorithm in terms of the computational complexity is identified by the fewest total number of arithmetic operations required to achieve an approximate result computed to some prescribed level of accuracy.

1.5 Numerical Accuracy

In the analysis of algorithms which solve numerical problems the accuracy of the computed results is another important criterion to distinguish between 'good' and 'bad' algorithms. The need to examine the accuracy of mathematical computations arises from the fact that a computer is a finite machine; it is capable of representing numbers only to a finite number of digit positions. As a result, most numbers, and even integers, if they are too long for the computer to represent exactly, are rounded, and so only a finite approximation to a number is stored in a computer. This implies that if a care is not exercised, some numerical algorithms implemented on a computer may produce approximations to the true results that are wildly inaccurate. The study of mathematical methods for solving problems numerically and ascertaining the bounds of errors in the results is a well developed field, known as numerical analysis. The foundations of modern mathematical error analysis were laid down in the celebrated paper of von Neumann and Goldstine (1947), although it was not until the mid-1950s to late 1950s that this area of study became really active.

One of the most important ideas in error analysis is that of backward error analysis. It was proposed by Givens (1954) and brilliantly utilized by Wilkinson (1963) in the analysis of the rounding errors in algebraic processes.

Applying backward analysis Wilkinson (1965) was able to give a complete *a priori* analysis of the solution of a set of linear algebraic equations by Gaussian elimination with pivoting. The question asked in backward error analysis is: What problem have we solved exactly and how far is this problem from the one we set out to solve? In the old-fashioned (pre-Wilkinson) forward error analysis the approach was: How much does the computed solution differ from the exact solution? The advantages of the backward error analysis have been firmly emphasized by many excellent results on the error analysis, and this error analysis is used today as a standard tool in the study of numerical processes.

1.6 Analysis of Algorithms

It is convenient to distinguish between the analysis of a particular algorithm and the analysis of a class of algorithms. In analysing a particular algorithm we would probably investigate its most important characteristics, such as the time complexity, that is, how many times each part of the algorithm is likely to be executed, and the space complexity, that is, how much computer storage space it is likely to need. In the analysis of a class of algorithms, on the other hand, we study the entire family of algorithms which solve a particular problem and seek to identify the 'best possible' algorithm of the family. If such an algorithm has not yet been found, then attempts are made to establish bounds on the computational complexity of the algorithms in the class.

Analyses of the first type have been made since the earliest days of computer programming. Analyses of the second type were, with a few exceptions, not embarked upon until much later. It is easy to see that analyses of the second type are far more powerful and useful since they study many algorithms at the same time. It has always been a dream of mankind to find the best (in mathematical terms, optimal) ways of solving the many problems with which we are faced. Computer algorithms is just another area of human activity where optimal ways to solve the problems are eagerly sought. Instead of studying every algorithm that is devised, it is obviously better to prove once and for all that a particular algorithm is the 'best possible'. However, it so happens that many of the chosen algorithms turn out to be the 'best' only in a very narrow sense; very small changes in the definition of the optimal criteria can significantly affect the choice of the 'best possible' algorithm. Consider, for example, the following problem: Compute x^n in the fewest possible number of multiplications. This problem was first raised by Arnold Scholtz in 1937. Let us consider x^{31}. It cannot be calculated in fewer than eight multiplications, i.e.

$$x, x^2, x^4, x^8, x^{16}, x^{24}, x^{28}, x^{30}, x^{31}.$$

It can however be done in only six arithmetic operations if division is allowed:

$$x, x^2, x^4, x^8, x^{16}, x^{32}, x^{31}.$$

Other examples of the dependence of complexity estimates on underlying assumptions are the different estimates of Pan (1966), Motzkin (1955), and Winograd (1967) on the number of multiplications necessary to evaluate a polynomial, and the different estimates of Kluyev and Kokovkin-Shcherbak (1965), Strassen (1969) and Miller (1973) on the number of arithmetic operations necessary to solve n linear equations in n unknowns.

Generally, analyses of the second type are difficult and relatively few results have been obtained to date. The problem of computing x^n with fewest multiplications, for instance, is far from being solved.

Another example is that of sorting. The need to sort a given group of elements into some prescribed order arise quite frequently in programming practice. Sorting algorithms have been studied intensively for a number of years. However, very few optimal parameters have yet been established, and even then only for problems of quite small size. For instance, for sorting algorithms which are based on comparisons of the elements, the exact value of the optimal number of comparisons, $C(n)$, is known only for $n \leqslant 12$ and $n = 20, 21$.

The difficulty in performing an analysis for a whole group of algorithms which solve the problem or a class of problems is the reason why such analysis has not superseded analysis of the individual algorithms. The study of individual algorithms is very important in practice since it can be applied to measure all the relevant factors about the performance of a particular algorithm.

1.7 Bounds on Complexity

Complexity analysis among other things is concerned with obtaining upper and lower bounds on the performance of algorithms or classes of algorithms that solve various problems. The existence of complexity bounds for the known algorithms can serve as a basis for classifying the problems.

There are those problems for which complexity bounds can be determined, using appropriate theorems, without the actual construction of an algorithm for the problems. Any proofs of the theorems in such a case are thus non-constructive and do not give any indication as to whether or how an algorithm solution can be obtained. A discussion of some complexity theorems of this kind, such as those concerned with some finite combinatorial problems, can be found in Rabin (1974).

For other problems lower bounds on complexity have been derived, but none of the available algorithms is known to attain the bounds. A notable example of this type is matrix multiplication, where a minimum bound of $O(n^2)$ is easily visualized but the fastest known methods are of $O(n^{2.496})$. This gap suggests that the bound $O(n^2)$ is not sharp enough.

Another group of problems is such that their lower bounds on computational complexity are known, the algorithms which attain these bounds can be constructed, but these algorithms are numerically unstable. Fast methods of matrix multiplications belong to this category. In fact using the modern approach of interpreting the operation of matrix multiplication as the computation of bilinear forms, Miller (1973) has shown that for any algorithm that multiplies two $n \times n$ matrices using only addition, subtraction and multiplication, to be strongly stable in the sense of the ideas drawn from backward error analysis, the algorithm has to be essentially equivalent to the usual algorithm, and hence must involve $O(n^3)$ operations.

Finally, there are problems for which lower bounds on complexity are known and the algorithms which attain these bounds can be built and the algorithms are numerically stable.

In this category, for example, belong the generally available numerical methods which are stable and attain well-known bounds of complexity. One particularly remarkable method of this class is the Fast Fourier Transform (FFT) algorithm for computing the discrete finite Fourier transforms of functions.

1.8 Models of Computation

Algorithms and their complexity are normally studied under assumption of a specific (precisely defined) model of a computing device that is capable of executing the algorithm in hand. A basic step in a computation has also to be precisely defined. Perhaps the most important motivation for formal models of computation is the desire to discover the inherent computational difficulty of various problems.

Unfortunately, there is no one computational model which is suitable for all situations, and so for each problem one selects an appropriate model which for the given situation reflects as accurately as possible the actual computation process on a real computer.

This approach is adopted in the present text. In each chapter we outline the main features of the computational model suitable for the problem in hand. The models vary from chapter to chapter because of the wide range of the problems considered. However one basic property that justifies and guides our choice of models is the fact that every model is defined in such a way that any computational process which can be carried out using such a model can also be carried out by a real computer.

In other words, we assume that none of our computational models possess such features that are not available on a modern real computer.

In terms of optimality, the goal of complexity analysis is to show that in order to solve a certain problem or class of problems computationally, one requires a certain number of operations of a certain type. This is a very difficult objective; for the majority of practical problems we still have to rely on experience to judge the goodness of an algorithm. However, in some cases useful results have been established. For example, we can now strictly bound from below the number of arithmetic operations required to perform certain calculations and in the course of the book we shall present some basic results of this nature. For instance, we shall show in the chapter on polynomial evaluation that the evaluation of an nth degree polynomial requires at least n multiplications and the same number of additions, unless certain conceptually different conditions are imposed on forming the polynomial coefficients. In other cases, indeed, it is possible to show that for some problems we shall never find optimal algorithms because they simply do not exist. For example, the problem of numerical differentiation is inherently unstable and so no good computational methods can be expected to exist at all. In solving many a problem a human touch will always be needed to check the meaningfulness of any computational model that may be constructed for the purpose of solving the problem.

2

Evaluation of Polynomials

The evaluation of a polynomial is one of the most widely encountered operations in computing. Indeed, some users of the extensive computer program libraries may not realize how ubiquitous this operation is, not only in contexts directly associated with the evaluation of polynomials, but also in problems of computing transcendental or more complex algebraic expressions. More often than not a polynomial of some kind is used to approximate a trigonometric or a logarithmic expression, or the ratio of polynomials. Evaluation of such functions as the sine, cosine, exponential and logarithmic functions represent familiar examples.

The problem of efficient and accurate numerical evaluation of a polynomial has received considerable attention already in the 1950s when first realization of the computer power came about. In this chapter we shall explore major results of these studies, and illustrate some general ideas that are involved in the analysis of numerical algorithms. The operation we consider is that of computing the value of a polynomial expression

$$P_n(x) = a_0 + a_1 x^{n-1} + \ldots + a_n \tag{2.1}$$

for a particular value, say, x_0, of the argument x. We shall refer to this as evaluation of the polynomial at a fixed point and it will be assumed throughout this chapter, with the exception of the final section, that both the argument and the coefficients are real numbers.

2.1 Polynomial Evaluation Algorithms

A polynomial may be expressed in a variety of ways. Each of these forms represents the polynomial exactly but where numerical evaluation is concerned the various forms may have different properties. We now examine various ways of evaluating a polynomial, depending on the form in which it is presented.

The Power Form

The most common form of a polynomial is the power form given by equation (2.1). To evaluate this polynomial term by term requires $(2n-1)$ multiplications and n additions. (On a computer the operations of addition and subtraction are of equal complexity, and so the term 'additions' is used here to embrace any combination of additions and subtractions.)

To obtain a more efficient evaluation algorithm we can rewrite equation (2.1) in the so-called *Horner,* or *nested,* form as

$$P_n(x) = (\ldots(a_0x + a_1)x + \ldots + a_{n-1})x + a_n. \tag{2.1.1}$$

This immediately suggests the following evaluation algorithm.

Algorithm powerform

```
V[n] := a[0]
for k := n - 1 downto 0 do
    V[k] := x × V[k+1] + a[n-k]
enddo
polyvalue := V[0]
```

This algorithm requires n multiplications and n additions and is the most frequently used algorithm.

The Root Product Form

The root product form is a polynomial given by

$$P_n(x) = a_0 \prod_{i=1}^{n} (x - \gamma_i), \text{ where } P_n(\gamma_i) = 0. \tag{2.1.2}$$

The roots γ_i's can be real or complex, though for our purposes here we may assume the form with all the roots real.

Important cases of polynomials in this form arise, for example, in statistics. The computational algorithm based on (2.1.2) is implemented as follows.

Algorithm rootform

```
V[n] := a[0]
for k := n - 1 downto 0 do
    V[k] := (x - gamma[k+1]) × V[k+1]
enddo
polyvalue := V[0]
```

Evaluation of the root product form requires n multiplications and n additions, the same as the Horner algorithm's.

The Newton Form

The Newton polynomial form is given by

$$P_n(x) = c_0(x-\beta_n)(x-\beta_{n-1})\dots(x-\beta_1) + c_1(x-\beta_{n-1})\dots(x-\beta_1) + c_{n-1}(x-\beta_1) + c_n. \quad (2.1.3)$$

The computational algorithm based on (2.1.3) may be given as follows.

Algorithm newtonform

```
V[n] := c[0]
for k := n − 1 downto 0 do
    V[k] := (x − beta[k + 1]) × V[k + 1] + c[n − k]
enddo
polyvalue := V[0]
```

The algorithm requires n multiplications and $2n$ additions.

This polynomial form arises in the problem of interpolating a function given in tabular form. The form does not occur directly but is arrived at after some manipulation on the direct interpolation polynomial.

Assume, for example, that the Lagrange polynomial is used to interpolate function $f(x)$. In the case of the three-point interpolation over ($f(x_k)$, x_k, $k = 0, 1, 2$), the Lagrange polynomial is

$$P_2^{(L)}(x) = f_0 \frac{(x-x_1)(x-x_2)}{(x_0-x_1)(x_0-x_2)} + f_1 \frac{(x-x_0)(x-x_2)}{(x_1-x_0)(x_1-x_2)} + f_2 \frac{(x-x_0)(x-x_1)}{(x_2-x_0)(x_2-x_1)}, \quad (2.1.4)$$

where $P_2^{(L)}(x)$ denotes the polynomial of degree 2.

This polynomial is easily obtained but its form is computationally uneconomical, if directly used for the evaluation. Indeed, for general n, as in $P_2^{(L)}(x)$, there are required $(2n^2 + n - 1)$ multiplications, $(n + 1)$ divisions and $(2n^2 + n + 2)$ additions. A computationally simpler form of this polynomial can be obtained using the following approach.

In the above example, in addition to $P_2^{(L)}(x)$ we consider also $P_1^{(L)}(x)$, where

$$P_1^{(L)}(x) = f_0 \frac{x-x_1}{x_0-x_1} + f_1 \frac{x-x_0}{x_1-x_0}.$$

Now examine the difference

$$P_2^{(L)}(x) - P_1^{(L)}(x) = f_0 \frac{(x-x_1)}{(x_0-x_1)}\left(\frac{x-x_2}{x_0-x_2} - 1\right) + f_1 \frac{(x-x_0)}{(x_1-x_0)}\left(\frac{x-x_2}{x_1-x_2} - 1\right) + f_2 \frac{(x-x_0)(x-x_1)}{(x_2-x_0)(x_2-x_1)}.$$

Here $[P_2^{(L)}(x) - P_1^{(L)}(x)]$ is a second degree polynomial in x, which vanishes at $x = x_0$ and x_1, and must therefore be a multiple of $(x - x_0)(x - x_1)$.

We therefore set

$$P_2^{(L)}(x) - P_1^{(L)}(x) = \alpha_2(x - x_0)(x - x_1)$$

that is,

$$P_2^{(L)}(x) = \alpha_2(x - x_0)(x - x_1) + P_1^{(L)}(x). \tag{2.1.5}$$

Similarly, since $P_0^{(L)}(x) = f_0$, we have

$$P_1^{(L)}(x) - P_0^{(L)}(x) = f_0 \frac{x - x_1}{x_0 - x_1} + f_1 \frac{x - x_0}{x_1 - x_0} - f_0$$

$$= f_0\left(\frac{x - x_1}{x_0 - x_1} - 1\right) + f_1 \frac{x - x_0}{x_1 - x_0}.$$

This polynomial vanishes when $x = x_0$, and so we can set

$$P_1^{(L)}(x) - P_0^{(L)}(x) = \alpha_1(x - x_0),$$

giving

$$P_1^{(L)}(x) = \alpha_1(x - x_0) + P_0^{(L)}(x). \tag{2.1.6}$$

Substituting (2.1.6) into (2.1.5), we find

$$P_2^{(L)}(x) = \alpha_2(x - x_0)(x - x_1) + \alpha_1(x - x_0) + \alpha_0, \tag{2.1.7}$$

the Newton form of the polynomial.

The Orthogonal Form

The orthogonal polynomial form is given by

$$P_n(x) = \sum_{k=0}^{n} b_k Q_k(x), \tag{2.1.8}$$

where the orthogonal polynomials, $\{Q_i(x), i = 0, 1, \ldots, n\}$, satisfy the recurrence

$$Q_{i+1}(x) = (A_i x + B_i)Q_i(x) - C_i Q_{i-1}(x),$$

with $A_i \neq 0$, $Q_0(x) = 1$, $Q_{-1}(x) = 0$,
and where A_i, B_i and C_i may depend on i but not on x.

An implementation of the orthogonal form algorithm is given below.

Algorithm orthogonalform

```
V[n] := b[n]
V[n-1] := (A[n-1] × x + B[n-1]) × V[n] + b[n-1]
```

```
for k := n − 2 downto 0 do
    V[k] := (A[k] × x + B[k]) × V[k+1] − C[k] × V[k+2] + b[k]
enddo
polyvalue := V[0]
```

This scheme requires $(3n-1)$ multiplications and $(3n-1)$ additions.

Among the classical orthogonal polynomials, the Chebyshev polynomials, $T_n(x)$, turn out to be the best choice on the grounds of efficiency of evaluation. In this case (2.1.8) becomes

$$P_n(x) = \sum_{k=0}^{n} b_k T_k(x), \tag{2.1.9}$$

and we have $B_i = 0, C_i = 1$ for $i \geqslant 0$, $A_i = 2$ for $i \geqslant 1$, and $A_0 = 1$.

One computational algorithm based on (2.1.9) is due to Clenshaw (1955) and is given below.

Algorithm clenshawform

```
Y := 2 × x
V[n] := b[n]
V[n − 1] := Y × V[n] + b[n − 1]
for k := n − 2 downto 1 do
    V[k] := Y × V[k+1] − V[k+2] + b[k]
enddo
polyvalue := x × V[1] − V[2] + b[0]
```

The evaluation of a general polynomial as a weighted sum of Chebyshev polynomials using algorithm *clenshawform* requires $(n+1)$ multiplications and $2n$ additions.

Another computational algorithm based on (2.1.9) was suggested by Bakhvalov (1971) and is as follows.

Algorithm bakhvalovform

```
D[0] := b[0] − b[2]/2
for k := 1 to n − 2 do
    D[k] := (b[k] − b[k+2])/2
enddo
D[n − 1] := b[n − 1]/2
V[n] := b[n]/2
Y := 2 × x
V[n − 1] := Y × V[n] + D[n − 1]
for k := n − 2 downto 0 do
    V[k] := (Y × V[k+1] − V[k+2]) + D[k]
enddo
polyvalue := V[0]
```

The algorithm requires $(n+1)$ multiplications, $(n+1)$ divisions by a factor of 2, and $3n-1$ additions. Algorithm *bakhvalovform* is less efficient than algorithm *clenshawform* in terms of the number of arithmetic operations, but for certain polynomials its use may be preferable to that of *clenshawform* on the grounds of numerical accuracy.

In the for-loop for computing V_k in algorithms *clenshawform* and *bakhvalovform* parameter k assumes value from $n-2$ downto 0. This 'back to front' ordering of the computational sequence is dictated by the considerations of numerical accuracy. When adding a large sequence of numbers on a computer under conditions of normalized floating-point arithmetic, it is desirable to add the smaller numbers first, since at each step of the computation the local round-off error is proportional to the partial accumulated sum. For 'smooth' polynomials the coefficients b_k in algorithm *clenshawform* have a tendency to decrease with k increasing and it is therefore desirable to have the order of accumulation of the summation terms as given in the algorithms. In the case of other algorithms discussed similar orderings for the computational sequences were used in the interests of a uniform presentation.

2.2 Preprocessing the Coefficients

Of the computational algorithms based on different polynomial forms the Horner algorithm requires for a polynomial of degree n the smallest number, n, of each multiplications and additions. The question may be raised however whether or not it is possible to devise an 'improved' form which will require fewer additions and/or fewer multiplications than the Horner algorithm.

The idea of establishing lower bounds on the number of arithmetic operations required to evaluate a polynomial was originally introduced by Ostrowski (1954). He showed that at least n multiplications and n additions are necessary to evaluate nth degree polynomial for $n \leqslant 4$, a crucial point in his proof being the underlying assumption that the polynomial coefficients were not in any way artificially transformed for the purpose of minimizing the number of multiplications and additions. Since then the result has been proved true under the same assumption for all non-negative values of n. However, in 1955, Motzkin introduced the notion of 'preprocessing' the polynomial coefficients for the purpose of designing evaluation algorithms which may require fewer multiplications and/or additions than, say, the Horner algorithm. Preprocessing of the coefficients clearly involves additional arithmetic operations, but if the polynomial is to be evaluated at many points, the overall savings on the number of arithmetic operations become significant.

To illustrate the idea of coefficient preprocessing we consider a simple example of a fourth degree polynomial which follows from the form proposed by Motzkin and described by Todd (1955).

Consider the following identity:

$$\begin{aligned} &a_0x^4 + a_1x^3 + a_2x^2 + a_3x + a_4 \\ &\quad \equiv a_0[(x(x+\alpha_0)+\alpha_1)(x(x+\alpha_0)+x+\alpha_2)+\alpha_3] \\ &\quad \equiv a_0x^4 + a_0(2\alpha_0+1)x^3 + a_0(\alpha_1+\alpha_2+\alpha_0(\alpha_0+1))x^2 \\ &\qquad + a_0((\alpha_1+\alpha_2)\alpha_0+\alpha_1)x + a_0(\alpha_1\alpha_2+\alpha_3). \end{aligned} \tag{2.2.1}$$

Equating coefficients of corresponding powers of x and solving the resulting equations, we obtain:

$$\alpha_0 = \frac{a_1 - a_0}{2a_0}, \qquad \alpha_1 = \frac{a_3}{a_0} - \alpha_0 \frac{a_2}{a_0} + \alpha_0^2(\alpha_0 + 1),$$

$$\alpha_2 = \frac{a_2}{a_0} - \alpha_0(\alpha_0 + 1) - \alpha_1, \qquad \alpha_3 = \frac{a_4}{a_0} - \alpha_1\alpha_2. \tag{2.2.2}$$

Thus, if we first compute α_i, $i = 0, \ldots, 3$ the evaluation algorithm becomes:

Algorithm motzkinform

$V[1] :=$ $(x + \text{alpha}[0]) \times x$
$V[2] :=$ $(V[1] + \text{alpha}[1]) \times (V[1] + x + \text{alpha}[2])$
$V[3] :=$ $V[2] + \text{alpha}[3]$
polyvalue $:=$ $\text{alpha}[0] \times V[3]$

The algorithm entails three multiplications and five additions, whereas the standard Horner algorithm would require four multiplications and four additions.

On a digital computer, multiplication is usually a more complex operation than addition, so even this modest saving is worthwhile if the polynomial is to be evaluated many times.

For polynomials of higher degree, the savings due to preprocessing their coefficients become steadily more significant as n increases. For instance, a polynomial of degree six

$$P_6(x) = a_0x^6 + \ldots + a_6$$

may be expressed in the form

$$F_6(x) = [\,(((x + \alpha_0)x + \alpha_1)(x + \alpha_2) + \alpha_3 + (x + \alpha_0)x + \alpha_1 + \alpha_4) \times (((x + \alpha_0)x + \alpha_1)(x + \alpha_2) + \alpha_3) + \alpha_5\,]\,a_0,$$

and solved using the Knuth algorithm:

Algorithm knuthform

$V[1] :=$ $(x + \text{alpha}[0]) \times x + \text{alpha}[1]$
$V[2] :=$ $V[1] \times (x + \text{alpha}[2]) + \text{alpha}[3]$
$V[3] :=$ $(V[2] + V[1] + \text{alpha}[4]) \times V[2] + \text{alpha}[5]$
polyvalue $:=$ $\text{alpha}[0] \times V[3]$

This requires four multiplications and seven additions as opposed to the Horner algorithm with six operations of each, multiplication and addition.

The operation of computing the parameters $\alpha_0, \ldots, \alpha_r$ in terms of the coefficients $a_0, \ldots, a_n$ is known as *preprocessing* the coefficients, and the polynomial $F_n(x)$ expressed in terms of the α_i, as *the polynomial with preprocessed coefficients.* In distinction to the polynomial forms considered earlier, polynomials with preprocessed coefficients do not arise 'naturally' but are obtained in an artificial way, with the sole purpose of facilitating their fast evaluation.

2.3 Optimality of the Evaluation Algorithms

The different polynomial evaluation algorithms yield different number of arithmetic operations required. One can then ask whether a minimum number of arithmetic operations can be determined that will suffice to evaluate a polynomial, given the choice of its forms. The following theorem answers one such question.

Theorem 2.3.1

Any algorithm to evaluate a polynomial $P_n(x)$ without preprocessing the coefficients involves at least n multiplications and at least n additions.

Proof of the theorem has been given by several authors, see for example, Belaga (1961), Pan (1966), Reingold and Stocks (1972) and Knuth (1973).

It has also been shown that the Horner algorithm is an optimal algorithm and the only one that optimizes both the number of additions and the number of multiplications (Borodin, 1971). We shall now show that the Horner algorithm is optimal in the number of multiplications under assumption that no preprocessing of the coefficients is made. However, first a specified computational model for the polynomial evaluation is in order. We shall assume the so called straight-line computation model which is defined as a sequence of statements $s_1, \ldots, s_r$. Each statement is of the form

$$\lambda_i = \lambda_j \circ \lambda_k, \qquad 0 \leqslant j, k < i, \tag{2.3.1}$$

where '$\circ$' stands for an operation of addition, subtraction or multiplication, and λ_i, λ_j, λ_k are the parameters and variables associated with the particular problem. In our case each λ_j and λ_k denotes one of the following: (a) the polynomial coefficient, a_k, $k = 0, 1, \ldots, n$; (b) the argument, x; (c) a previous variable in the straight-line program, where $j, k < \mathrm{i}$; (d) an arbitrary rational constant, and denotes a new variable in the straight-line program. We call step λ_i in (2.3.1) an 'active step in a_m' if at least one of λ_j and λ_k is either a_m (or a linear function of a_m, i.e. $\beta a_m + \lambda$, where β and γ are real constants, and $\beta \neq 0$) or an active step in a_m, and the other is *not* a constant.

As an example of a straight-line program to evaluate a polynomial, consider:

$$P_5(x) = ((((a_0 x + a_1)x + a_2)x + a_3)x + a_4)x + a_5.$$

Set $\lambda_{-6} = a_5, \quad \lambda_{-5} = a_4, \quad \lambda_{-4} = a_3, \quad \lambda_{-3} = a_2$
$\lambda_{-2} = a_1, \quad \lambda_{-1} = a_0, \quad \lambda_0 = x,$

and define a sequence, $\lambda_1, \ldots, \lambda_{10}$, by

$$\lambda_1 = \lambda_0 \times \lambda_{-1} \quad \lambda_2 = \lambda_1 + \lambda_{-2} \quad \lambda_3 = \lambda_2 \times \lambda_0$$
$$\lambda_4 = \lambda_3 + \lambda_{-3} \quad \lambda_5 = \lambda_4 \times \lambda_0 \quad \lambda_6 = \lambda_5 + \lambda_{-4}$$
$$\lambda_7 = \lambda_6 \times \lambda_0 \quad \lambda_8 = \lambda_7 + \lambda_{-5} \quad \lambda_9 = \lambda_8 \times \lambda_0 \qquad \lambda_{10} = \lambda_9 + \lambda_{-6}.$$

Then the sequence $\{\lambda_k, k = 1, \ldots, 10\}$ defines the Horner straight-line program for evaluating a general polynomial of degree 5.

The steps λ_1, λ_3, λ_5, λ_7 and λ_9 are the multiplication steps active in a_0; the steps λ_2, λ_4, λ_6, λ_8 and λ_{10} are the addition steps active in a_1, a_2, a_3, a_4, a_5, respectively.

Now, we wish to establish the fact that in order to evaluate a polynomial of degree n,

$$P(x; a_0, \ldots, a_n) = a_0x^n + a_1x^{n-1} + \ldots + a_{n-1}x + a_n,$$

using the computational model (2.3.1), at least n multiplications are required which actively involve $\{a_k, k = 0, \ldots, n\}$.

To do that we prove a more general theorem.

Theorem 2.3.2

Let $\mathbf{U} = \{u_{ij}\}, 0 \leqslant i \leqslant m, 0 \leqslant j \leqslant n, n \leqslant m$, be an $(m+1) \times (n+1)$ real matrix of rank $n+1$, and $\mathbf{V} = (v_0, v_1, \ldots, v_m)$ a real vector. Then any straight-line program which computes

$$P(x; a_0, \ldots, a_n) = \sum_{i=0}^{m} (u_{i0}a_0 + u_{i1}a_1 + \ldots + u_{in}a_n + v_i)x^i$$
$$= (\mathbf{Ua} + \mathbf{V})^T\mathbf{x}, \tag{2.3.2}$$

where $\mathbf{a} = (a_0, \ldots, a_n)$ and $x = (x^0, x, x^2, \ldots, x^m)$

are column vectors, involves at least n multiplication steps active in $\{a_k, k = 0, \ldots, n\}$.

Here the matrix $\mathbf{U}$ and vector $\mathbf{V}$ are chosen arbitrarily, and $P(x, \mathbf{a})$ can be thought of as a general polynomial in the sense that it is a function not only of x but of the inputs $\mathbf{a}$ as well, while the straight-line program and its variables, λ_i, λ_j, λ_k, will generally depend on the chosen $\mathbf{U}$ and $\mathbf{V}$. The proof establishes the fact that no matter how $\mathbf{U}$ and $\mathbf{V}$ are chosen, any straight-line program, which computes $P(x, \mathbf{a})$, will use at least n multiplication steps active in a_ks.

Proof. The theorem holds for $n = 0$ since in this case we have

$$P(x, a_0) = (u_{00}a_0 + v_0) + (u_{10}a_0 + v_1)x + \ldots + (u_{m0}a_0 + v_m)x^m, \tag{2.3.3}$$

and for $m = 0, 1, \ldots$ (recall that by definition $m \geqslant n$), the straight-line program, whatever form it may take, will include a multiplication step which multiplies a_0 by a constant u_{00}.

Next, for $n = 1$ we have

$$P(x; a_0, a_1) = \sum_{i=0}^{m} (u_{i0}a_0 + u_{i1}a_1 + v_i)x^i$$

and, in particular, for $m = n = 1$:

$$P(x; a_0, a_1) = (u_{00}a_0 + u_{01}a_1 + v_1) + (u_{10}a_0 + u_{11}a_1 + v_1)x. \tag{2.3.4}$$

The rank of matrix

$$\mathbf{U} = \begin{bmatrix} u_{00} & u_{01} \\ u_{10} & u_{11} \end{bmatrix}$$

is by definition equal to 2; this means that at least three of the entries u_{ij} are non-zero values. It follows that at least one of u_{10} and u_{11} is non-zero and therefore the evaluation of $P(x; a_0, a_1)$ in (2.3.4) involves at least one multiplication step which is active either in a_0 or in a_1.

Now consider a polynomial straight-line program which computes $P(x; a_0, a_1, \ldots, a_n)$ for $n > 1$. Let $\lambda_{i_q} = \lambda_{j_q} \times \lambda_{k_q}$ be the first multiplication step active in one of $a_0, \ldots, a_n$. Without loss of generality, we may assume that λ_{j_q} involves a_n. Thus, λ_{j_q} has the form $\beta a_n + \alpha$, where β and α are real constants, $\beta \neq 0$. Now change step i_q to be $\lambda_{i_q} = \varkappa \times \lambda_{k_q}$ where $\varkappa$ is an arbitrary real number. Add a further step to calculate

$$\lambda = a_n = (\varkappa - \alpha)/\beta.$$

These new steps involve only additions and constant multiplication by suitable new constants.

Finally, replace a_n everywhere in the straight-line program by the new variable λ. The new straight-line program calculates

$$R(x; a_0, \ldots, a_{n-1}) = P\left(x; a_0, \ldots, a_{n-1}, \frac{\varkappa - \alpha}{\beta}\right),$$

and has one less multiplication step active in $\{a_k, k = 0, \ldots, n\}$, since a_n has been eliminated.

By virtue of (2.3.2), the polynomial R may be expressed as

$$\begin{aligned} R(x; a_0, \ldots, a_{n-1}) &= \sum_{i=0}^{m} \left(u_{i0}a_0 + u_{i1}a_1 + \ldots + u_{in-1}a_{n-1} + u_{in}\frac{\varkappa - \alpha}{\beta} + v_i\right)x^i \\ &= (\mathbf{U}_1\mathbf{a}_1 + \mathbf{V}_1)^{\mathrm{T}}\mathbf{x}, \end{aligned} \tag{2.3.5}$$

where

$$\mathbf{U}_1 = \begin{bmatrix} u_{00}u_{01} \ldots u_{0n-1} \\ u_{10}u_{11} \ldots u_{1n-1} \\ \vdots \\ u_{m0}u_{m1} \ldots u_{mn-1} \end{bmatrix}, \quad \mathbf{a}_1 = \begin{bmatrix} a_0 \\ a_1 \\ \vdots \\ a_{n-1} \end{bmatrix}, \quad \mathbf{V}_1 = \begin{bmatrix} u_{0n}\dfrac{x-\alpha}{\beta} + v_0 \\ u_{1n}\dfrac{x-\alpha}{\beta} + v_1 \\ \vdots \\ u_{mn}\dfrac{x-\alpha}{\beta} + v_m \end{bmatrix}, \quad \mathbf{x} = \begin{bmatrix} x^0 \\ x^1 \\ \vdots \\ x^m \end{bmatrix}.$$

It follows from (2.3.5) that R satisfies the general form (2.3.2) assumed by theorem 2.3.2, and by induction on n, its straight-line program involves at least $n-1$ multiplication steps active in $\{a_k, k=0, \ldots, n-1\}$. Hence the original straight-line program which computes P has at least n multiplication steps active in $\{a_k, k=0, \ldots, n\}$.
This completes the proof.

The optimality of the Horner algorithm in terms of the number of multiplications follows as a special case of theorem 2.3.2, when the matrix **U** is set as the identity matrix, **I**, and the vector **V** as zero-vector.

Since the Horner alogorithm is an optimal algorithm for evaluation of a polynomial with 'natural' coefficients and since this algorithm requires n multiplications to evaluate a polynomial of degree n, it follows that n multiplications is the minimum number of these operations which is needed by any algorithm to evaluate a polynomial with 'natural' coefficients. The proof of an optimal number of additions is developed in the way similar to that for the multiplications.

2.4 Belaga Theorems

Lower bounds on the number of arithmetic operations required to evaluate a polynomial with preprocessed coefficients are given by the following theorems.

Theorem 2.4.1

For any polynomial $P_n(x)$ of degree n, there exists a computational evaluation scheme that requires $\lfloor (n+1)/2 \rfloor + 1$ multiplications and $n+1$ additions.

Theorem 2.4.2

No evaluation scheme exists with less than $\lfloor (n+1)/2 \rfloor$ multiplications or with less than n additions.

It has not yet been shown whether an evaluation scheme exists with exactly $\lfloor (n+1)/2 \rfloor$ multiplications and n additions. Thoerems 2.4.1 and 2.4.2 were formulated by Belaga (1958). The first of the theorems is proved by a constructive approach, whereby an evaluation scheme is developed with the number of multiplications and additions as stated in the theorem. We shall explore this scheme in greater detail. Theorem 2.4.2 establishes lower bounds on the number of multiplications and additions, under the assumption that some preprocessing of the coefficients is allowed without cost. Proofs of the theorem have been elaborated mainly by Winograd (1970b) who in the course of proof uses some deep results from topology, and by Knuth (1969) who argues the facts using the so-called 'degrees of freedom' from number theory. The reader might wish to consult a somewhat simpler proof of Reingold and Stocks (1972) which is based on the notion of algebraic independence of the preprocessed parameters. It is felt, however, that a careful treatment of the proof of theorem 2.4.2 is beyond the scope of the present text and an interested reader is referred to the sources of information just mentioned. We shall now turn to the scheme for evaluation of a polynomial that has been suggested by Belaga. It is a general scheme which evaluates any polynomial of degree $n \geqslant 4$ with prior preprocessing of the coefficients and is given as:

Algorithm belaga

```
while n ⩾ 4 do
    V[1] := (x + alpha[1]) × x
    V[2] := (V[1] + x + alpha[2]) × (V[1] + alpha[3]) + alpha[4]
    for k := 3 to n div2 do
        V[k] := V[k − 1] × (V[1] + alpha[2k − 1]) + alpha[2k]
    enddo
    if n div2 > 0 then polyvalue := alpha[n + 1] × x × V[ndiv2]
    else polyvalue := alpha[n + 1] × V[ndiv2]
    endif
enddo
```

We can see that the Belaga computational scheme requires $\lfloor (n+1)/2 \rfloor + 1$ multiplications and $2\lfloor n/2 \rfloor + 1$ additions.

A method for obtaining preprocessed parameters α_i in the algorithm in terms of $(a_0, \ldots, a_n)$

$$\text{where } P_n(x) = \sum_{k=0}^{n} a_k x^{n-k},$$

was given by Cheney (1962).

Algorithm *belaga* gives

$$P_n(x) = a_0 x^n + a_1 x^{n-1} + \ldots + a_n = \begin{cases} \alpha_{n+1} V_m \\ \alpha_{n+1} x V_m \end{cases}, \quad m = n \textbf{ div} 2. \tag{2.4.1}$$

Without restricting the generality of the method we may assume that $a_0 = \alpha_{n+1} = 1$, and consider

$$\begin{aligned} V_m &= x^{2m} + a_1 x^{2m-1} + \ldots + a_{2m} \\ &= (x^{2m-2} + b_1 x^{2m-3} + b_2 x^{2m-4} + \ldots + b_{2m-2})(x^2 + \alpha_1 x + \alpha_{2m-1}) \\ &\qquad + \alpha_{2m}. \end{aligned} \tag{2.4.2}$$

Algorithm cheney

equate like powers of x in (2.4.2) *to obtain* $a_1 = \alpha_1 + b_1$
set $\alpha_1 = (a_1 - 1)/m$
obtain $b_1 = ((m-1)a_1 + 1)/m$.//This solution satisfies the requirement that the higher term coefficient in the polynomial is unity.//
while more α_j **do**
write $b_2 = a_2 - \alpha_1 b_1 - \alpha_{2m-1}$. //At this point α_{2m-1} is unknown.//
write $b_k = a_k - \alpha_1 b_{k-1} - \alpha_{2m-1} b_{k-2}$ *for* $k = 3, \ldots, 2m-2$
solve $a_{2m-1} = \alpha_{2m-1} b_{2m-3} + \alpha_1 b_{2m-2}$ *for* α_{2m-1}
substitute the value of α_{2m-1} *into the expressions for* $b_{2m-2}, \ldots, b_2$ *and obtain their values*
compute $\alpha_{2m} = a_{2m} - \alpha_{2m-1} b_{2m-2}$//We have now values for α_1, α_{2m-1} and α_{2m}.//
replace in (2.4.2) m by $m-1$ *and solve the new system to obtain* α_{2m-2} and α_{2m-3}
enddo

Since the Cheney scheme is valid for any polynomial of degree n, it serves as a constructive proof of theorem 2.4.1.

Example 2.4.1

This is evaluation of the Belaga parameters for a power form polynomial.

Consider $P_7(x) = x^7 - x^6 + 8x^5 - 4x^4 + 6x^3 + 2x^2 - 5x + 1$.

Using (2.4.1) and (2.4.2) we can write

$$\begin{aligned} V_3 &= x^6 - x^5 + 8x^4 - 4x^3 + 6x^2 + 2x - 5 \\ &= (x^4 + b_1 x^3 + b_2 x + b_4)(x^2 + \alpha_1 x + \alpha_5) + \alpha_6 \end{aligned}$$

with $c_1 = -1$, $c_2 = 8$, $c_3 = -4$, $c_4 = 6$, $c_5 = 2$, $c_6 = -5$.
We have

(c_1) $-1 = \alpha_1 + b_1$. Set $\alpha_1 = -\frac{2}{3}$, then $b_1 = -\frac{1}{3}$.

(c_2) $8 = \alpha_5 + \alpha_1 b_1 + b_2$, giving $b_2 = 7.778 - \alpha_5$.

(c_3) $-4 = \alpha_5 b_1 + \alpha_1 b_2 + b_3$, giving $b_3 = 1.185 - 0.333\alpha_5$.

(c_4) $6 = \alpha_5 b_2 + \alpha_1 b_3 + b_4$, giving $b_4 = 6.790 - 8\alpha_5 + \alpha_5^2$.

(c_5) $2 = \alpha_5 b_3 + \alpha_1 b_4$, giving $\alpha_5^2 - 6.519\alpha_5 + 6.527 = 0$, and $(\alpha_5)_1 = 5.284$, $(\alpha_5)_2 = 1.236$.

(c_6) $-5 = \alpha_5 b_4 + \alpha_6$, giving two values for α_6, $(\alpha_6)_1 = 34.952$ and $(\alpha_6)_2 = -3.059$, corresponding to $(\alpha_5)_1$ and $(\alpha_5)_2$, respectively.

At this stage two sets of the parameters values have been obtained:

(i) $\alpha_6 = 34.952$, $\alpha_5 = 5.284$, $\alpha_1 = -0.667$,
$b_1 = -0.333$, $b_2 = 2.494$, $b_3 = -0.576$, $b_4 = -7.561$;

and

(ii) $\alpha_6 = -3.059$, $\alpha_5 = 1.236$, $\alpha_1 = -0.667$,
$b_1 = -0.333$, $b_2 = 6.542$, $b_3 = 0.773$, $b_4 = -1.570$.

For set (i) we obtain

$$V_2 = x^4 - 0.333x^3 + 2.494x^2 - 0.576x - 7.561$$
$$= (x^2 + d_1x + d_2)(x^2 + \alpha_1x + \alpha_3) + \alpha_4.$$

The system for obtaining the values for α_3, and α_4 is as follows:

(b_1) $-0.333 = \alpha_1 + d_1$, giving $d_1 = \frac{1}{3} = 0.333$.

(b_2) $2.494 = \alpha_3 + \alpha_1 d_1 + d_2$, giving $d_2 = 2.716 - \alpha_3$.

(b_3) $-0.576 = \alpha_1 d_2 + \alpha_3 d_1$, giving $\alpha_3 = 1.235$.
We then get $d_2 = 1.481$, and
$\alpha_2 = d_2 = 1.481$.

(b_4) $-7.561 = \alpha_3 d_2 + \alpha_4$, giving $\alpha_4 = -9.390$.

This completes calculations for set (i).
For set (ii) we similarly obtain

$$V_2 = x^4 - 0.333x^3 + 6.542x^2 + 0.773x - 1.570$$
$$= (x^2 + g_1x + g_2)(x^2 + \alpha_1x + \alpha_3) + \alpha_4.$$

The values of α_3 and α_4 are obtained by solving the following system:

(b_1) $-0.333 = \alpha_1 + g_1$, giving $g_1 = \frac{1}{3} = 0.333$.

(b_2) $6.542 = \alpha_3 + \alpha_1 g_1 + g_2$, giving $g_2 = 6.764 - \alpha_3$.

(b_3) $0.773 = \alpha_3 g_1 + \alpha_1 g_2$, giving $\alpha_3 = 5.282$.
We thus find $g_2 = 1.481$, and
$\alpha_2 = g_2 = 1.481$.

(b_4) $-1.570 = \alpha_3 g_2 + \alpha_4$, giving $\alpha_4 = -9.398$.

This completes calculations for set (ii).
Two sets of the Belaga parameters as shown:

α_1	α_2	α_3	α_4	α_5	α_6	α_7
$-\frac{2}{3}$	1.481	1.235	−9.390	1.236	−3.059	1
$-\frac{2}{3}$	1.481	5.282	−9.398	5.284	34.952	1

give two Belaga algorithms for $P_7(x)$:

$$V_1 = (x - \tfrac{2}{3})x,$$
$$V_2 = (V_1 + x + 1.481)(V_1 + 1.235) - 9.390,$$
$$V_3 = V_2(V_1 + 1.236) - 3.059,$$
$$P_7(x) = xV_3,$$

and

$$V_1 = (x - \tfrac{2}{3})x,$$
$$V_2 = (V_1 + x + 1.481)(V_1 + 5.282) - 9.398,$$
$$V_3 = V_2(V_1 + 5.284) + 34.952,$$
$$P_7(x) = xV_3.$$

As the example shows Belaga parameters for a given polynomial are not unique. The other more serious practical flaw in the Belaga algorithm arises from the fact that some of the α_j may be complex for a polynomial with real 'natural' coefficients.

As an example of such a case consider the polynomial

$$P_6(x) = x^6 + x^5 + x^4 + x^3 + x^2 + x + 1.$$

The Belaga coefficients of this polynomial are:

$$\alpha_1 = 0 \qquad \alpha_2 = 0 \qquad \alpha_3 = 1 - q$$
$$\alpha_4 = 0 \qquad \alpha_5 = q \qquad \alpha_6 = 1 \qquad \alpha_7 = 1,$$

where

$$q = \tfrac{1}{2}(1 \pm i\sqrt{3}),$$

and the Belaga algorithm is given as

$$V_1 = x^2,$$
$$V_2 = (V_1 + x)(V_1 + (1 - q)),$$
$$P_6(x) = V_3 = V_2(V_1 + q) + 1.$$

The Pan Form

Another algorithm for an efficient evaluation of a polynomial with preprocessed coefficients has been devised by Pan(1959). The algorithm holds for any polynomial of degree $n \geqslant 5$, and is implemented as follows.

Algorithm panform

```
while n ≥ 5 do
    W[0] := (x + lambda[0]) × x
    W[1] := W[0] + x
```

```
V[0] := x
for k := 1 to ndiv4 do
    V[k] := V[k − 1] × ((W[0] + lambda[4k − 3]) × (W[1]
            + lambda[4k − 2]) + lambda[4k − 1]) + lambda[4k]

enddo
case polyvalue of

    n = ndiv4 + 1 : lambda[n] × V[ndiv4]
    n = ndiv4 + 2 : lambda[n] × x × V[ndiv4] + lambda[n − 1]
    n = ndiv4 + 3 : lambda[n] × (V[ndiv4] × (W[0]
                    + lambda[n − 2]) + lambda[n − 1]
    n = ndiv4 + 4:  lambda[n] × x × (V[ndiv4] × (W[0]
                    + lambda[n − 3]) − lambda[n − 2])
                    + lambda[n − 1]
endcase
enddo
```

The Pan algorithm requires $[n/2] + 2$ multiplications and $n + 1$ additions. For n odd, this gives the same number of arithmetic operations as the Belaga algorithm, and for n even, it gives one extra multiplication.

The Pan parameters are, again, not uniquely determined. For example. Rice (1965) gives the following three variants of the Pan form for the polynomial

$$x^7 - 2x^5 + x^3 - 0.1x$$

λ_0	λ_1	λ_2	λ_3	λ_4	λ_5	λ_6	λ_7
−1/3	−1.15	−0.481	0.179	0.223	−0.035	0.0078	1
−1/3	−0.268	−0.481	−0.264	0.018	−0.917	0.016	1
−1/3	−0.804	−0.481	−0.087	−0.043	−0.381	−0.016	1

However, the great advantage of the Pan form over the Belaga form is that for a polynomial with real 'natural' coefficients, all Pan parameters are always real.

Rabin and Winograd (1971) have shown how to achieve $\lceil n/2 \rceil + O(\log_2 n)$ multiplication/division operations and $n + O(n)$ additions using only rational preprocessing of coefficients. Eve (1964) has also suggested schemes for rational preprocessing of coefficients.

A non-uniqueness of the preprocessed coefficients may at times be considered as a useful attribute since it gives extra freedom of choice among available computational schemes in specific cases. Winograd(1973) gives an example of such a polynomial:

$$\frac{1}{81}x^4 + \frac{5}{81}x^3 + \frac{4}{27}x^2 + \frac{11}{27}x + \frac{5}{9}$$

for evaluation of which along with the traditional Horner scheme,

$$\left(\left(\left(\frac{1}{81}x+\frac{5}{81}\right)x+\frac{4}{27}\right)x+\frac{11}{27}\right)x+\frac{5}{9},$$

we can consider using one of the following preprocessed schemes:

$$\text{(Todd–Motzkin)}\quad \frac{1}{81}((x(x+2)+21)(x(x+2)+x-15)+360),$$

$$\text{(Rabin–Winograd)}\left(\frac{1}{3}x\left(\frac{1}{3}x-\frac{2}{3}\right)+\frac{85}{81}\right)\left(\frac{1}{3}x\left(\frac{1}{3}x-\frac{2}{3}\right)+x-\frac{59}{81}\right)+\frac{8660}{6561},$$

$$\text{(Rabin–Winograd)}\left(\frac{1}{3}x\left(\frac{1}{3}x+\frac{1}{3}\right)+\frac{25}{27}\right)\left(\frac{1}{3}x\left(\frac{1}{3}x+\frac{1}{3}\right)+\frac{1}{3}x-\frac{1}{27}\right)+\frac{430}{729}.$$

Here an inspection of the Todd–Motzkin expression shows that if the expression is used to evaluate a polynomial for small x, then some significant digits can be lost due to the fact that the scheme's constants are large. Since the evaluation algorithms of this type might well be used for building up, say the elementary function procedures, such as $\sin x$, where it is quite common to call the procedure for small x the Todd–Motzkin scheme, will prove unsuitable in practice. On the other hand the use of the Rabin–Winograd schemes reduces the rounding error problems because of their evenly balanced coefficients.

2.5 Numerical Solution and Conditioning of the Problem

The problem of assessing the accuracy of computed results depends critically on the exact way in which numbers are represented within the computer.

For our needs, we shall distinguish between mathematical (or round-off) errors and physical (or inherent) errors in the data. The round-off errors are usually small and what is very important can be kept under control, in the sense that is is usually possible to specify upper bounds for the absolute values of such errors. The physical errors, on the other hand, may be quite large and are outside our control. We must first define the various types of error with which we shall be concerned.

If x is the true value of the number, and X an approximation to the true value, then we define

$$e = X - x \qquad \text{as the } \textit{error} \tag{2.5.1}$$

$$|e| = |X - x| \qquad \text{as the } \textit{absolute error} \text{ and} \tag{2.5.2}$$

$$\varepsilon = \frac{e}{x} = \frac{X-x}{x} \qquad \text{as the } \textit{relative error;} \tag{2.5.3}$$

we can also write $X = x + e$ and

$$X = x(1+\varepsilon). \tag{2.5.4}$$

Floating-point Number Representation

In the computer numbers are stored in the so-called floating-point (fl) form, i.e.

$$x = 2^b a \tag{2.5.5}$$

where b (the exponent) is a positive or negative integer (or zero) and a (the mantissa) is a fraction in the range $-\frac{1}{2} \geqslant a > -1$ or $\frac{1}{2} \leqslant a < 1$, and of fixed length t digits. If we assume that each arithmetic operation is carried out internally to double precision and the result subsequently rounded to t significant digits, then denoting the floating-point digital representation of the number by the symbol fl or by the bar, we can write

$$\bar{x} = \text{fl}(x) = x(1 + \varepsilon), \quad |\varepsilon| \leqslant 2^{-t}, \tag{2.5.6}$$

where $|\varepsilon|$ denotes the absolute value of the relative error. The results given below are due to Wilkinson (1963).

Assuming that the input data is read in exactly, we have

$$\text{fl}(x_1 \pm x_2) = (x_1 \pm x_2)(1 + \varepsilon), \qquad |\varepsilon| \leqslant 2^{-t} \tag{2.5.7}$$

$$\text{fl}(x_1 x_2) \quad = (x_1 x_2)(1 + \varepsilon), \qquad |\varepsilon| \leqslant 2^{-t} \tag{2.5.8}$$

$$\text{fl}(x_1/x_2) \quad = (x_1/x_2)(1 + \varepsilon), \qquad |\varepsilon| \leqslant 2^{-t} \tag{2.5.9}$$

Next, we shall state bounds for the errors involved in the calculation of extended products, sums and inner-products, since these will be used in various analyses throughout the text.

For the extended product $x_1 x_2 \ldots x_n$, we have

$$\text{fl}(x_1 x_2 \ldots x_n) = x_1 x_2 \ldots x_n(1 + E), \tag{2.5.10}$$

where

$$|E| < (n - 1)2^{-t_1} \tag{2.5.11}$$

and t_1, for compactness of presentation, is defined by the relation

$$2^{-t_1} = (1.06)2^{-t}$$

so that

$$t_1 = t - 0.08406.$$

Much the same result may be obtained for extended sequences of multiplication and divisions. We have

$$\text{fl}(x_1 x_2 \ldots x_m / y_1 y_2 \ldots y_n) = (x_1 x_2 \ldots x_m / y_1 y_2 \ldots y_n)(1 + E) \tag{2.5.12}$$

where

$$|E| < (m + n - 1)2^{-t_1} \tag{2.5.13}$$

It may be observed from (2.5.11) and (2.5.13) that for n (and $n + m$) appreciably smaller than 2^t, the computed results, given by (2.5.10) and (2.5.12), have a low relative error, E.

This observation is not, however, valid for extended sequences of additions, and here we have

$$\mathrm{fl}(x_1 + x_2 + \ldots + x_n) = x_1(1 + \eta_1) + x_2(1 + \eta_2) + \ldots + x_n(1 + \eta_n), \tag{2.5.14}$$

where

$$|\eta_r| < (n + 1 - r)2^{-t_1}, r = 1, \ldots, n. \tag{2.5.15}$$

Finally, for the inner product with accumulation we have

$$\mathrm{fl} \sum_{i=1}^{n} x_i y_i = \left[\sum_{i=1}^{n} x_i y_i (1 + e_i) \right] (1 + e) \tag{2.5.16}$$

where

$$|e_r| < \tfrac{3}{2}(n + 2 - r)2^{-2t_2},\ 2^{-t_2} = (1.06)2^{-2t} \tag{2.5.17}$$
$$|e| < 2^{-t}$$

Conditioning of a Problem

A computer representation of any number consists of a finite number of digits only. This limited accuracy in the representation of numbers gives rise to a phenomenon known as the numerical instability. Numerical instability is a feature of a particular computational scheme. It manifests itself in a growth of round-off errors during the computation. A computational scheme which may generate an uncontrolled growth of round-off errors is said to be numerically unstable, and a computational scheme which controls the growth of round-off errors is described as numerically stable.

Another concept which relates to the fact that for some forms there occurs a much higher loss of accuracy during evaluation than one would normally expect is that of ill-conditioning. Ill-conditioning is associated with a particular formulation of the mathematical problem. It exists independently of round-off effects and its nature is unaffected by various schemes for the control of round-off errors. Suppose that the Horner form of a polynomial which has been obtained as a mathematical model describing a certain physical process, has one (or more) coefficient, a, which is uncertain. By uncertainty in this context we mean that the coefficient is known only to lie within some interval $(a_h - \varepsilon_1, a_h + \varepsilon_2)$. The value of the polynomial, p, taken at some fixed point will then lie in the corresponding interval $(p - \eta_1, p + \eta_2)$. The quantities η_1 and η_2 are functions of ε_1 and ε_2 and hence one of the two possible situations occurs: (i) for 'small' ε_1 and ε_2, the quantities η_1 and η_2 will be 'small' as well; (ii) for 'small' ε_1 and ε_2, the quantities η_1 and η_2 will be very 'large'. (We use the terms 'small' and 'large' in the relative sense, hence the quotation marks.)

In case (i) we say that the particular polynomial form, i.e. the Horner form in our example, 'defines the polynomial with the accuracy of the coefficients'. Computationally, it means that provided a stable computational algorithm is used, a value of the polynomial may be computed with a degree of accuracy comparable to the accuracy of the coefficients. In case (ii) we say that the particular polynomial form does not define the polynomial with the accuracy of its coefficients. We also say that this polynomial form is ill-conditioned. For such a form, however carefully a computational evalutation algorithm may be chosen, it cannot alleviate the effects of ill-conditioning in the form. There is as yet no general theory about the circumstances which give rise to ill-conditioning in various problems, and each problem must be analysed individually.

Ill-conditioning of a polynomial form is a relative concept in the sense that here one is concerned with the question of how different the computed result is from what one would expect it to be, allowing normal round-off effects. A quantitative measure of ill-conditioning of a polynomial form at a point relates the number of arithmetic operations required by a particular algorithm and the number of digits used in a floating-point arithmetic to the number of accurate significant digits in the polynomial value (Rice, 1965).

A well-conditioned polynomial form $P(x)$ is characterized by the property that small errors incurred in the coefficients and in the argument x_0, result in a polynomial value $P(x_0)$ which is different from the value $P(x_0)$ by the amount comparable in size to the errors in the coefficients and in the x_0. In this context the Chebyshev and the root product forms have been shown to be well-conditioned. In Table 2.5.1 a brief summary is given of the conditioning of various polynomial forms.

Numerical stability of the polynomial evaluation algorithms has been studied by several authors. The Horner algorithm, in particular, has been examined very thoroughly (see, for example, Fox and Mayers (1968)). Other algorithms for polynomial evaluation have been analysed by Mesztenyi and Witzgall (1967), Hart (1968), Bakhvalov (1971), Newbery (1974). In Table 2.5.1 the results of these studies are summarized.

No computational algorithm based on an ill-conditioned form will accurately evaluate the polynomial since ill-conditioning is a function of the way in which a problem is formulated.

One may seek a different better-conditioned polynomial form instead but this is a very delicate problem as a particular polynomial form may be ill-conditioned not only in terms of its evaluation at a specified argument but also in terms of the problems associated with the transformation of one polynomial form into another. For example, the form may be ill-conditioned in terms of its root-finding. However if the evaluation algorithm based on a particular polynomial form is numerically unstable, while the form itself is well-conditioned, then it may be preferable to transform the form in question to some other form for which a stable evaluation algorithm is available. For example, in some cases it is advantageous to transform the conventional power form into an orthogonal form, e.g. the Chebyshev form, and use this to evaluate the polynomial. Ideally, of course, one should first ensure that the chosen polynomial form is well-conditioned and then select an evaluation algorithm that is numerically stable.

Table 2.5.1 Summary of the conditioning and numerical stability results for various polynomial forms

Conditioning of the form	Polynomial form	Stability of the algorithm
May be ill-conditioned	Horner	May be unstable
Always well-conditioned	Chebyshev	Both algorithms are stable
Comparatively well-conditioned	Root product	Always stable
Well-conditioned variants of Newton's form are obtainable, given a standard Newton's form, which displays symptoms of ill-conditioning. This however at a price of involving extra parameters and extra operations as compared with the standard form	Newton	Stability depends on the choice of the interpolation points. More or less equal distribution of these points over the interval of evaluation will usually give satisfactory stability. The interpolation points can also be chosen so as to guarantee maximum stability
May be ill-conditioned, by empirical evidence	Belaga	May be very unstable
Conjecture: If the polynomial is such that its Belaga and/or Pan coefficients are small then it is well-conditioned	Pan	May be extremely unstable

The cases when the algorithm based on the Chebyshev polynomial form will not perform better than the Horner algorithm has been studied by Newbery (1974). He has shown that when the Horner coefficients of the polynomial are of uniform sign or of strictly alternating sign, then the Horner algorithm, in terms of the upper bound on the computational error, may indeed be better—or, at least not worse—than the algorithm based on the Chebyshev form. Bakhvalov (1971) has given an instance when it is advantageous to transform a given Horner polynomial form into a Chebyshev form.

2.6 Error Analysis of Evaluation Algorithms

In this section we assume well-conditioned polynomial forms and for such forms present major results of the error analysis. Consider the following polynomial:

$$P_n(x) = \sum_{k=0}^{n} a_{n-k}x^k = \sum_{k=0}^{n} b_k T_k(x) = Q_n(x) \tag{2.6.1}$$

It is important to bear in mind that in the form $Q_n(x)$ x is given in the interval $[-1, 1]$ since the Chebyshev polynomials, $T_k(x)$, are defined in this interval only. The transformation when considering the form $Q_n(x)$ does not result in any loss of generality and may be readily carried out, cf. Newbery (1974).

We assume, further, that the a_{n-k}, b_k and x have round-off errors only, that is

$$\begin{aligned} \bar{a}_{n-k} &= \text{fl}(a_{n-k}) = a_{n-k}(1+\varepsilon_{n-k}), \\ \bar{b}_k &= \text{fl}(b_k) = b_k(1+\varepsilon_k), \\ \bar{x} &= \text{fl}(x) = x(1+\delta), \end{aligned} \tag{2.6.2}$$

where $|\varepsilon_{n-k}|$, $|\varepsilon_k|$, $|\delta|$ do not exceed 2^{-t}.

Let the exact value $P(x)$ of the polynomial with coefficients a_r at the point x be

$$P(x) = \sum_{k=0}^{n} a_{n-k}x^k. \tag{2.6.3}$$

and the exact value $\bar{P}(x)$ of the polynomial with coefficients $\bar{a}_r$ at the point x be given by

$$\bar{P}(x) = \sum_{k=0}^{n} \bar{a}_{n-k}x^k. \tag{2.6.4}$$

If $\bar{a}_{n-k} = a_{n-k}(1+\varepsilon_{n-k})$ and $|\varepsilon_{n-k}| \leqslant \varepsilon < 2^{-t}$ then the difference

$$\bar{P}(x) - P(x) = \sum_{k=0}^{n} a_{n-k}\varepsilon_{n-k}x^k,$$

giving

$$|\bar{P}(x) - P(x)| \leqslant \varepsilon\Sigma|a_{n-k}||x^k|. \tag{2.6.5}$$

For all polynomials this bound can be attained by taking $\varepsilon_{n-k} = \varepsilon\text{sign}(a_{n-k}x^k)$.

Now, let the exact value $P(\bar{x})$ of the polynomial with coefficients a_r at the point $\bar{x} = x(1+\delta)$ be given by

$$P(\bar{x}) = \sum_{k} a_{n-k}x^k(1+\delta)^k. \tag{2.6.6}$$

Then the error

$$P(\bar{x}) - P(x) = \sum_{k} k\delta a_{n-k}x^k, \tag{2.6.7}$$

which is obtained by retaining in the expression for $P(\bar{x})$ the first order terms only. We have

$$|P(\bar{x}) - P(x)| \leqslant \sum_{k} k\varepsilon|a_{n-k}||x^k| \quad \text{when } |\delta| \leqslant \varepsilon < 2^{-t},$$

and the bound can be attained only if all $a_{n-k}x^k$ are of the same sign. Note that in this case,

$$\frac{|P(\bar{x})-P(x)|}{|P(x)|}\leqslant n\varepsilon, \tag{2.6.8}$$

and hence the result then has a low relative error.

The computed value $\bar{P}(\bar{x})=\mathrm{fl}(\bar{P}(\bar{x}))$ obtained using floating-point arithmetic for the polynomial with coefficients $\bar{a}_r$ at the point $\bar{x}$ is given by

$$\bar{P}(\bar{x})=\mathrm{fl}(\bar{P}(\bar{x}))=\sum_k \bar{a}_{n-k}\bar{x}^k(1+\varepsilon_k)$$

with $|\varepsilon_k|\leqslant(2k+1)2^{-t}$ to first order.

Hence to first order

$$\begin{aligned}|\bar{P}(\bar{x})-P(x)| &\leqslant \Sigma|a_{n-k}||x^k|(3k+1)2^{-t}\\ &\leqslant (3n+1)2^{-t}\Sigma|a_{n-k}||x^k|\end{aligned} \tag{2.6.9}$$

with the same assumptions about $\bar{a}_r$ and $\bar{x}$ as above.

We notice from (2.6.9) that for all polynomials and all x this bound is merely $(2n+1)$ times as large as the bound resulting from the errors in the coefficients alone. However, while the error bound in the case of the errors in the coefficients alone was attainable the bound given by (2.6.9) is rarely if ever attainable.

We now follow a more detailed analysis of round-off error accumulation in various evaluation algorithms. First consider the Horner algorithm. By virtue of assumptions (2.6.2), we have

$$\bar{V}_n=\mathrm{fl}(a_0)=a_0(+\varepsilon_0) \tag{2.6.10}$$

and $\bar{V}_k=\mathrm{fl}(\bar{x}\bar{V}_{k+1}+\bar{a}_{n-k})$

$$\begin{aligned}&=[x(1+\delta)\bar{V}_{k+1}(1+\alpha_k)+a_{n-k}(1+\varepsilon_{n-k})](1+\beta_k)\\ &=x\bar{V}_{k+1}(1+\delta)(1+\alpha_k)(1+\beta_k)+a_{n-k}(1+\varepsilon_{n-k})(1+\beta_k).\end{aligned} \tag{2.6.11}$$

Hence to first order (and using T_k to denote $\bar{V}_k-V_k$):

$$\begin{aligned}\bar{V}_k=V_k+T_k &= x\bar{V}_{k+1}+xV_{k+1}(\delta+\alpha_k+\beta_k)\\ &\quad +a_{n-k}+a_{n-k}(\varepsilon_{n-k}+\beta_k)\\ &=(xV_{k+1}+a_{n-k})+xT_{k+1}+xV_{k+1}(\delta+\alpha_k+\beta_k)\\ &\quad +a_{n-k}(\varepsilon_{n-k}+\beta_k),\end{aligned}$$

wherefrom

$$T_k=xT_{k+1}+xV_{k+1}(\delta+\alpha_k+\beta_k)+a_{n-k}(\varepsilon_{n-k}+\beta_k),$$

and thus

$$|T_k|\leqslant|x||T_{k+1}|+(3|x||V_{k+1}|+2|a_{n-k}|)\varepsilon. \tag{2.6.12}$$

Here T_k denotes the absolute error in the computed value of V_k, and, as usual,

$|\delta|$, $|\alpha_k|$, $|\beta_k|$ and $|\varepsilon_{n-k}|$ are the errors, each less than or equal to $\varepsilon \leqslant 2^{-t}$, $k = n-1, \ldots, 0$.

The estimate (2.6.12) shows that the Horner algorithm is sensitive to the magnitude of x. However, the effect of error accumulation in the Horner process is relatively insignificant, while the power form representation itself is inherently ill-conditioned. Thus if one takes the function $f(x) = (x-b)(x-2b)\ldots(x-20b)$ with $b = 0.9987632$, say, there will not be a set of a_{n-k} on, say, an eight-digit decimal computer for which even the exact value of $\Sigma a_{n-k}x^k$ gives a good representation of $f(x)$. This surprising fact is of vital importance.

Next, for the root product algorithm we have:

$$
\begin{aligned}
\bar{V}_n &= \mathrm{fl}(a_0) = a_0(1+\varepsilon_0) \\
\bar{V}_k &= \mathrm{fl}([\bar{x} - \bar{\gamma}_{k+1}]\,\bar{V}_{k+1}) \\
&= \mathrm{fl}([x(1+\delta) - \gamma_{k+1}(1+w_{k+1})]\,(V_{k+1}+T_{k+1})) \\
&= [x(1+\delta) - \gamma_{k+1}(1+w_{k+1})]\,(1+\beta_k)(V_{k+1}+T_{k+1})(1+\alpha_k) \\
&= x(V_{k+1}+T_{k+1})(1+\delta)(1+\beta_k)(1+\alpha_k) - \gamma_{k+1}(V_{k+1}+T_{k+1}) \\
&\quad \times (1+w_{k+1})(1+\beta_k)(1+\alpha_k)
\end{aligned} \tag{2.6.13}
$$

Proceeding as before, by first retaining only linear error terms in (2.6.13) and then considering the differences between the exact and computed values of V_j's we get for the errors

$$
\begin{aligned}
|T_n| &= |\bar{V}_n - V_n| \leqslant |a_0|\,\varepsilon \\
|T_k| &= |\bar{V}_k - V_k| \leqslant |x - \gamma_{k+1}|\,|T_{k+1}| + 3\,|(x-\gamma_{k+1})V_{k+1}|\,\varepsilon
\end{aligned} \tag{2.6.14}
$$

where $\varepsilon \leqslant 2^{-t}$.

Further, using

$$|V_k| = |x - \gamma_{k+1}|\,|V_{k+1}| \tag{2.6.15}$$

we can write

$$|T_k| \leqslant 3\,|V_k|\,\varepsilon + \frac{|V_k|}{|V_{k+1}|}\,|T_{k+1}|, \tag{2.6.16}$$

and for the relative error we get

$$\frac{|T_k|}{|V_k|} \leqslant \frac{|T_{k+1}|}{|V_{k+1}|} + 3\varepsilon \tag{2.6.17}$$

We see that the algorithm *rootform* based on the root product form is completely stable.

It should however be kept in mind that although theoretically transformation into the root product form is always possible, finding the roots may be an unstable process itself, notwithstanding availability of a whole arsenal of techniques for polynomial root searching, and thus, the transformation may be of little practical help.

For the Newton algorithm a similar analysis gives the following bounds on the errors

$$|T_n| = |\bar{V}_n - V_n| \leqslant |a_0|\varepsilon$$
$$|T_k| = |\bar{V}_k - V_k| \leqslant |x - \beta_{k+1}|\,|T_{k+1}| + (4|(x-\beta_{k+1})V_{k+1}| + 2|a_{n-k}|)\varepsilon \tag{2.6.18}$$

where $\varepsilon \leqslant 2^{-t}$.

Bounds obtained show that the numerical stability of this evaluation algorithm depends on the choice of the interpolation points, β_k. More or less equal distribution of the β_k over the interval of evaluation will usually give satisfactory stability. The points β_k can also be chosen so at to guarantee maximum stability. If there are no zeros in the interval of evaluation, the stability will be complete.

The upper bound on the round-off error accumulated in the Chebyshev–Clenshaw algorithm may be derived using the following results due to Newbery (1974):

$$Q_n(x) = xV_1 - V_2 + b_0 \tag{2.6.19}$$

and

$$|\bar{V}_1 - V_1| \leqslant \frac{\sigma\left(\sum_k |\bar{b}_k|\right) n[1 + (1 + 2|x|)n(n+1)/2]}{1 - \sigma(1 + 2|x|)n(n+1)/2} \tag{2.6.20}$$

$$|\bar{V}_2 - V_2| \leqslant \frac{\sigma\left(\sum_k |\bar{b}_k|\right)(n-1)[1 + (1 + 2|x|)n(n-1)/2]}{1 - \sigma(1 + 2|x|)n(n-1)/2} \tag{2.6.21}$$

where $\sigma = \varepsilon(2+\varepsilon)$, $|\varepsilon| \leqslant 2^{-t}$.

We proceed using relations (2.6.19)–(2.6.21):

$$T_0 = |\bar{V}_0 - V_0| = |\mathrm{fl}(\bar{x}\bar{V}_1 - \bar{V}_2 + \bar{b}_0) - (xV_1 - V_2 + b_0)| \tag{2.6.22}$$

where

$$\begin{aligned}\bar{V}_0 &= \mathrm{fl}(\bar{x}\bar{V}_1 - \bar{V}_2 + \bar{b}_0)\\ &= \mathrm{fl}(x(1+\delta)(V_1+T_1) - (V_2+T_2) + b_0(1+\varepsilon_0))\\ &= [[x(1+\delta)(V_1+T_1)(1+\alpha_0) - (V_2+T_2)](1+\beta_0) + b_0(1+\varepsilon_0)]\\ &\quad\times(1+\xi_0)\\ &= x(V_1+T_1)(1+\delta)(1+\alpha_0)(1+\beta_0)(1+\xi_0) - (V_2+T_2)(1+\beta_0)\\ &\quad\times(1+\xi_0) + b_0(1+\varepsilon_0)(1+\xi_0)\end{aligned} \tag{2.6.22}$$

Using arguments similar to the Horner algorithm analysis, we derive

$$|T_0| \leqslant |x-1|\,|T| + (4|xV_1| + 2|V_2| + 2|b_0|)\varepsilon. \tag{2.6.23}$$

Recalling that here $-1 \leqslant x \leqslant 1$, we conclude that the Chebyshev–Clenshaw algorithm is always numerically stable. A similar analysis can be readily carried out for the Chebyshev–Bakhvalov algorithm.

Bakhvalov (1971) has given an upper bound on the round-off error for the Chebyshev–Bakhavalov algorithm:

$$|\bar{Q}_n(\bar{x}) - Q_n(x)| \leqslant Cq(x,n)\left(\sum_k |\bar{b}_k|\right)\log n\varepsilon \tag{2.6.24}$$

where $q(x,n) = \min\left\{\frac{1}{\sqrt{(1-x^2)}}, n\right\}, \quad -1 \leqslant x \leqslant 1,.$

and C is an absolute constant.

Using similar techniques and after somewhat tedious arithmetic manipulations the following error estimates for the Belaga algorithm may be given:

$$|T_1| = |\bar{V}_1 - V_1| \leqslant 4(|x|^2 + |\alpha_1 x|)\varepsilon, \tag{2.6.25}$$

$$\begin{aligned}|T_2| = |\bar{V}_2 - V_2| \leqslant{} & (2|V_1| + |x| + |\alpha_2| + |\alpha_3|)|T_1| + (5|V_1|^2 + 6|xV_1| \\ & + 5|\alpha_2 V_1| + 6|\alpha_3 V_1| + 7|\alpha_3 x| + 6|\alpha_2\alpha_3| + 2|\alpha_4|)\varepsilon,\end{aligned} \tag{2.6.26}$$

$$\begin{aligned}|T_{k+1}| = |\bar{V}_{k+1} - V_{k+1}| \leqslant{} & (|V_1| + |\alpha_{2k+1}|)|T_k| + (3|V_k V_1| + 4|V_k||x|^2 \\ & + 4|V_k||\alpha_1 x| + 4|V_k\alpha_{2k+1}| + 2|\alpha_{2k+2}|)\varepsilon\end{aligned} \tag{2.6.27}$$

where $\varepsilon \leqslant 2^{-t}$.

From (2.6.27) it follows that if a well-conditioned Belaga polynomial has 'small' coefficients, then the numerical stability of the Belaga algorithm is highly sensitive to the magnitude of x, since the error, T_k, grows as

$$|T_k| \leqslant P_2(|x|)|T_{k-1}| + A\varepsilon, \quad \varepsilon \leqslant 2^{-t}, \tag{2.6.28}$$

where $P_2(x)$ is a quadratic polynomial, parameter A depends on x and the Belaga coefficients as in (2.6.27).

For the Pan algorithm the following error estimates yield:

$$|E_0| = |\bar{W}_0 - W_0| \leqslant 4(|x|^2 + |\lambda_0 x|)\varepsilon, \tag{2.6.29}$$

$$|E_1| = |\bar{W}_1 - W_1| \leqslant 1.25E_0 + 2|x|\varepsilon. \tag{2.6.30}$$

Then, noting that

$$\bar{V}_0 = \mathrm{fl}(x) = x(1+\delta), |\delta| \leqslant 2^{-t},$$

an estimate for the error of the general term is shown to be:

$$|T_k| = |\bar{V}_k - V_k| \leqslant P_4(|x|)|T_{k-1}| + B\varepsilon, \qquad \varepsilon \leqslant 2^{-t}. \tag{2.6.31}$$

Here $P_4(x)$ is a fourth degree polynomial and parameter B is a function of x and the Pan coefficients.

From (2.6.31) we again can see that the Pan algorithm is extremely sensitive to the magnitude of x and may be very unstable indeed.

2.7 Evaluation of the Derivatives of a Polynomial

The need to evaluate the derivatives of a polynomial at a specific point arises in many contexts, a typical example being that of finding the roots of polynomial equations. The usual procedure in such cases is to use Horner's algorithm repetitively and we shall now show that to evaluate all the derivatives in this way requires $\frac{1}{2}n(n+1)$ multiplications and the same number of additions.

Given a polynomial of degree n,

$$P_n(x) = a_0x^n + a_1x^{n-1} + \ldots + a_n$$
$$= (\ldots((a_0x + a_1)x + a_2)x + \ldots + a_{n-1})x + a_n,$$

$n-1$ brackets

for the first derivative we write

$$\begin{aligned} P_n'(x) &= na_0x^{n-1} + (n-1)a_1x^{n-2} + \ldots + a_{n-1} \\ &= [a_{n-1} + x(a_{n-2} + x(a_{n-3} + \ldots + x(a_1 + a_0x)\ldots))] \\ &\quad + x[a_{n-2} + x(a_{n-3} + \ldots + x(a_1 + a_0x)\ldots)] \\ &\quad + x^2[a_{n-3} + x(a_{n-4} + \ldots + x(a_1 + a_0x)\ldots)] \\ &\quad + \ldots + x^{n-2}[a_1 + a_0x] + x^{n-1}[a_0]. \end{aligned} \tag{2.7.1}$$

In (2.7.1), each of the expressions in the square brackets is computed as an intermediate result in the Horner process.

Denoting the values in the square brackets by b_i, we can write

$$P_n'(x) = b_0x^{n-1} + b_1x^{n-2} + \ldots + b_{n-3}x^2 + b_{n-2}x + b_{n-1}$$
$$= (\ldots((b_0x + b_1)x + b_2)x + \ldots + b_{n-2})x + b_{n-1}.$$

We see that all the b's have been computed while evaluating the polynomial itself and a further application of the same procedure suffices to evaluate the derivative. The process can, obviously, be extended to compute in turn the higher derivatives, $P_n^{(i)}(x)$ for $i = 2, \ldots, n$.

Hence the algorithm which evaluates both a polynomial and its first derivative may be expressed as follows:

Algorithm evaluate

```
V[n] := a[0]
for k := n − 1 to 0 do
    V[k] := x × V[k + 1] + a[n − k]
enddo
W[n − 1] := x × V[n] + V[n − 1]
for k := n − 2 to 1 do
    W[k] := x × W[k + 1] + V[k − 1]
enddo
derivative := W[1]
polyvalue := V[0]
```

The algorithm uses $2n-1$ operations of each, multiplication and addition;

of these $n-1$ are required for the evaluation of the derivative. The algorithm is readily extended for evaluation of all derivatives of $P_n(x)$ and this would call for

$$n+(n-1)+(n-2)+\ldots+1=n(n+1)/2$$

of each, multiplications and additions.

Kirkpatrick (1972) has shown that provided the operation of division is excluded, a minimum of at least $2n-1$ additions is required to evaluate both a polynomial of degree n and its first derivative. The minimum number of multiplications, however, does not need to be as large as $2n-1$ and in fact can be reduced to $n+1$; see Borodin (1973).

An alternative and a more economical way to compute the derivatives has been proposed by Shaw and Traub (1974). They consider the so called normalized derivatives which are defined as $P_n^{(s)}(x)/s!$, where $P_n(x)$ is, as usual, a polynomial of degree n. The algorithm computes all the normalized derivatives of a polynomial in a total of $3n-2$ operations of multiplication and division and requires the same number of additions as in the 'repeated Horner' algorithm

Let $P_n(x)=\sum_{i=0}^{n} a_{n-i}x^i$,

then the Shaw–Traub algorithm is given as:

Algorithm shaw_traub

```
for i := 0 to n − 1 do
      V^{-1}[i] := a[i+1] × x^{n-i-1}
enddo
for j := 0 to n do
      V^j[j] := a[0] × x^n
enddo
for j := 0 to n − 1 do
      for i := j+1 to n do
            V^j[i] := V^{j-1}[i−1] + V^j[i−1]
      enddo
enddo
```

In the algorithm $V^j[i]$ are given as

$$V^j[i]=x^{n-i+j}\sum_{k=j}^{i}\binom{k}{j}a_{i-k}x^{k-j},$$

the $\binom{k}{j}$ being binomial coefficients.
Thus

$$V^j{}_n=x^j\sum_{k=j}^{n}\binom{k}{j}a_{n-k}x^{k-j}=\frac{P^{(j)}{}_n(x)}{j!}x^j,\quad j=0,1,\ldots,n-1,$$

where $P_n{}^{(0)}(x)=P_n(x)$.

To estimate the number of arithmetic operations involved note that $x^2, \ldots, x^n$ are obtained by $n-1$ multiplications and $a_0x^n, \ldots, a_{n-1}x$ are obtained by further n multiplications.

Hence the $V^j[n]$ requires $2n-1$ multiplications. The normalized derivatives

$$\frac{P_n^{(j)}(x)}{j!} = \frac{V_n^j}{x^j} \quad j = 1, \ldots, n-1 \tag{2.7.2}$$

can then be calculated in $n-1$ divisions.

(Note that $\dfrac{P_n^{(n)}(x)}{n!} = a_0$ and so need not be computed.)

Thus the algorithm *shaw_traub* and (2.7.2) yield point values of all the normalized derivatives as well as the polynomial itself in $3n-2$ multiplications and divisions, and $n(n+1)/2$ additions.

It has also been shown that the computation of the

$$V_n^j = \frac{x^j P^{(j)}{}_n(x)}{j!}, \quad j = 0, \ldots, n$$

in $2n-1$ multiplications as accomplished in the above algorithm optimizes the number of multiplications required by any algorithm for solving the problem; see Borodin and Munro (1975).

Wozniakowski (1974a) has shown that algorithm *shaw_traub* is numerically stable. In particular, he proved that under conditions of normalized floating-point arithmetic, the computed values of the V_i^j's are given by

$$\bar{V}_i^j = \mathrm{fl}(V_i^j) = x^{n-i+j} \sum_{k=j}^{i} \binom{k}{j} a_{i-k}(1 + \eta_{i,k}^j)x^{k-j},$$

where

$$|\eta_{i,-1}^{-1}| \leqslant (n-i-1)\varepsilon, \qquad i = 0, \ldots, n-1$$

$$|\eta_{j,j}^j| \leqslant n\varepsilon, \qquad j = 0, \ldots, n$$

$$|\eta_{i,k}^j| \leqslant \{2k + 1 - \delta_{i,k} - j + (n-i+j)\}\varepsilon, \qquad j = 0, \ldots, n; \quad i = j+1, \ldots, n; \quad k = j, \ldots, i$$

and $\delta_{i,k}$ denotes the Kronecker symbol.

2.8 Evaluation of a Polynomial with a Complex Argument

We now turn to the problem of evaluating a polynomial

$$P(z) = \sum_{k=0}^{n} a_{n-k} z^k$$

for a complex value of the argument z_0. The coefficients a_{n-k} are assumed to be real. The operations of addition and multiplication of complex numbers can be reduced to a sequence of arithmetic operations on real numbers in accordance with the following scheme: a real and a complex number are added in one addition, two complex numbers are added in two additions; the product of a real and a complex numbers requires two multiplications; the product of two complex numbers requires four multiplications and two additions, or three multiplications and five additions.

In the scheme, the operation of computing the product of two complex numbers in 3 real multiplications and five additions is less obvious than the rest of the scheme. The operation, however, readily follows from the identity (here $\mathrm{i} = \sqrt{-1}$)

$$(a + \mathrm{i}b)(c + \mathrm{i}d) = a(c + d) - (a + b)d + \mathrm{i}(a(c + d) + b - a)c)$$

$$\equiv \mathrm{ac} - bd + \mathrm{i}((a + b)(c + d) - ac - bd).$$

The identity was first suggested by Winograd (1970a).

An algorithm for evaluation of $P(z)$ at a fixed point $z = z_0$ is based on the following polynomial representation

$$P(z) = \sum_{k=0}^{n} a_{n-k} z^k = g(z)(z - z_0) + V_n, \tag{2.8.1}$$

where

$$g(z) = V_0 z^{n-1} + V_1 z^{n-2} + V_2 z^{n-3} + \ldots + V_{n-1}$$

and V_n is a constant.

The process of obtaining (2.8.1) is referred to as synthetic division of $P(z)$ by $(z - z_0)$, and $g(z)$ is called the reduced degree polynomial. From (2.8.1) it follows that provided the coefficients of $g(z)$ and the constant V_n are known, the polynomial value $P(z_0) = V_n$.

The evaluation algorithm for (2.8.1) is:

Algorithm complex_evaluate_one

$V[0] = a[0]$
for $k := 1$ **to** n **do**
$\quad V[k] := V[k-1] \times z[0] + a[k]$
enddo
polyvalue $:= V[0]$

It requires either $4n - 2$ multiplications and $3n - 2$ additions or $3n - 1$ multiplications and $6n - 5$ additions and is essentially the Horner algorithm. An alternative procedure for evaluating $P(z)$ is based on division of the polynomial by a quadratic factor, i.e. synthetic division of the form

$$P(z) = \sum_{k=0}^{n} a_{n-k} z^k = g(z) d(z) + V_{n-1} z + V_n, \tag{2.8.2}$$

where the divisor is

$$d(z) = (z - z_0)(z - \bar{z}_0)$$

and the quotient is

$$g(z) = V_0 z^{n-2} + V_1 z^{n-3} + \ldots + V_{n-2}.$$

Let point z_0 at which $P(z)$ is evaluated be expressed as $z_0 = x_0 + iy_0$. Equating like powers of z in (2.7.2) we obtain

$$\begin{aligned} a_0 &= V_0, \\ a_1 &= V_1 - 2x_0 V_0, \\ a_k &= V_k - 2x_0 V_{k-1} + (x_0^2 + y_0^2) V_{k-2}, \qquad k = 2, \ldots, n-2, \\ a_{n-1} &= -2x_0 V_{n-2} + (x_0^2 + y_0^2) V_{k-3} + V_{n-1}, \\ a_n &= (x_0^2 + y_0^2) V_{n-2} + V_n. \end{aligned} \tag{2.8.3}$$

Hence the following evaluation algorithm can be developed.

Algorithm complex_evaluate_two

```
V[0] := a[0]
p := x[0] + x[0]
q := x[0]^2 + y[0]^2
V[1] := V[0] × p + a[1]
for k := 2 to n − 1 do
    V[k] := V[k − 1] × p − V[k − 2] × q + a[k]
enddo
V[n] := − V[n − 2] × q + a[n]
polyvalue := V[n] × z[0] + V[n − 1]
```

The algorithm requires $2n + 2$ multiplications and $2n + 1$ additions, and is due to Knuth (1965). It is an improvement over the Horner-like algorithm for $n \geqslant 3$.

To evaluate a polynomial with complex coefficients at a complex argument, algorithm *complex_evaluate_one* requires either $4n$ multiplications and $4n$ additions, or $3n$ multiplications and $7n - 3$ additions. Knuth's algorithm *complex_evaluate_two* becomes less interesting with either $4n + 10$ operations of each, multiplication and addition, or with $4n + 9$ multiplications and $4n + 13$ additions.

2.9 Comments

The concept of preprocessing the coefficients was conceived in the 1950s. At the time it contributed significantly to the development of the analysis of algorithms. The design of the polynomial evaluation schemes which use the preprocessed coefficients manifested a breakthrough in scientific endavours to bring about efficient evaluation schemes by means of conceptually new representation forms. Since then the concept has found wide applications in many computation fields. The method of preprocessing is generally useful wherever one has the problem of continually re-evaluating a set of expressions where some of the parameters remain fixed.

We have concentrated on efficient evaluation of a polynomial at a given point. Other problems involving polynomials include efficient evaluation of a polynomial at more than one point, e.g. in a process of finding polynomial roots, polynomial interpolation, polynomial division, polynomial multiplication. The latter is closely related to the problem of efficient multiplication of matrices; for example, given an $m \times n$ matrix $\mathbf{X} = (x_{ij})$ and an $n \times r$ matrix $\mathbf{Y} = (y_{ij})$, their product $\mathbf{Z} = \mathbf{XY}$ means that

$$\mathbf{Z} = (z_{ik}), \quad \text{where } z_{ik} = \sum_{1 \leqslant j \leqslant n} x_{ij} y_{jk}, \quad 1 \leqslant i \leqslant m, \quad 1 \leqslant k \leqslant r.$$

This equation can be regarded as the computation of mr simultaneous polynomials in $mn + nr$ variables. The problem is then analysed with view to design an economic computation scheme. Fiduccia (1971) has developed a technique for multiplying an $m \times m$ matrix by two vectors in about $3m^2/2$ multiplications rather than $2m^2$. The algorithm may be used to evaluate a polynomial at any two points in $3n/2 + 0(\sqrt{n})$ multiplications. In

$$P_n(x) = \sum_{i=0}^{n} a_{n-i} x^i$$

without loss of generality assume that n is a perfect square. Write

$$\mathbf{A} = \begin{bmatrix} a_{n-1} & a_{n-2} & \cdots & a_{n-\sqrt{n}} \\ a_{n-(\sqrt{n}+1)} & & & \cdots \\ \vdots & & & \vdots \\ \cdots & & & a_0 \end{bmatrix}$$

and then if $\mathbf{X}_1$ and $\mathbf{X}_2$ are the points of evaluation,

$$\mathbf{X} = \begin{bmatrix} x_1 & x_2 \\ x_1^2 & x_2^2 \\ \vdots & \vdots \\ x_1^{\sqrt{n}} & x_2^{\sqrt{n}} \end{bmatrix}$$

$\mathbf{X}$ is found in $2\sqrt{n} - 2$ multiplications. Then $\mathbf{Y} = \mathbf{AX} = (y_{ij})$ may be found in roughly $3n/2$ more multiplications and

$$P_n(x_j) = a_n + y_{1j} + \sum_{k=2}^{\sqrt{n}} y_{kj} x_j^{(k-1)\sqrt{n}}, \ j = 1, 2$$

is computed in $2\sqrt{n} - 2$ more. The entire process takes $3n/2 + 0(\sqrt{n})$ multiplications.

However, the algorithm is not generally suitable as an efficient evaluation scheme because it requires the number of additions such that more than $4n$ total arithmetics are needed ($4n$ arithmetics suffices for two applications of the Horner scheme). The point of the algorithms like Fiduccia's is to illustrate that the number of specific arithmetics can nearly always be reduced but only at a price of expending more of some other type of operations. In many applications such pay-off situations are carefully examined in order to decide on the algorithm most efficient for the specified efficiency criteria.

Exercises

1. Given the polynomial $P_4(x) = a_0x^4 + a_1x^3 + a_2x^2 + a_3x + a_4$ and the Motzkin–Todd form, $P_4(x) = M_4(x) = a_0[[x(x+\alpha_0) + \alpha_1][x(x+\alpha_0) + \alpha_1 + x + \alpha_2] + \alpha_3]$, find the expressions for the preprocessed parameters α_0, α_1, α_2 and α_3 in terms of the coefficients a_0, a_1, a_2, a_3 and a_4, and write down the straight-line program for the Motzkin–Todd form, which would require three multiplications and five additions.
2. Knuth's form of the $P_4(x)$ is given as $P_4(x) = K_4(x) = a_0[[x(x+\alpha_0) + \alpha_1][x(x+\alpha_0) + \alpha_1 - x + \alpha_2] + \alpha_3]$. Find the expressions for the preprocessed parameters $\alpha_0, \alpha_1, \alpha_2$ and α_3 in terms of the coefficients a_0, a_1, a_2, a_3 and a_4. The two forms, $M_4(x)$ and $K_4(x)$, when used to evaluate the same $P_4(x)$ may give, in terms of the accuracy, significantly different computed results. Suggest some reasons, or perhaps an example, bearing this observation.
3. For the polynomial

$$P_5(x) = a_0x^5 + a_1x^4 + a_2x^3 + a_3x^2 + a_4x + a_5$$

Knuth has suggested the following form

$$P_5(x) = V_5(x) = [[[(x+\alpha_0)^2 + \alpha_1](x+\alpha_0)^2 + \alpha_2](x+\alpha_3) + \alpha_4)]\alpha_5,$$

where α_0 is obtained as a root of a certain cubic equation. Derive the cubic equation. Assuming that the cubic equation has only one real root (otherwise the form is not uniquely determined) obtain the expression for the preprocessed parameters in terms of the polynomial coefficients.
4. For the polynomial $P_6(x)$ of degree six the Pan form is given by

$$P_6(x) = A_6(x) = [[[x(x+\alpha_0) + \alpha_1 - x + \alpha_3][x(x+\alpha_0) + \alpha_1 + x + \alpha_2] + \alpha_4][x(x+\alpha_0) + \alpha_1] + \alpha_5]\alpha_6.$$

Verify that for this form the following holds:

$$\alpha_0 = a_1/3a_0 \text{ and}$$

$$\alpha_1 = (a_5 - \alpha_0 a_4 + \alpha_0^2 a_3 - \alpha_0^3 a_2 + 2\alpha_0^5)/(a_3 - 2\alpha_0 a_2 + 5\alpha_0^3),$$

then obtain the expressions for the remaining Pan parameters in terms of $a_0, a_1, \ldots, a_6$.

5. The polynomial $x^2 + y^2 + z^2 + t^2 - 2(xz - yt)$ can obviously be evaluated with seven multiplications. By suitable factoring show how the evaluation can be accomplished with only two multiplications.
6. Consider the polynomial $P_n(x) = \sum_{k=0}^{n} a_{n-k}x^k$ with real coefficients and a real argument x. Let the time complexity function $F(n)$ of an algorithm to evaluate $P_n(x)$ be the number of multiplications required to evaluate $P_n(x)$ for any set of $\{a_{n-k}\}$. Suppose it is known that $a_{n-k} = 0$ for all odd/even k. Construct an algorithm to take advantage of this restriction and analyse its complexity.
7. Certify that any algorithm which uses only $+$, $-$ and $\times$, and computes $\sum_{k=0}^{n} a_{n-k}x^k$ requires at least n additions.
8. Write a program which evaluates the polynomial $P_n(x) = \sum_{k=0}^{n} a_{n-k}x^k$ and all its derivatives by a repetitive use of the Horner scheme.
9. Prove that the algorithm of exercise 8 requires $n(n+1)/2$ operations of each multiplication and addition.
10. Estimate the total number of arithmetic operations required by each algorithm *complex_evaluate_one* and *complex_evaluate_two* in the case of a polynomial with complex coefficients, evaluated at a real value of the argument.
11. Give a straight-line program for the algorithm *shaw_traub* for $n = 6$.

3

Iterative Processes

3.1 An Iterative Process

The idea of *iteration* (from the Latin *iteratio* = repetition) is one of the most important concepts used in the numerical methods for solving mathematical problems. It is also called *successive approximation*. Let us illustrate the idea on the example of solving an equation of the form:

$$x = F(x). \tag{3.1.1}$$

We assume that the function $F(x)$ is differentiable.

In an iterative method we start with some initial approximation x_0, which for the majority of equations may be quite crude, and thereafter form a sequence of numbers using the following rule:

$$x_1 = F(x_0), \quad x_2 = F(x_1), \; \ldots \tag{3.1.2}$$

Each calculation of the type

$$x_{n+1} = F(x_n) \tag{3.1.3}$$

is called an *iteration*.

If the sequence of numbers obtained converges to a limiting value α, then

$$\lim_{n \to \infty} F(x_n) = F(\alpha). \tag{3.1.4}$$

Hence, $x = \alpha$ satisfies the equation $x = F(x)$.

We may say that with n increasing, the number x_n tends to the root α. An algorithm based on an iterative process is assumed to be completed when some prescribed accuracy for the approximation to the root is achieved.

In this chapter we shall be concerned with iterative methods for computing approximate solutions of the non-linear equation

$$f(x) = 0 \tag{3.1.5}$$

where attention is restricted to functions of a real variable that are real, single-valued, and possess a certain number of continuous derivatives in the neighbourhood of a real root, α, of (3.1.5).

One important group of non-linear equations is the set of polynomial equations, and iterative methods constitute a powerful group of methods for determining the roots of polynomials. One frequently needs to determine both the real and complex roots of a polynomial or to determine all the roots simultaneously, e.g. when stability problems are considered. Also, polynomial equations have a special structure as compared with general non-linear equations. These considerations call for methods which are specially suitable for solving polynomial equations and a number of special techniques have in fact been developed which are not suitable for the general equations. These specialized techniques will not be considered here. The evaluation of complex and multiple roots of a general non-linear equation and the solution of systems of non-linear equations will not be discussed here either.

Bisection Method

Probably the most simple procedure for approximating a real root of an equation is the following bisection method. Let a and b be two points such that $f(a)f(b) < 0$. Then the equation $f(x) = 0$ has at least one real root in (a, b). For simplicity, let us assume that $f(a) < 0$, $f(b) > 0$. A new approximation is now obtained by computing $(a + b)/2$ and finding the sign of $f((a + b)/2)$. If $f((a + b/2) = 0$, a root has been found, otherwise the following changes are made:

$$\text{if } f\left(\frac{a+b}{2}\right) > 0 \quad \text{then} \quad \frac{a+b}{2} \quad \text{replaces } b,$$

$$\text{if } f\left(\frac{a+b}{2}\right) < 0 \quad \text{then} \quad \frac{a+b}{2} \quad \text{replaces } a.$$

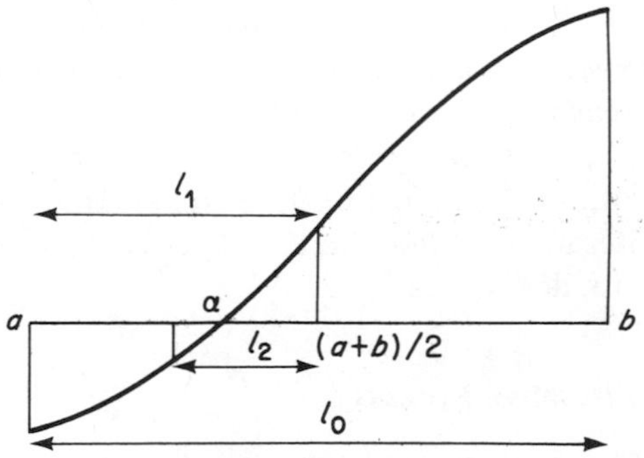

Figure 3.1.1 An illustration of the bisection method. l_0, l_1, l_2 ... is a sequence of diminishing intervals, $l_k = |b - a|/2^k$, containing the root α

The process is then repeated. We can see that the root will either be found by this procedure or be known to lie on an interval of length $|b-a|/2^m$ after the bisection operation has been applied m times. We may then take the approximate value of the root to be the midpoint of this last interval; this estimate can have a maximum error of $\frac{1}{2}|b-a|/2^m$, and to achieve this accuracy we need just m evaluations of $f(x)$.

In Fig. 3.1.1 an illustration is given of bisection method, and an implementation of the method is given below.

Algorithm bisection (root)

```
    initialize leftend and rightend
    initialize found and stop
    repeat
        root := (leftend + rightend)/2
        if f(root) = 0 then found := true
        else
            if f(leftend) × f(root) < 0
            then rightend := root
            else leftend := root
            endif
            compute new stop
        endif
    until (found or stop)
    if stop then root := (leftend + rightend)/2
```

In the algorithm the variables *leftend* and *rightend* are the left-hand and the right-hand endpoints of the root interval, respectively, and *found* and *stop* are stopping criteria variables. It is interesting to observe that in this procedure the accuracy to which the root can be found is limited only by the accuracy with which $f(x)$ can be evaluated.

Bisection method is guaranteed to *converge*. However since no use is made of the structure of $f(x)$—it is necessary only to be able to decide on the sign of $f(x)$—one can expect that a large number of function evaluations would be required to get a reasonably accurate result. At each iteration the gain in accuracy is one binary digit. Since $10^{-1} \approx 2^{-3.3}$, on average one decimal digit is gained in 3.3 iterations. It is most likely then that the convergence could be speeded up if the iterative method were to take account of the structure of $f(x)$, and possibly its derivatives.

Convergence of an Iterative Process

We shall study iterative methods of the general form

$$x_{n+1} = \phi(x_n, x_{n-1}, \ldots, x_{n-m+1}), \tag{3.1.6}$$

where ϕ is known as the *iteration function*.

Convergence of an iterative method is all important in the application of the method and is studied by means of error analysis. Provided the method converges, the next question is: How fast does the method converge? One way to interpret this question is to ask how many iterations it takes for the method to achieve the result of prescribed accuracy.

Definition

Let $x_0, x_1, x_2, \ldots$ be a sequence of values obtained using (3.1.6) which converges to α. Let further e_n be the error in the iterate x_n, such that $e_n = |x_n - \alpha|$.

If there exists a number p and a constant $C \neq 0$ such that

$$\lim_{n \to \infty} \frac{e_{n+1}}{e_n^{\,p}} = C, \tag{3.1.7}$$

then p is called the order of convergence of the sequence and C its asymptotic error constant.

For $p = 1$, 2, or 3 the convergence is said to be linear, quadratic, or cubic, respectively. The corresponding iteration function $\phi(x_n, x_{n-1}, \ldots, x_{n-m+1})$ is then said to be of order p.

The order p may be considered as a suitable measure of the improvement in accuracy of the approximation to the root which is gained at one iteration.

Assume, for example, that the magnitude of C in (3.1.7) is unity. Let x_n agree with the exact solution, α, to γ significant figures. Then, for n sufficiently large, x_{n+1} approximates α to $p\gamma$ significant figures. It is obvious that the improvement in accuracy will be most rapid when p is large and C small.

3.2 The Newton–Raphson Method: Convergence

In the bisection method shown in Fig. 3.1.1 the root is the point of intersection of the curve and the x axis. An approximation to the root can be obtained by replacing the function curve with a suitable straight line whose intersection with the x axis is easily computed. Starting with an arbitrary initial approximation to α, say, x_0, we then compute an iterative sequence $x_0, x_1, x_2, \ldots$.

This way different iterative methods can be derived depending on how the direction of the straight line is chosen. One choice for the direction of the straight line is that of the tangent to the curve at the current root estimate, as shown in Fig. 3.2.1. The iterative method based on this idea is the celebrated Newton–Raphson method. It is defined as follows.

$$x_{n+1} = \phi(x_n) = x_n - \frac{f(x_n)}{f'(x_n)}, \quad x_0 \text{ arbitrary}. \tag{3.2.1}$$

An algorithmic form of the method is given below.

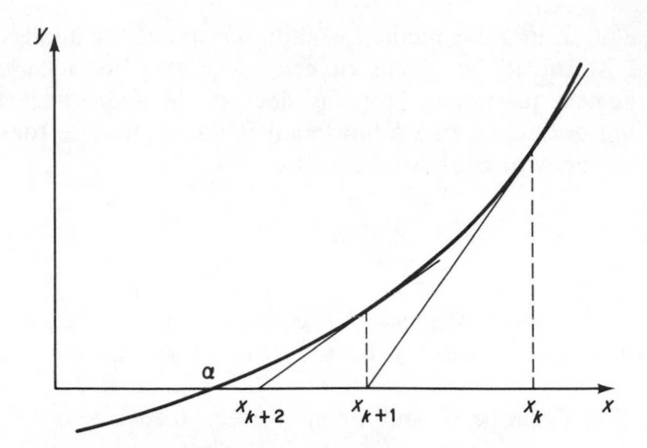

Figure 3.2.1 The Newton–Raphson method for a convex function

Algorithm newton (root)

initialize root
initialize stop
while not *stop* **do**
compute f(root)
compute f' (root)
root := root − f(root)/f' (root)
compute new step
enddo

To study the converge properties of this method, we asume that $f(x)$ is twice differentiable, and we keep in mind our earlier general assumption that the root α is simple, that is $f'(\alpha) \neq 0$, from which it follows that $f'(x) \neq 0$ for all x in some vicinity of the root α.

We now establish the relation between the errors e_n and e_{n+1} under the above assumptions.

The Taylor series expansion about x_n gives

$$0 = f(\alpha) = f(x_n) + (\alpha - x_n)f'(x_n) + \tfrac{1}{2}(\alpha - x_n)^2 f''(\xi), \quad \xi \in (x_n, \alpha).$$

It follows that

$$\alpha - x_{n+1} = \frac{f(x_n)}{f'(x_n)} + \alpha - x_n = -\frac{1}{2}(\alpha - x_n)^2 \frac{f''(\xi)}{f'(x_n)}. \tag{3.2.2}$$

Thus

$$e_{n+1} = \frac{1}{2} e_n{}^2 \frac{f''(\xi)}{f'(x_n)} \tag{3.2.3}$$

and if $x_n \to \alpha$, we have

$$\frac{e_{n+1}}{e_n^2} \to \frac{1}{2}\frac{f''(\alpha)}{f'(\alpha)}. \tag{3.2.4}$$

Formula (3.2.4) shows that e_{n+1} is approximately proportional to e_n^2, that is provided the Newton–Raphson method converges its convergence is of order 2.

Let us now study the conditions under which the method converges. Assume that I is a vicinity of the root α, such that

$$\frac{1}{2}\left|\frac{f''(y)}{f'(x)}\right| \leqslant M \qquad \text{for all } x, y \in I \tag{3.2.5}$$

If $x_n \in I$ then $|e_{n+1}| \leqslant Me_n^2$, which can be written as

$$|Me_{n+1}| \leqslant (Me_n)^2.$$

Assume that $|Me_0| < 1$ and that the interval $[\alpha - e_0, \alpha + e_0]$ is contained in I. By induction it can be shown that $x_n \in I$ for all n and that

$$|e_n| \leqslant \frac{1}{M}(Me_0)^{2^n} \qquad \text{for } n > 0. \tag{3.2.6}$$

It follows that the Newton–Raphson method converges provided

$$|Me_0| = M|x_0 - \alpha| < 1,$$

that is provided that the initial approximation, x_0, is chosen sufficiently close to α.

Another reason for a possible non-convergence of the method is the situation when at some stage in the iteration process two iterates, x_k and x_{k+1}, will be computed alternately, without any further progress in approaching the root. This case is demonstrated in Fig. 3.2.2.

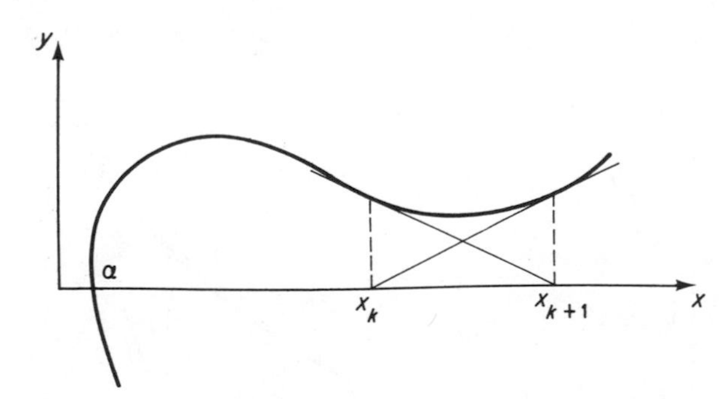

Figure 3.2.2 A case of non-convergence of the Newton–Raphson method. The iterations after $(k+1)$st will produce two points, x_k and x_{k+1}, alternately, without any further progress in approaching root α

Secant Method: Convergence

If in the Newton–Raphson method we replace the derivative $f'(x_n)$ by the difference ratio

$$\frac{f(x_n) - f(x_{n-1})}{x_n - x_{n-1}},$$

another iterative method is obtained, that is

$$x_{n+1} = x_n - f(x_n)\frac{x_n - x_{n-1}}{f(x_n) - f(x_{n-1})}, \quad f(x_n) \neq f(x_{n-1}). \tag{3.2.7}$$

The method is called the secant method and is one of the oldest methods known for the solution of $f(x) = 0$. After some period of neglect, its use has been revived since the method is particularly convenient for use on a computer. The method is illustrated in Fig. 3.2.3. Here x_{n+1} is determined as the absissa of the intersection between the secant through $(x_{n-1}, f(x_{n-1}))$ and $(x_n, f(x))$ and the axis.

To study the relation between the errors at two consecutive steps in the secant method, consider the Newton interpolation formula for the points x_n and x_{n-1}:

$$f(x) = f(x_n) + (x - x_n)\frac{f(x_n) - f(x_{n-1})}{x_n - x_{n-1}} + \frac{1}{2}(x - x_{n-1})(x - x_n)f''(\xi), \tag{3.2.8}$$

where $\xi \in$ the smallest interval containing x, x_{n-1}, x_n.

The secant method is obtained by ignoring the remainder term in this formula, and choosing x_{x+1} so as to satisfy the equation

$$0 = f(x_n) + (x_{n+1} - x_n)\frac{f(x_n) - f(x_{n-1})}{x_n - x_{n-1}}, \tag{3.2.9}$$

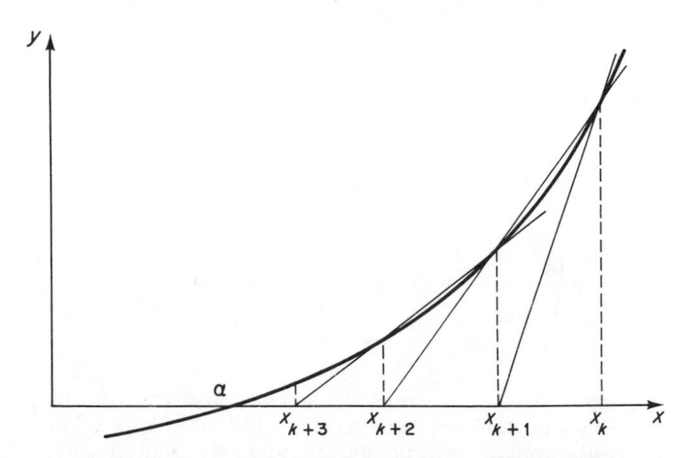

Figure 3.2.3 The secant method for a convex function

Setting $x = \alpha$ in (3.2.8) and using equation (3.2.9) we obtain

$$(\alpha - x_{n+1}) \frac{f(x_n) - f(x_{n-1})}{x_n - x_{n-1}} + \frac{1}{2}(\alpha - x_n)(\alpha - x_{n-1}) f''(\xi) = 0, \tag{3.2.10}$$

since $f(\alpha) = 0$. By virtue of the mean value theorem we have

$$\frac{f(x_n) - f(x_{n-1})}{x_n - x_{n-1}} = f'(\xi'), \tag{3.2.11}$$

where $\xi' \in$ the smallest interval containing x_{n-1} and x_n. It follows that

$$e_{n+1} = \left| \frac{f''(\xi)}{2f'(\xi')} \right| e_n e_{n-1}. \tag{3.2.12}$$

The secant method converges for sufficiently good initial values, x_0 and x_1, provided that $f'(\alpha) \neq 0$ and $f(x)$ is twice continuously differentiable.

Assuming now that the secant method converges, for very large n, we may take $\xi \simeq \alpha$, $\xi' \sim \alpha$, and hence

$$|e_{n+1}| \approx C |e_n| |e_{n-1}|, \quad \text{where } C = \left| \frac{1}{2} \frac{f''(\alpha)}{f'(\alpha)} \right|. \tag{3.2.13}$$

Here the constant C is the same as in Newton–Raphson's method. To establish the order of convergence, p, of the secant method we make the following conjecture:

$$|e_{n+1}| \approx K |e_n|^p,$$
$$|e_n| \approx K |e_{n-1}|^p.$$

Substituting the proposed relation between the errors into (3.2.13) we obtain

$$K |e_n|^p \approx C |e_n| K^{-1/p} |e_n|^{1/p}.$$

This relation is valid only if $p = 1 + (1/p)$, that is if $p = \frac{1}{2}(1 \pm \sqrt{5})$, and if $C = K^{1 + (1/p)} = K^p$.

We thus have

$$|e_{n+1}| \approx C^{1/p} |e_n|^p \quad \text{for } p = \tfrac{1}{2}(1 \pm \sqrt{5}). \tag{3.2.14}$$

Now, the value $p = \frac{1}{2}(1 - \sqrt{5}) = -0.618\ldots$ implies that the error of the current iteration, $|e_{n+1}|$, depends inversely on the error of the preceding iteration, $|e_n|$. Since the error $|e_n|$, in general, is very small, for example it is much smaller than unity, its reciprocal, $|e_n|^{-1}$ may be large, which in turn implies that the error $|e_{n+1}|$ will grow uncontrollably with n increasing. This contradicts the assumptions on which the proof is based, that the method is convergent. Hence, the proof is valid only for $p = \frac{1}{2}(1 + \sqrt{5})$.

We conclude that

$$|e_{n+1}| \approx C^{1/p} |e_n|^p, \quad p = \tfrac{1}{2}(1 + \sqrt{5}) = 1.618\ldots, \; n \gg 1.$$

We shall confine ourselves to this heuristic discussion of the order of

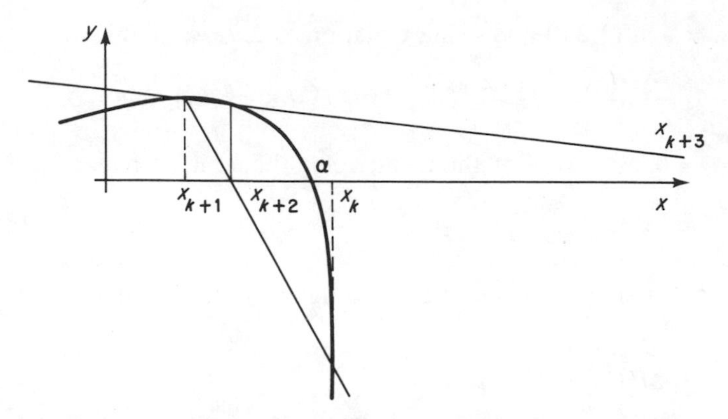

Figure 3.2.4 An example of the divergent secant method

convergence; a rigorous proof of the above result has been given by Ostrowski (1966). We wish once again to emphasize that a good initial approximation to the root is essential for the secant method to converge. It is not difficult to find examples of functions for which the method would give a divergent sequence of approximations. One such example is shown in Fig. 3.2.4.

Method of False Position (*Regula Falsi*)

A simple modification of the Secant method will produce a method which is always convergent. Suppose that the two initial approximations can be chosen so that the two function values have opposite signs. It is possible then to generate a sequence of values which always possesses this property. The value of x_2 is found as the intersection between the chord joining $f(x_0)$ and $f(x_1)$, and the x-axis. Then choosing x_2 and x_i, where i is either 0 or 1, so as to make $f(x_2)f(x_i) < 0$, the above procedure is repeated to obtain x_3 and so on. The method of false position is given as

$$x_{n+1} = x_n - f(x_n)\,\frac{x_n - x_{n-1}}{f(x_n) - f(x_{n-1})}, \quad \text{where } f(x_n)f(x_{n-1}) < 0.$$

An illustration of the method is shown in Fig. 3.2.5.

The penalty involved in the modification of the method is to sacrifice the high rate of convergence of the secant method. As soon as an interval is reached on which the function is convex or concave, thereafter one of the end-points of this interval is always retained and this feature slows down convergence to first order. This is, of course, the price one would expect to have to pay for a guaranteed convergence.

Various modifications of the *regula falsi* basic scheme which lead to a considerable enhancement in speed of convergence without the guarantee of convergence being lost, are discussed in an excellent paper by Jarratt (1973).

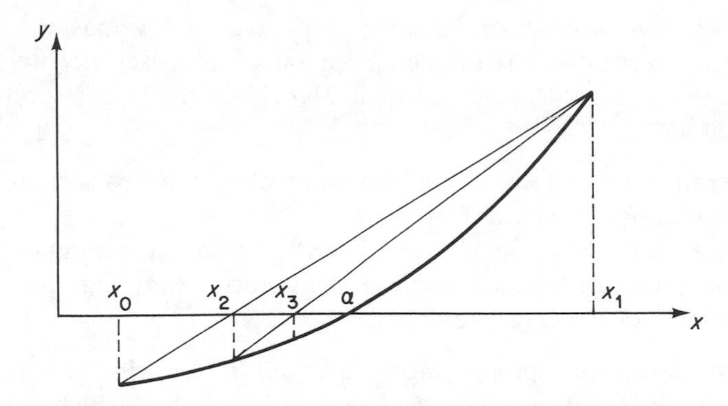

Figure 3.2.5 The method of false position for a convex function

3.3 Formulation of the Optimality Problem

Using the order p as a measure of the relative efficiency of an iterative method (e.g. p shows how fast the method converges) one intuitively feels that the Newton–Raphson method will be 'faster' than the secant method in terms of the number of iterations required to arrive at a given accuracy of the root value. Simple examples show this to be true at least for some problems (e.g. the well-separated simple roots of well-conditioned polynomials). It is not difficult, however, to find examples which contradict the above conjecture. Consider the problem of finding a root of $g(x)=0$ where

$$g(x)=x-\cos\left(\frac{0.785-x\sqrt{(1+x^2)}}{1+2x^2}\right).$$

To compute the root $x=0.97989961$ requires 8 iterations by the Newton–Raphson method and 5 iterations by the Secant method. (As a matter of interest, it takes 5 iterations to compute the same root by the Traub method, which is given in Section 3.7). In our case the initial approximations were taken as $x_0=0$, $x_1=0.1$ (the Secant method needs two starting values).

Such 'unexpected behaviour' of iterative methods is explained by the fact that the order of a method is a property local to the neighbourhood of the root. It therefore measures how good the method is when it is near the convergence, while the total number of iterations depends on the initial approximations to the root.

It is also important to realize that the time taken to execute one iteration on a digital computer varies from method to method. Among other things, this time depends on the number of function evaluations per iteration and on the number of arithmetic operations necessary to combine the function values to form the iterative formula. It, thus, makes more sense to compare the computational efficiency of various methods by measuring how much computation must be done to compute an approximation to the root to a prescribed numerical accuracy. Efficiency of the root-finding methods can be discussed in terms of (i) the number of iteration steps required versus the numerical accuracy achieved; and (ii) the enhancement of accuracy per iteration step versus the iteration 'cost'. The 'optimal' algorithm may then be thought of as that for either

(a) the iterative method which would guarantee a prescribed accuracy after the fewest number of iteration steps; or
(b) the iterative method which would need the minimum amount of work, e.g. the number of arithmetic operations per iteration step, required to produce an approximation of a prescribed accuracy.

These two forms of optimal model are not necessarily contradictory or independent; indeed if one could find an iterative method which combined the least possible amount of work per iteration and the fewest number of iteration steps to give a numerical approximation of prescribed accuracy, then such a method would clearly be ideal. The optimal requirements formulated above, however, prompt the observation that any reasonable efficiency measure of an iterative method should be a function both of the parameters set by condition (a) and of those set by condition (b). This suggests that an efficiency measure should be maximized for these parameters.

The complete optimization problem for the universe of root searching iterative techniques has yet to be solved. We shall now study some recent development in this area.

First, it is convenient to distinguish two groups of iterative methods. One group consists of those methods that are designed to solve root-finding problems where enough is known about the root location, to enable one to choose starting values such that the convergence of the method adopted would not be likely to present any difficulties. Then with every new iteration step, one 'marches' closer and closer to the root until sufficiently accurate approximation is obtained.

The Newton–Raphson and secant methods are examples of this group.

In cases where the *a priori* information on the location of the root is poor, some guarantee of the convergence of the method becomes of paramount importance. This has led to the development of a number of algorithms which are relatively insensitive to the choice of starting values. Most of these so called 'robust' methods are based on the technique of searching until two values of x are found for which $f(x)$ possesses opposite signs. The root is then bracketed between these values, and the iterations are repeated so that the brackets are preserved. The bisection algorithm of Section 3.1 is an example of such a method.

For methods in this group, provided the tolerance ε is known, an estimate of the upper bound can be given of the number of iterations required to achieve an adequate solution. For instance, in bisection method, starting with some initial root interval $[a, b]$, one repeatedly halves it. After n iterations, it is easy to see that the root will lie in an interval of length $|b-a|/2^n$. We deduce that the method guarantees an approximation to the root, with a tolerance $\pm\varepsilon$, after $n = \log_2|(b-a)/\varepsilon|$ iterations. Clearly this does not take into account round-off errors which may make the computed result somewhat worse. We shall refer to such an estimate of the number of iterations as the guaranteed convergence number of the iterative method. Efficiency of different methods may then be compared in terms of the guaranteed convergence number.

We shall now investigate further methods for finding a zero of a function defined on an interval.

3.4 Iterative Methods with Guaranteed Convergence

Bisection and the method of false position are two examples of iterative methods suitable for finding a zero of a function $f(x)$ defined in an interval $[a, b]$ and such that $f(a)f(b) < 0$. Using these basic techniques several hybrid algorithms have been developed which for the majority of problems converge faster than either of the two basic methods. One such method is due to Dekker (1969). The method is shown in Fig. 3.4.1.

The sequence of computations corresponding to the case shown in Fig. 3.4.1 is as follows.

Initially the root is known to lie in $[a, b]$.

Set $x_0 = a$ and $x_1 = b$.

//The root interval is $[x_0, x_1]$.//

Compute $f((x_0 + x_1)/2)$.

Compute $x_2 = x_1 - \dfrac{f(x_1)(x_1 - x_0)}{f(x_1) - f(x_0)}$

Compute $f(x_2)$.

//The new root interval is $[x_0, x_2]$,

since $|f(x_2)| \leqslant |f((x_0 + x_1)/2)|$ and $f(x_2)f(x_0) < 0$.//

Compute $f((x_0 + x_2)/2)$

Compute $x_3 = x_2 - \dfrac{f(x_2)(x_2 - x_1)}{f(x_2) - f(x_1)}$

Compute $f(x_3)$.

//The new root interval is $[x_0, (x_0 + x_2)/2]$,

since $f((x_0 + x_2)/2) \leqslant f(x_3)$ and

$f((x_0 + x_2)/2)f(x_0) < 0$.//

etc.

The general iteration step in Dekker's method involves two function evaluations, one at the midpoint of the current root interval and the other at the cross-section point (as in secant method). The new endpoint is chosen so as to achieve the largest reduction of the root interval. The following algorithm implements Dekker's method.

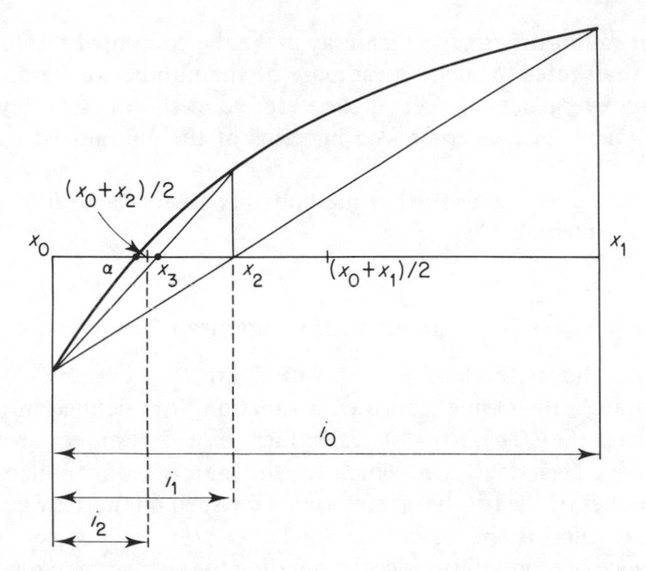

Fig. 3.4.1. Dekker's zero-finding method for a non-linear convex function. At each iteration the function is evaluated at two points, the midpoint of the current root interval and the cross-section point (as in secant method), and the new endpoint is chosen so as to achieve maximum reduction of the root interval

Algorithm dekker (root)

> *initialize leftend and rightend*
> *initialize stop*
> **while not** *stop* **do**
>> *compute* $f((leftend + rightend)/2)$
>>
>> *compute* $root := rightend - \dfrac{f(rightend) \times (rightend - leftend)}{f(rightend) - f(leftend)}$
>>
>> *compute* $f(root)$
>> **if** $ABS(f(root)) \leqslant ABS(f((leftend + rightend)/2))$
>> **then**
>>> **if** $f(root) \times f(leftend) < 0$
>>> **then** *rightend* := *root*
>>> **else** *leftend* := *root*
>>> **endif**
>>
>> **else**
>>> **if** $f((leftend + rightend)/2) \times f(leftend) < 0$

```
            then rightend := (leftend + rightend)/2
            else leftend := (leftend + rightend)/2
            endif
        endif
        compute new stop
    enddo
    root := (leftend + rightend)/2
```

The order of convergence of Dekker's algorithm is 1.618, and according to Brent (1971a), the algorithm performs much better than bisection on well-behaved functions, such as polynomials of moderate degree with well-separated zeros. In fact, the algorithm has proved satisfactory in most practical cases. However, the algorithm's guaranteed convergence number is $n = |(b-a)/\varepsilon|$ which compares unfavourably with the bisection method. Indeed Brent gives an example when this full number of iterations is required. Such slow convergence may occur in problems where, at each iteration, the point computed using linear interpolation (in secant method) has to be taken as the next approximation to the root. It may then happen that at each step, the length of the root interval is reduced by no more than an amount equal to ε, the prescribed absolute tolerance. The number of iterations required to achieve an approximation to the root within this prescribed tolerance is then given by the relation

$$|b - a - n\varepsilon| \leqslant \varepsilon,$$

giving in turn

$$n = \left|\frac{b-a}{\varepsilon}\right|. \tag{3.4.1}$$

Brent has suggested a modification to the Dekker algorithm which ensures that a bisection is done at least once in every $2\log_2|(b-a)/\varepsilon|$ consecutive iterations with the result that the modified algorithm is never much slower than bisection.

Another modification of the Dekker algorithm due to Brent (1971b) combines linear interpolation and inverse quadratic interpolation with bisection. This modified algorithm retains the same order of convergence as Dekker's algorithm and guarantees convergence to the root within tolerance ε in $n = [\log_2|(b-a)/\varepsilon| + 1]^2 - 2$ iterations.

In 1974 Bus and Dekker developed a new algorithm for finding the zero of a function within a given interval which uses both bisection and rational approximation technique. Iterative methods in which a rational function is fitted through previously computed values were first studied by Jarratt and Nudds (1965) and Jarratt (1966a, 1966b). For example, a useful practical technique may be obtained using the three-parameter bilinear form

$$g(x) = \frac{x - r}{px + q} \tag{3.4.2}$$

such that $g(x_\nu) = f(x_\nu)$, $\nu = i,\ i-1,\ i-2$, where x_ν are previously computed approximations to the root.

Using bisection and a rational function of form (3.4.2), Bus and Dekker derived an algorithm within the order of convergence equal to 1.84 and a guaranteed convergence number equal to $n = 4 \log_2 |(b-a)/\varepsilon|$.

We would also like to mention an algorithm due to Krautstengel (1968). It is based on the use of linear interpolation (*regula falsi*) and linear extrapolation (secant method) and is illustrated in Fig. 3.4.2. The class of functions to which the algorithm may usefully be applied is restricted by rather severe conditions which require that depending on whether the function is convex or concave on $[a, b]$, one of the following conditions holds

$$\text{(i)} \quad \frac{f(b)}{b-a} \leqslant f'(b),$$

$$\text{(ii)} \quad -\frac{f(a)}{b-a} \leqslant f'(a). \tag{3.4.3}$$

The illustration shown in Fig. 3.4.2 is given for a convex function that satisfies

$$f(a) < 0,\ f(b) > 0,$$
$$f'(x) > 0,\ f''(x) > 0,\ x \in [a, b]$$

$$\text{and } -\frac{f(a)}{b-a} \leqslant f'(a).$$

It is shown by Krautstengel that when used on functions that satisfy conditions (3.4.3), the algorithm converges with the order $1 + \sqrt{2}$. The guaranteed

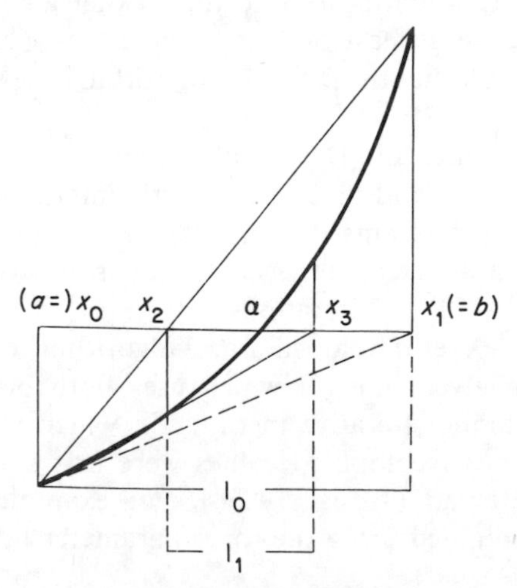

Figure 3.4.2 The Krautstengel method for a convex function

Table 3.4.1 Parametric characteristics of root finding methods with guaranteed convergence

Method	Order of convergence	Guaranteed convergence number
Bisection	1	$\log_2 \left\lvert \frac{b-a}{\varepsilon} \right\rvert$
Secant and *regula falsi*	1.618	$\left\lvert \frac{b-a}{\varepsilon} \right\rvert$
Dekker's (bisection + secant)	1.618	$\left\lvert \frac{b-a}{\varepsilon} \right\rvert$
Dekker–Brent (bisection + secant; with bisection ensured in every $2 \log_2 \left\lvert \frac{b-a}{\varepsilon} \right\rvert$ consecutive iterations)	1.618	$\log_2 \left\lvert \frac{b-a}{\varepsilon} \right\rvert$
Brend–Dekker (bisection + secant + inverse quadratic interpolation)	1.618	$\left[\log_2 \left\lvert \frac{b-a}{\varepsilon} \right\rvert + 1\right]^2 - 2$
Bus–Dekker (bisection + rational approximation)	1.84	$4 \log_2 \left\lvert \frac{b-a}{\varepsilon} \right\rvert$

convergence number of the algorithm has not yet been determined. Table 3.4.1 gives a brief summary of the characteristics of the robust root finding methods discussed.

Optimality of Root Finding Algorithms with Guaranteed Convergence

We have seen that in root finding methods that are suitable for functions which change sign in the root interval, the basic iteration step consists of calculating a value of $f(x)$ at some point in the interval and comparing it with the previously obtained value or values. With such methods, therefore, the number of iteration steps is usually the same as the number of function evaluations required. Furthermore, the guaranteed convergence of the method gives the maximum number of funciton evaluations required to achieve an *a priori* prescribed numerical accuracy of the approximation to the root. This number may therefore be considered as one possible criterion with respect to which an 'optimal' method can be formulated. That is to say, the 'optimal' method can be identified by the minimum value of the maximum number of function evaluations required to obtain an approximation to the root within a fixed tolerance ε and starting with a fixed root interval. This problem belongs to the general class of optimization problems known as minimax problems and we shall refer to it as the *guaranteed convergence minimax problem*. It still awaits complete solution.

Another optimization problem can be formulated as follows: Assuming that the initial root interval and the number of function evaluations are given, find the iterative method which computes the best numerical approximation to the root. This latter problem can also be described as that of minimizing the maximum length of the interval which can be guaranteed to include the desired zero, after a fixed number of function evaluations, performed sequentially. It thus is another minimax problem and we shall refer to it as the *numerical accuracy minimax problem*.

It has been shown that bisection is an optimal algorithm in the sense of the numerical accuracy minimax problem, for finding zeros of functions which change sign in the root interval, see Brent (1971a). For functions which satisfy the Lipschitz condition with a known constant K, i.e.

$$|f(x') - f(x'')| \leqslant K|x' - x''|, \quad x', x'' \in [a, b] \tag{3.4.4}$$

the optimality of bisection has been proved by Sukharev (1976) and Chernous'ko (1968).

Assuming the root interval $[a, b]$ and $f(a)f(b) < 0$, an optimal algorithm is given in the form:

$$b_0 = b$$

$$a_0 = b,$$

$$x_1 = \frac{a_0 + b_0}{2},$$

$$a_k = \begin{cases} a_{k-1}, & \text{if } f(x_k) > 0, \\ x_k - \dfrac{f(x_k)}{K}, & \text{if } f(x_k) < 0, \end{cases} \tag{3.4.5}$$

$$b_k = \begin{cases} x_k - \dfrac{f(x_k)}{K}, & \text{if } f(x_k) > 0, \\ b_{k-1}, & \text{if } f(x_k) < 0, \end{cases}$$

$$x_{k+1} = \frac{a_k + b_k}{2}, \qquad k = 1, 2, \ldots .$$

The current root intervals are computed using condition (3.4.4).

Sukharev (1976) has also shown that for the class of functions which satisfy condition (3.4.4), the bisection method given by (3.4.5) is optimal in the sense of the so-called W-criterion. The W-criterion is defined as

$$W(X^r, Y^r) = \min_{x \in [a,b]} \max_{f \in L, f(X^r) = Y^r} |f(x)| \tag{3.4.6}$$

where X^r and Y^r denote the sets $\{x_i, i=0, 1, \ldots, r\}$
and $\{y_i = f(x_i), i=0, 1, \ldots, r\}$ respectively,
and '$f \in L, f(X^r) = Y^r$' reads as 'all the functions which satisfy the Lipschitz condition and such that $f(x_i) = y_i$, $i = 0, 1, \ldots, r$'.

A 'W-optimal' root-finding algorithm guarantees that after n iteration steps there will be found a point, x_{n+1}, at which the absolute value of the function will not exceed some tolerance w, and also that there exists no other algorithm which, with the same number of iteration steps, would guarantee the point, x_{n+1}, at which the absolute value of the function would not exeed any $w' < w$.

However, bisection does not yield optimality conditions for various suitably restricted classes of allowable functions. For example, for convex functions, Bellman and Dreyfus (1961) (see also Gross and Johnson 1959), have developed an algorithm which is 'better' than bisection in the sense of the numerical accuracy minimax problem.

3.5 Complexity Parameters and Measures of Efficiency

We now proceed to study computational complexity of the general class of iterative methods given by (3.1.6). We shall define an optimal root finding process as a method with the 'lowest cost' iteration step. The iterative methods will be considered under assumption of a good initial approximation to the root. We shall follow the work of Traub and his co-workers. Consider the two iterative methods.

The Newton–Raphson iteration step involves the following operations:

Evaluate $f(x_{i-1})$

Evaluate $f'(x_{i-1})$

Compute $x_i = x_{i-1} - f(x_{i-1})/f'(x_{i-1})$.

In one iteration, given the value x_{i-1}, one function and one derivative evaluation is carried out, and the data obtained is 'combined' to compute the value x_i.

The secant iteration step consists of the following:

Evaluate $f(x_{i-1})$

Compute $x_i = x_{i-1} - f(x_{i-1}) \dfrac{x_{i-1} - x_{i-2}}{f(x_{i-1}) - f(x_{i-2})}$

In one iteration, given two iterates, one function evaluation is carried out and then the data is 'combined' to compute x_i.

In order to relate explicitly the iteration function ϕ to the 'cost' of an iteration, we introduce the following parameters:

- e the number of function evaluations per iteration, i.e. f, f', etc.,
- m the number of iterates, except the immediately preceding one, that occur in the iteration function, i.e. x_{i-2}, etc.

Denote the iteration function by $\phi_{e,m}$.
Thus for the Newton-Raphson method we have

$\phi_{2,0}$ since $e = 2, \quad m = 0,$

and for secant method $\phi_{1,1}$ since $e = 1$, $m = 1$.

The well-known efficiency measures based on the order p and the complexity parameter e are

$$E_1 = p/e, \quad \text{Traub's efficiency index,} \tag{3.5.1}$$

$$E_2 = p^{1/e}, \quad \text{Ostrowski's efficiency index.} \tag{3.5.2}$$

These are asymptotic measures; the efficiency which they measure is achieved when the number of iterations tends to infinity. The higher the value of E_i, the more efficient (asymptotically) is the iterative method.
For the Newton–Raphson method we have $E_1 = 1$, $\quad E_2 = 1.414$,
and for secant method $E_1 = 1.62$, $E_2 = 1.62$.

In terms of the two measures secant method is (asymptotically) faster than the Newton–Raphson's. The indexes E_1 and E_2 are applicable only to the vicinity of the root. They are not always adequate for measuring performance of a method in a real situation. The algorithms of identical theoretical efficiency, say, in terms of E_2, may not perform equally well in practice. Two reasons are possible for this to happen: the number of function evaluations, e, treats the function and its derivatives alike, which may be not appropriate in some problems; also different methods have different iteration step formulae. For example, Newton–Raphson's method requires only two operations (one division and one addition) per iteration, the secant method needs five operations (one division, one multiplication and three additions/subtractions), while Traub's method (see Section 3.7) needs seven arithmetic and one shifting operations. The effect of these differences, however, is frequently masked by the deferring costs of evaluating f, f', etc. Therefore a more practical measure of efficiency should be a function of

(a) the number of function and any needed derivatives(s) evaluations per iteration and the cost of these evaluations (problem cost),
(b) the cost of forming the iterative formula (combinatory cost),
(c) the order of convergence, p,
(d) size of the asymptotic error constant.

Feldstein and Firestone (1969) have suggested one such measure:

$$E_3 = p^{1/H}, \quad \text{where } H = e\left(1 + \frac{A}{B}\right) \tag{3.5.3}$$

with p and e as before. Here, A is the number of arithmetic operations, exclusive of function evaluations, required to complete one step of the algorithm, B is the number of arithmetic operations required to evaluate f and any needed derivatives.

Another efficiency measure of similar type is due to Paterson (1972),

$$E_4 = \frac{1}{M} \log_2 p \tag{3.5.4}$$

where M is the number of divisions and multiplications required to compute $\phi_{e,m}$.

Kung and Traub (1973) have proposed an efficiency measure which, as opposed to E_1, takes account of the number of arithmetic operations required. The measure is as follows:

$$E_5 = \sum_i \frac{\log_2 p}{n_i V(f^{(i)}) + c(\phi)} \tag{3.5.5}$$

where n_i is the number of evaluations of $f^{(i)}$ used in $\phi_{e,m}$, $V(f^{(i)})$ is the number of arithmetic operations for one evaluation of $f^{(i)}$, $c(\phi)$ is the minimum number of arithmetic operations, required to combine the $f^{(i)}$ to form $\phi_{e,m}$ by any procedure.

Note here, particularly in relation to measure E_5, that different expressions of an iterative formula may not all be equally good numerically—there may be some conflict between efficiency and, for example, numerical stability of the formula. Numerical stability of iterative methods is discussed in greater detail in Section 3.10. In relation to quantities E_3, E_4, and E_5 certain asymptotic

Table 3.5.1 Asymptotic estimates of efficiency measures for different iterative methods when applied to compute a simple real root of an nth degree polynomial

Method / Efficiency measure	$E_3 = p^{1/II}$	$E_4 = \frac{1}{M} \log_2 p$	$E_5 = \sum_i \frac{\log_2 p}{n_i V(f^{(i)}) + c(f)}$
Secant	$\sqrt{1.618}$	$\frac{1}{1.44n}$	$\frac{1}{2.87n}$
Newton–Raphson's	$\sqrt{2}$	$\frac{1}{2n}$	$\frac{1}{4n}$
Traub's (formula (3.7.3))	$\sqrt[3]{4}$	$\frac{1}{1.5n}$	$\frac{1}{3n}$
Jarratt's (formula (3.7.10))	$\sqrt[3]{4}$	$\frac{1}{1.5n}$	$\frac{1}{3n}$
Fifth-order (formula (3.7.14))	$\sqrt[4]{5}$	$\frac{1}{1.7}$	$\frac{1}{3.4n}$

estimates may be given in terms of the size of the problem. For example, for polynomial equations such estimates are obtained in terms of the degree of the polynomial, n. Assuming that the polynomial and its derivatives are evaluated using the Horner algorithm which requires n operations each of multiplication and addition, we obtain for large n the asymptotic estimates for E_3, E_4 and E_5 as shown in Table 3.5.1.

3.6 One-point Iterative Methods

The iteration steps of the Newton–Raphson and secant methods show that at the ith iteration, the function and its derivative are evaluated at one point only, x_{i-1}. We refer to these evaluations as new data; other information, for example, the value $f(x_{i-2})$ is available in the computer memory for use in forming the current iterate, x_i. If the iterative method uses the 'old' information then this information is 'passively' combined to form x_i. Iterative methods of this nature are called one-point methods with memory, such as secant method, or without memory, such as Newton–Raphson's method. The theory of one-point methods is developed below using the concepts of direct and inverse interpolation.

Direct Polynomial Interpolation

Let $x_i, x_{i-1}, \ldots, x_{i-m}$ be $m+1$ approximates to a zero of the function $f(x)$.

Let $Q_{e,m}$ be a polynomial of degree $e(m+1)-1$ whose first $e-1$ derivatives agree with first $e-1$ derivatives of $f(x)$ at these points, i.e.

$$Q_{e,m}^{(s)}(x_{i-j}) = f^{(s)}(x_{i-j}), \quad j=0, 1, \ldots, m; \; s=0, \ldots, e-1. \tag{3.6.1}$$

Define the next approximant, x_{i+1}, as the root of $Q_{e,m}$, i.e.

$$Q_{e,m}(x_{i+1}) = 0. \tag{3.6.2}$$

If $\phi_{e,m}$ is the iteration function that maps x_i into x_{i+1} it can be obtained from the polynomial $Q_{e,m}$.

The process is then repeated for the set $x_{i+1}, x_i, \ldots, x_{i-m+1}$, and so on.

The majority of computationally useful root finding techniques fall into two groups; those whose iterative formulae require values of the function f only, and those whose iterative formulae require values of both the function f and its derivative f' (processes which use higher derivatives than f' are relatively uncommon in practice, though some examples will be given later). We give below some examples of how iterative methods of either type can be derived using direct polynomial interpolation.

Example 3.6.1

Let $e=1$, $m=1$.

(The iteration function is generated using direct using direct interpolation at two points.) We have

$$Q_{1,1}(x_i) = f(x_i)$$
$$Q_{1,1}(x_{i-1}) = f(x_{i-1}),$$

and

$$Q_{1,1}(x) = f(x_i)\,\frac{x - x_{i-1}}{x_i - x_{i-1}} + f(x_{i-1})\,\frac{x - x_i}{x_{i-1} - x_i}. \tag{3.6.3}$$

Setting

$Q_{1,1}(x_{i+1}) = 0$, we get

$$x_{i+1} = x_i - f(x_i)\,\frac{x_i - x_{i-1}}{f(x_i) - f(x_{i-1})} = \phi_{1,1},$$

which is the secant method.

Example 3.6.2

Let $e = 1$, $m = 2$.
(The iteration function is generated using direct interpolation at three points.)
We have

$$Q_{1,2}(x_i) \quad = f(x_i)$$
$$Q_{1,2}(x_{i-1}) = f(x_{i-1})$$
$$Q_{1,2}(x_{i-2}) = f(x_{i-2})$$

and

$$\begin{aligned} Q_{1,2}(x) \quad &= f(x_i)\frac{(x - x_{i-1})(x - x_{i-2})}{(x_i - x_{i-1})(x_i - x_{i-2})} \\ &\quad + f(x_{i-1})\,\frac{(x - x_i)(x - x_{i-2})}{(x_{i-1} - x_i)(x_{i-1} - x_{i-2})} \\ &\quad + f(x_{i-2})\,\frac{(x - x_i)(x - x_{i-1})}{(x_{i-2} - x_i)(x_{i-2} - x_{i-1})} \end{aligned} \tag{3.6.4}$$

Setting $Q_{1,2}(x_{i+1}) = 0$ we get an iterative method which is known as Muller's method:

$$x_{i+1} = x_i - \frac{2f(x_i)}{w + \{w^2 - 4f(x_i)f[x_i, x_{i-1}, x_{i-2}]\}^{1/2}}$$

where $w = f[x_i, x_{i-1}] + (x_i - x_{i-1})f[x_i, x_{i-1}, x_{i-2}]$.

Here $f[x_i, x_{i-1}]$ denotes the first divided difference

$$\frac{f(x_i) - f(x_{i-1})}{x_i - x_{i-1}}$$

and $f[x_i, x_{i-1}, x_{i-2}]$, the second divided difference

$$\frac{f[x_i, x_{i-1}] - f[x_{i-1}, x_{i-2}]}{x_i - x_{i-2}}.$$

The method was originally expounded by Muller in 1956 and is particularly suitable for solving high-order polynomial equations, although it can be applied equally well to transcendental equations. It is used to find both real and complex roots. Muller's was the first published work in which the direct interpolation was proposed as a tool for developing iterative methods. The order of convergence of Muller's method is 1.839 and its discussion can be found in several good textbooks on numerical computation, for example, Conte and de Boor (1972).

Example 3.6.3

Let $e = 2$, $m = 0$.

(The iteration function is generated using direct interpolation at one point.) Hence we have

$$Q_{2,0}(x_i) = f(x_i)$$
$$Q'_{2,0}(x_i) = f'(x_i)$$

and

$$Q_{2,0}(x) = f(x_i) + (x - x_i)f'(x_i). \tag{3.6.5}$$

Setting $Q_{2,0}(x_{i+1}) = 0$ we get

$$x_{i+1} = x_i - \frac{f(x_i)}{f'(x_i)} = \phi_{2,0},$$

which is the Newton–Raphson formula. We note that although $e(m+1)$ values are necessary to start the process, thereafter, only one new function value per iteration is required.

The use of direct interpolating polynomials has a number of practical disadvantages. For example, to obtain any iterate, x_{i+1}, one has to solve a polynomial of degree $e(m+1) - 1$. This makes the solution non-unique as we don't know which of the zeros of the polynomial we get. The rule must be such that the point x_{i+1} is determined uniquely by the points $x_i, x_{i-1}, \ldots, x_{i-m}$; that is, there must be a function which maps $x_i, \ldots, x_{i-m}$ into x_{i+1}, Also, some or all the zeros of the polynomial may be complex, and although the fact that a new approximation may be predicted as being complex is acceptable in general for polynomial equations, it is a nuisance in the case where only real roots are of interest.

Inverse Polynomial Interpolation

To avoid the need for solving a polynomial equation, an interpolation of F, the inverse of f, at the points $y_i = f(x_i)$, $y_{i-1} = f(x_{i-1})$, $\ldots$, $y_{i-m} = f(x_{i-m})$, can be carried out instead.

Let f' be non-zero (i.e. the root α is simple) and let $f^{(q)}$ be continuous in an interval J. Let f map J into K. Then f has an inverse F and $F^{(q)}$ is continuous on K. There exists a unique polynomial $R_{e,m}$ of degree $e(m+1)-1$ such that

$$R_{e,m}^{(s)}(y_{i-j}) = F^{(s)}(y_{i-j}), \quad j = 0, \ldots, m; \quad s = 0, \ldots, e-1. \tag{3.6.6}$$

Then, evaluating the interpolation polynomial at zero determines the point x_{i+1} uniquely, i.e. $x_{i+1} = R_{e,m}(0)$. The function that maps $x_i, x_{i-1}, \ldots, x_{i-m}$ into x_{i+1} is denoted by $\Psi_{e,m}$.

We now give examples of how iterative methods can be arrived at using inverse polynomial interpolation.

Example 3.6.4

Let $e = 1$, $m = 1$, that is the iteration function is generated using inverse polynomial interpolation over two points. We write

$$R_{1,1}(y) = F(y_i)\,\frac{y - y_{i-1}}{y_i - y_{i-1}} + F(y_{i-1})\,\frac{y - y_i}{y_{i-1} - y_i} \tag{3.6.7}$$

where $\quad y = f(x), \quad x = F(y)$.

An iterative method is obtained as

$$x_{i+1} = R_{1,2}(0) = -x_i\,\frac{f(x_{i-1})}{f(x_i) - f(x_{i-1})} - x_{i-1}\,\frac{f(x_i)}{f(x_{i-1}) - f(x_i)}$$

or

$$x_{i+1} = x_i - f(x_i)\,\frac{x_i - x_{i-1}}{f(x_i) - f(x_{i-1})},$$

which is the Secant method again.

Example 3.6.5

Let $e = 1$, $m = 2$, tht is the iteration function is generated using inverse polynomial interpolation over three points.

Write

$$R_{1,2}(y) = F(y_i)\,\frac{(y - y_{i-1})(y - y_{i-2})}{(y_i - y_{i-1})(y_i - y_{i-2})} + F(y_{i-1})\,\frac{(y - y_i)(y - y_{i-2})}{(y_{i-1} - y_i)(y_{i-1} - y_{i-2})}$$

$$+ F(y_{i-2})\,\frac{(y - y_i)(y - y_{i-1})}{(y_{i-2} - y_i)(y_{i-2} - y_{i-1})}, \tag{3.6.8}$$

giving

$$x_{i+1} = R_{1,2}(0) = x_i\,\frac{f(x_{i-1})f(x_{i-2})}{(f(x_i) - f(x_{i-1}))(f(x_i) - f(x_{i-2}))}$$

$$+ x_{i-1}\,\frac{f(x_i)f(x_{i-2})}{(f(x_{i-1}) - f(x_i))(f(x_{i-1}) - f(x_{i-2}))}$$

$$+ x_{i-2}\,\frac{f(x_i)f(x_{i-1})}{(f(x_{i-2}) - f(x_i))(f(x_{i-2}) - f(x_{i-1}))}. \tag{3.6.9}$$

Expression (3.6.9) can be presented in the form:

$$x_{i+1} = x_i - \frac{f(x_i)}{f[x_i, x_{i-1}]} + \frac{f(x_i)f(x_{i-1})}{f(x_i) - f(x_{i-2})}\left[\frac{1}{f[x_i, x_{i-1}]} - \frac{1}{f[x_{i-1}, x_{i-2}]}\right]. \tag{3.6.10}$$

The order of convergence of the method is 1.839.

Example 3.6.6

We now deduce an algorithm for obtaining the iteration function $\Psi_{e,m}$ using inverse polynomial interpolation at one point. To do this we first observe that

$$\alpha = F(0),$$

since $x \equiv F(f(x))$, $x \in J$, and $y \equiv f(F(y))$, $y \in K$. (3.6.11)

By the Taylor formula:

$$\alpha = F(0) = F(y) + \sum_{k=1}^{r} (-1)^k \frac{y^k}{k!} F^{(k)}(y)$$

$$+ (-1)^{r+1} \frac{y^{r+1}}{(r+1)!} F^{(r+1)}(\eta), \quad 0 \leqslant \eta \leqslant y$$

we obtain

$$\alpha = x + \sum_{k=1}^{r} (-1)^k \frac{F^{(k)}(f(x))}{k!} [f(x)]^k + (-1)^{r+1} \frac{F^{(r+1)}(\eta)}{(r+1)!} [f(x)]^{r+1}. \tag{3.6.12}$$

For simplicity, we denote

$$F^{(k)}(f(x)) \equiv g_k(x) \text{ and } \Psi_{r,0}(x) = x + \sum_{k=1}^{r} (-1)^k \frac{g_k(x)}{k!} [f(x)]^k. \tag{3.6.13}$$

Equation $x = \Psi_{r,0}(x)$ has the root $x = \alpha$, since

$$\Psi_{r,0}(\alpha) = \alpha + \sum_{k=1}^{r} (-1)^k \frac{g_k(\alpha)}{k!} [f(a)]^k = \alpha.$$

Thus, setting

$$x_{i+1} = \Psi_{r,0}(x_i), \quad i = 0, 1, \ldots, \quad x_0 \in J, \tag{3.6.14}$$

we obtain an iterative method.

If x_0 is chosen close enough to the root α, then the sequence $\{x_i\}$ converges to α, as there exists an interval containing the point α such that for any point x of this interval the following holds:

$$|\Psi'(x)| \leqslant C < 1. \tag{3.6.15}$$

It follows that for the convergence of the sequence $\{x_i\rangle$, it is only necessary for x_0 to belong to this interval.

The iteration function $\Psi_{r,0}(x)$ can be found in an explicit form in terms of $f(x)$ and its derivatives. By virtue of (3.6.11) we have:

$$\begin{aligned} &F'(f(x))f'(x) = 1 \\ &F''(f(x))[f'(x)]^2 + F'(f(x))f''(x) = 0 \\ &F'''(f(x))[f'(x)]^3 + 3F''(f(x))f'(x)f''(x) + F'(f(x))f'''(x) = 0, \end{aligned} \tag{3.6.16}$$

giving

$$\begin{aligned} &g_1(x)f'(x) = 1 \\ &g_2(x)[f'(x)]^2 + g_1(x)f''(x) = 0 \\ &g_3(x)[f'(x)]^3 + 3g_2(x)f'(x)f''(x) + g_1(x)f'''(x) = 0. \end{aligned} \tag{3.6.17}$$

From (3.6.17), the functions $g_1(x)$, $g_2(x)$, ... can be found successively.

We now give some examples of the use of inverse polynomial interpolation at one point.

Example 3.6.7

Let $e = 2$, $m = 0$.
From (3.6.13), (3.6.14) and (3.6.17), we have

$$x_{i+1} = \Psi_{2,0} = x_i - \frac{f(x_i)}{f'(x_i)},$$

and we again have the Newton–Raphson method.

Example 3.6.8

Let $e = 3$, $m = 0$.
We have

$$x_{i+1} = \Psi_{3,0} = x_i + (-1)g_1(x_i)f(x_i) + \frac{1}{2!} g_2(x_i)f^2(x_i),$$

where

$$g_1(x_i) = \frac{1}{f'(x_i)},$$

$$g_2(x_i) = -g_1(x_i)\frac{f''(x_i)}{[f'(x_i)]^2}.$$

It follows that

$$x_{i+1} = \Psi_{3,0} = x_i - \frac{f(x_i)}{f'(x_i)} - \left[\frac{f(x_i)}{f'(x_i)}\right]^2 \frac{f''(x_i)}{f'(x_i)}, \tag{3.6.18}$$

which is known as the Chebyshev method. The order of the method is 3.

The orders of convergence of the corresponding direct and inverse methods are the same. For both direct and inverse interpolation, the starting values, $x_0, x_1, \ldots, x_m$, must be available. One method for obtaining these starting values is to use $\phi_{e,1}$ followed successively by $\phi_{e,2}, \phi_{e,3}, \ldots, \phi_{e,m}$.

Derivative Estimated Iterative Methods

The term identifying this group of methods is due to Traub (1964) who has extensively explored the possibilities for constructing new methods by taking an iterative formula involving the derivatives and replacing either all or the highest derivatives by estimates, generated in some suitable fashion.

Take, for example, the Newton–Raphson formula

$$x_{i+1} = x_i - \frac{f(x_i)}{f'(x_i)}.$$

Using direct polynomial interpolation over two approximations to the root, x_i and x_{i-1}, the polynomial of degree 1 is set up, giving

$$P_1(x) = f(x_i)\frac{x - x_{i-1}}{x_i - x_{i-1}} + f(x_{i-1})\frac{x - x_i}{x_{i-1} - x_i}.$$

The value of $P_1'(x_i)$ is now taken as an estimate of $f'(x_i)$ and used to replace the derivative in the Newton–Raphson formula, giving

$$x_{i+1} = x_i - f(x_i)\frac{x_i - x_{i-1}}{f(x_i) - f(x_{i-1})},$$

which, obviously, is the secant formula.

Similarly, using interpolation over three approximations, x_i, x_{i-1}, x_{i-2}, the polynomial of degree 2 is set up, for which $P_2(x_{i-j}) = f(x_{i-j})$, $j = 0, 1, 2$. The value of $P_1'(x_i)$ is now taken as an estimate of $f'(x_i)$ and used to replace the derivative in the Newton–Raphson formula. We get a new iterative method given by

$$x_{i+1} = x_i - \frac{f(x_i)}{f[x_i x_{i-1}] + f[x_i, x_{i-2}] - f[x_{i-1}, x_{i-2}]} \tag{3.6.19}$$

where as before, $f[x_s, x_{s-1}]$ denotes the first divided difference.

The method (3.6.19) is one of the two new methods which were explored by Traub as a result of utilizing the idea of derivative estimation. He has also shown that the order of convergence and the asymptotic error constant of the method are precisely the same as those of the method of direct interpolation using a quadratic fit; the order is 1.839.

Inverse polynomial intrapolation can be employed in a similar fashion to estimate the derivative. Generally, the derivative estimates found using a direct and an inverse interpolating polynomial of the same degree are different. Thus, using inverse interpolation over three approximations, x_i, x_{i-1}, x_{i-2}, to estimate f' and then replacing the derivative in the Newton–Raphson formula by the estimate obtained lead to a new iterative formula given by

$$x_{i+1} = x_i - f(x_i)\left\{\frac{1}{f[x_i, x_{i-1}]} + \frac{1}{f[x_i, x_{i-2}]} - \frac{1}{f[x_{i-1}, x_{i-2}]}\right\} \tag{3.6.20}$$

which is the second of the two new methods obtained by Traub. The order of convergence of this method is again 1.839.

We now take the Chebyshev third order method:

$$x_{i+1} = x_i - \frac{f(x_i)}{f'(x_i)} - \left[\frac{f(x_i)}{f'(x_i)}\right]^2 \frac{f''(x_i)}{f'(x_i)}.$$

Using a two-point Hermite polynomial interpolation formula over two approximations, x_i and x_{i-1}, we obtain

$$H_3(x) = \frac{1}{(x_i - x_{i-1})^2}\left[\left(1 - 2\frac{x - x_i}{x_i - x_{i-1}}\right)(x - x_{i-1})^2 f(x_i)\right.$$

$$+ \left(1 - 2\frac{x - x_{i-1}}{x_{i-1} - x_i}\right)(x - x_i)^2 f(x_{i-1}) + (x - x_i)(x - x_{i-1})^2 f'(x_i)$$

$$\left. + (x - x_{i-1})(x - x_i)^2 f'(x_{i-1})\right] \tag{3.6.21}$$

Differentiating $H_3(x)$ twice gives an estimate of $f''(x_i)$ as

$$f''(x_i) = \frac{2}{x_i - x_{i-1}}\{2f'(x_i) + f'(x_{i-1}) - 3f]x_i, x_{i-1}]\} \tag{3.6.22}$$

while using inverse interpolation under similar conditions, one obtains

$$f''(x_i) = \frac{2}{f(x_i) - f(x_{i-1})}\left\{\frac{2}{f'(x_i)} + \frac{1}{f'(x_{i-1})} - \frac{3}{f[x_i, x_{i-1}]}\right\}. \tag{3.6.23}$$

Replacing the second derivative in Chebyshev's formula by the estimates given in (3.6.22) and (3.6.23) we obtain two different iterative formulae, each of order 2.73.

Optimality of One-point Iterative Methods

The order of an iterative method is considered a suitable measure of the improvement in accuracy gained in one iteration. Furthermore, the order of a one-point method is dependent on the number of 'new' function evaluations in one iteration, and this number includes the count of the evaluations of the function as well as its derivatives. Traub has observed that the efficiency of a one-point iterative method without memory is limited by the fact that such a method of order p depends on at least $p-1$ derivatives of f, and thus requires at least p 'new' function evaluations per iteration. In searching for the ways to improve efficiency of iterative methods, one attempts to make more efficient use of the 'old' information: What is the most that the data obtained at preceding iterations can do to enhance the accuracy at the current interation? Again, Traub has noted that the contribution of the 'old' data to the convergence is less than unity when quantitatively measured in terms of the number of required function evaluations per iteration. Subsequently Traub (1964) conjectured that

> an analytic one-point iterative method, with or without memory, which is based on p evaluations per iteration step is of order at most p.

This conjecture has been proved by Brent, Winograd and Wolfe (1973), and Kung and Traub (1974). Below we outline the proof following Brent *et al.*

Convergence of Polynomial Schemes

The theory of one-point iterative methods using the concept of polynomial interpolation shows that a one-point iterative method is obtained by replacing the function f at each iteration by a polynomal function Q which agrees with f in a specified way. As always, when one function is interpolated by another, a 'remainder' formula can be obtained for the error incurred. One such useful formula is due to Cauchy:

Let I be the smallest interval containing $(x_i, x_{i-1}, \ldots, x_{i-m})$.
For any x there is a ξ, in the smallest interval containing I and x, such that

$$f(x) - Q(x) = \frac{1}{n!}[f-Q]^{(n)}(\xi) \prod_{j=0}^{m} (x - x_{i-j})^e \tag{3.6.24}$$

where $n = e(m+1)$.

This formula leads to the derivation of upper bounds on the order of convergence as a function of e for polynomial interpolation schemes with fixed m. If Q_i is the interpolating polynomial agreeing with f at $x_i, \ldots, x_{i-m}$ for the first $(e-1)$ derivatives, then Q_i is of degree $n-1$, where $n = e(m+1)$, and so (3.6.24) gives

$$f(x) - Q_i(x) = \frac{1}{n!} f^{(n)}(\xi) \prod_{j=0}^{m} (x - x_{i-j})^e,$$

or

$$| Q_i(\alpha) | = \frac{1}{n!} | f^{(n)}(\xi) | \prod_{j=0}^{m} \varepsilon_{i-j}^{e}, \tag{3.6.25}$$

where $\varepsilon_k = | \alpha - x_k |$.

Since $Q_i(x_{i+1}) = 0$, the $| Q_i(\alpha) | = \varepsilon_{i+1} | Q_i'(\eta) |$ for some value of η in $[x_{i+1}, \alpha]$.

This is under assumption that in addition to $f'(\alpha) \neq 0$ (this assumption has been made at the beginning of the chapter), we have $f^{(n)}(\alpha) \neq 0$. Estimation of $| Q_i'(\eta) |$ and $| f^{(n)}(\xi) |$ reduces (3.6.25) to

$$\varepsilon_{i+1} = (C + \delta_i) \prod_{j=0}^{m} \varepsilon_{i-j}^{e}, \tag{3.6.26}$$

where

$$C = \left| \frac{f^{(n)}(\alpha)}{n! f'(\alpha)} \right| \quad \text{and} \quad \delta_i \to 0.$$

Taking logarithms, we can rewrite (3.6.26) as

$$\log_2 \varepsilon_{i+1} = \log_2 (C + \delta_i) + e \sum_{j=0}^{m} \log_2 \varepsilon_{i-j}. \tag{3.6.27}$$

Since with i increasing, $\log_2 \varepsilon_{i+1} \to \infty$, the term $\log_2 (C + \delta_i)$ becomes unimportant, and we can consider the difference equation

$$\log_2 \varepsilon_{i+1} = e \sum_{j=0}^{m} \log_2 \varepsilon_{i-j}.$$

We solve this equation in a normal way by first letting

$\log_2 \varepsilon_s = \lambda^s$, and then considering

$$\lambda^{i+1} = e \sum_{j=0}^{m} \lambda^{i-j},$$

which gives, after multiplying both sides by λ^{m-i},

$$\lambda^{m+1} - e \sum_{j=0}^{m} \lambda^{j} = 0. \tag{3.6.28}$$

Let λ be the unique real root of largest modulus of (3.6.28). For $m = 0, 1, 2, 3$ and a fixed integer value of e (e.g. $e = 2$) these roots are

$$\lambda = e, \ \lambda = e + 0.732, \ \lambda = e + 0.920, \ \lambda = e + 0.974.$$

Moreover, $\lambda \to e + 1$ as $m \to \infty$.

Now, we have

$$\lim_{k \to \infty} \frac{\log \varepsilon_{k+1}}{\log \varepsilon_k} = \lambda. \tag{3.6.29}$$

We shall now show that the order of convergence of the one-point iterative method defined by $Q_i(x_{i+1}) = 0$ is equal to λ.

By definition we have

$$\lim_{k \to \infty} \frac{|\alpha - x_{k+1}|}{|\alpha - x_k|} = \lim_{k \to \infty} \frac{\varepsilon_{k+1}}{\varepsilon_k{}^p} = C \neq 0.$$

For large k, we can write

$$\frac{\varepsilon_{k+1}}{\varepsilon_k{}^p} \approx C.$$

Taking logarithms, we obtain

$$\log_2 \varepsilon_{k+1} - p \log_2 \varepsilon_k = \log_2 C,$$

which can be rewritten as

$$\frac{\log_2 \varepsilon_{k+1}}{\log_2 \varepsilon_k} = p + \frac{\log_2 C}{\log_2 \varepsilon_k},$$

giving

$$\lim_{k \to \infty} \frac{\log_2 \varepsilon_{k+1}}{\log_2 \varepsilon_k} = p, \tag{3.6.30}$$

since $\log_2 C$ is a constant.
Comparing (3.6.29) and (3.6.30), we obtain $p = \lambda = e + 1$.

We may thus conclude that if an iterative method uses only the values of the function f and its derivatives up to the $(e - 1)$st inclusive (evaluated at arbitrarily many previous values of the argument), then the order of convergence of the method is less or equal to $e + 1$.

This result is often referred to as the basic optimality theorem for one-point iterative methods.

3.7 Multipoint Iterative Methods

The optimality theorem for one-point iterative methods establishes that the relation between the order of convergence of such methods and the number of function evaluations per iteration step is linear.

We may now pose a further question:

> Is it possible to construct an iterative process which would not be subject to these restrictions?

The concept of the so called multipoint iterative method (without memory or with memory, in the same sense as used to distinguish between two types of the one-point method) has been developed by Traub so as to give a positive answer to this question. A multipoint iterative method may be considered as one possible technique for increasing efficiency of the iterative methods.

The general idea of a multipoint method is based on the use of an appropriate combination of iterates. One simple example would be to use two successive Newton–Raphson iterates. The method obtained would be of fourth order and would require four function evaluations per iteration. We give below three further examples of multipoint iterative methods.

Example 3.7.1

Assume that the ith approximation to the root α has been computed.

Now let $z_i = x_i - \dfrac{f(x_i)}{f'(x_i)}$ and $x_{i+1} = z_i - \dfrac{f(z_i)}{f'(x_i)}$.

Combining the two equations we get

$$x_{i+1} = x_i - u_i - \frac{f(z_i)}{f'(x_i)}, \quad \text{where} \quad u_i = \frac{f(x_i)}{f'(x_i)}. \tag{3.7.1}$$

$\phi_{3,0} = x_i - u_i - \dfrac{f(z_i)}{f'(x_i)}$ is of order 3.

An illustration of the method is given in Fig. 3.7.1.

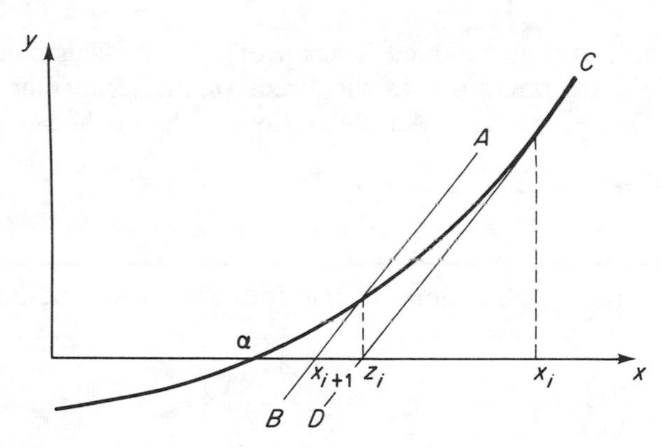

Fig. 3.7.1. An illustration of Example 3.7.1. The straight lines AB and CD are parallel

Example 3.7.2

Combine the Newton–Raphson method

$$z_i = x_i - \frac{f(x_i)}{f'(x_i)},$$

and the secant method,

$$x_{i+1} = z_i - f(z_i)\,\frac{z_i - x_i}{f(z_i) - f(x_i)},$$

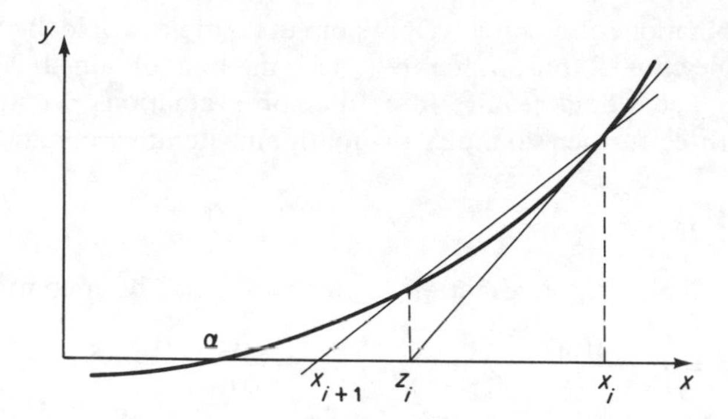

Fig. 3.7.2. An illustration of Example 3.7.2

to obtain

$$x_{i+1} = \phi_{3,0} = x_i - u_i + f(z_i)\frac{u_i}{f(z_i) - f(x_i)}$$

$$= x_i + \frac{u_i}{f(z_i) - f(x_i)}. \tag{3.7.2}$$

The iteration function obtained is again of order 3. This method may be visualized as the intersection with the x-axis of the secant line through the points $(x, f(x))$ and $(z, f(z))$. An illustration of the method is given in Fig. 3.7.2.

Example 3.7.3

Now consider the combination of the following two one-point iterative methods

$$z_i = x_i - \frac{f(x_i)}{f'(x_i)}, \qquad x_{i+1} = z_i - \frac{f(z_i)(z_i - x_i)}{2f(z_i) - f(x_i)}.$$

Here at every second step the derivative f' is replaced by a linear combination of $f(x)$ at two immediately preceding points.

The method is depicted in Fig. 3.7.3.

Combining the above two equations we obtain

$$x_{i+1} = \phi_{3,0} = x_i - u_i \frac{f(z_i) - f(x_i)}{2f(z_i) - f(x_i)} \tag{3.7.3}$$

We shall call (3.7.3) the Traub iterative method. The method's iteration step is as follows.

evaluate $f(x_i)$
evaluate $f'(x_i)$
compute $u_i = f(x_i)/f'(x_i)$

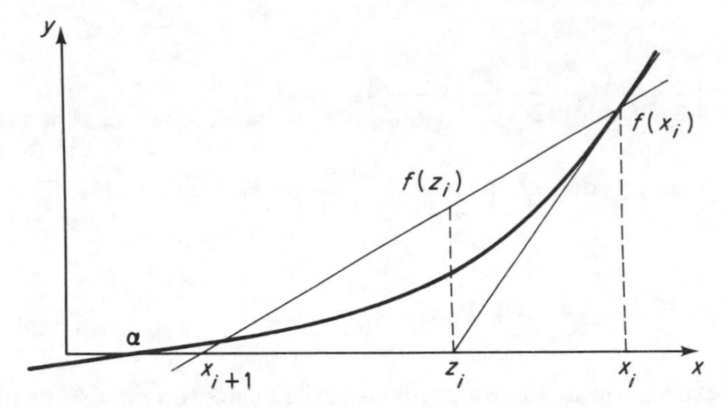

Fig. 3.7.3. An illustration of the Traub method

compute $z_i = x_i - u_i$
evaluate $f(z_i)$

compute $x_{i+1} = x_i - u_i \dfrac{f(z_i) - f(x_i)}{2f(z_i) - f(x_i)}$

Here we have $e = 3$, $m = 0$, and, hence, $\phi = \phi_{3,0}$; however, as we hall now see, the method is of order 4.

Write formula (3.7.3) as

$$x_{i+1} = x_i - (x_i - z_i)\,\frac{f(z_i) - f(x_i)}{2f(z_i) - f(x_i)}.$$

Let ε_i be the error such that $x_i = \alpha + \varepsilon_i$. Note that

$$z_i = \alpha + K\varepsilon_i^2 + M\varepsilon_i^3 + O(\varepsilon_i^4) \tag{3.7.4}$$

where $K = \dfrac{1}{2}\dfrac{f''}{f'}$ and $M = \dfrac{1}{3}\dfrac{f'''}{f'} - \dfrac{1}{2}\left(\dfrac{f''}{f'}\right)^2$.

We can write for (3.7.3):

$$\alpha + \varepsilon_{i+1} = \alpha + \varepsilon_i - (\alpha + \varepsilon_i - \alpha - K\varepsilon_i^2 - M\varepsilon_i^3)$$
$$\times \frac{f(\alpha + K\varepsilon_i^2 + M\varepsilon_i^3) - f(\alpha + \varepsilon_i)}{2f(\alpha + K\varepsilon_i^2 + M\varepsilon_i^3) - f(\alpha + \varepsilon_i)}. \tag{3.7.5}$$

Using the Taylor series expansion and some simple transformations, the relation (3.7.5) is brought to the form:

$$\varepsilon_{i+1} = \varepsilon_i - \varepsilon_i\,\frac{N}{D}, \tag{3.7.6}$$

where $N = -f' + \frac{1}{2}f''\varepsilon_i + (2Mf' - \frac{1}{6}f''')\varepsilon_i^2 + (\frac{1}{2}K^2 f'' - \frac{1}{24}f^{iv} - KMf' + \frac{1}{6}Kf''')\varepsilon_i^3 + \dots,$

and $D = -f' + \frac{1}{2}f''\varepsilon_i + (2Mf' - \frac{1}{6}f''')\varepsilon_i^2 + (K^2 f'' - \frac{1}{24}f^{iv})\varepsilon_i^3 + \dots.$

which gives the leading term as

$$\varepsilon_{i+1}=\left[\frac{1}{8}\left(\frac{f''}{f'}\right)^3-\frac{1}{12}\left(\frac{f''}{f'}\right)\left(\frac{f'''}{f'}\right)\right]\varepsilon_i^4+0(\varepsilon_i^5). \tag{3.7.7}$$

It is customary to denote the ratio $\dfrac{f^{(r)}(\alpha)}{f'(\alpha)}$ by F_r. Relation (3.7.7) can then be written as

$$\varepsilon_{i+1}=\left[\frac{1}{8}F_2^3-\frac{1}{12}F_2F_3\right]\varepsilon_i^4+O(\varepsilon_i^5),$$

where the expression in square brackets is the asymptotic constant of Traub's method. This completes the proof.

Further Examples of Multipoint Iterative Methods

A particular feature of the Traub multipoint method given by (3.7.3) lies in the fact that it is an iterative method whose order of convergence is higher than the number of function evaluations per iteration. The order of the method is $p=4$ but it requires only three function evaluations per iteration. In this respect, however, even more dramatic examples of multipoint methods may be given. Below we illustrate some families of multipoint methods that possess specific useful features.

Suppose that the equation $f(x)=0$ is such that the derivative f' can be rapidly evaluated compared with f itself (for example for a function defined as an integral). For problems of this kind a multipoint iterative method which minimizes the number of new function evaluations at the price of extra evaluations of its derivative(s) may be considered as more efficient. Traub (1964), Kogan (1967), and Jarratt (1966a, 1966c, 1969) have constructed multipoint iterative methods which utilize this feature.

Consider a family of methods defined by the general formula

$$\phi(x)=x-\sum_{j=1}^{s-1}\alpha_j^s\omega_j^s(x) \tag{3.7.8}$$

where

$$\omega_j^s(x)=f(x)/f'\left[x+\sum_{i=1}^{j}\beta_{j,i}^s\omega_i^s(x)\right],\quad j>1.$$

Expanding (3.7.8) about the root α and defining the ω_1^s in various ways, the free parameters, α_j^s and $\beta_{j,i}^s$, can be used to optimize, at least in the asymptotic sense, the order of convergence of a specific iterative method of form (3.7.8) for locating the simple roots of an arbitrary function f.

Traub considered in detail the simplified form

$$\phi(x)=x-\alpha_1\omega_1(x)-\alpha_2\omega_2(x)-\alpha_3\omega_3(x), \tag{3.7.9}$$

where

$$\omega_1 = u(x) = \frac{f(x)}{f'(x)},$$

$$\omega_2 = f/f'(x + \beta\omega_1),$$

$$\omega_3 = f/f'(x + \gamma_1\omega_1 + \gamma_2\omega_2).$$

He has shown that for $\alpha_3 = 0$ and appropriate choices of the parameters α_1, α_2 and β, it is possible to build a family of third order methods, each requiring one function and two derivative evaluations per iteration. Furthermore, by including the term $-\alpha_3\omega_3(x)$ fourth order formulae are possible to derive which require one function and three derivative evaluations per iteration step.

A further search for the efficient iteration functions based on (3.7.8) resulted in a rather remarkable method due to Jarratt which is given by (3.7.8)

$$x_{k+1} = \phi(x_k) = x_k - \frac{5}{8}\frac{f(x_k)}{f'(x_k)} - \frac{3}{8}f(x_k)\frac{f'(x_k)}{\left\{f'\left[x_k - \frac{2}{3}\frac{f(x_k)}{f'(x_k)}\right]\right\}^2}. \tag{3.7.10}$$

The method is of fourth order but requires one function and only two derivative evaluations.

Another simplified form of (3.7.8) is

$$\phi_r(x) = x - \frac{f(x)}{f'\left(x + \sum_{s=1}^{r} \beta_s(f(x))^s\right)}, \tag{3.7.11}$$

where $(f(x))^s$ denotes the sth power of $f(x)$.

Using this form leads to a third order iterative method due to Kogan and given by

$$x_{k+1} = \phi_1(x_k) = x_k - \frac{f(x_k)}{f'\left(x_k - \frac{f(x_k)}{2f'(x_k)}\right)}. \tag{3.7.12}$$

The process requires one function and two derivative evaluations per iteration.

Yet another group of interesting methods considered by Jarratt arises from the formula

$$\phi(x) = x - \frac{f(x)}{\alpha_1\omega_1(x) + \alpha_2\omega_2(x) + \alpha_3\omega_3(x)} \tag{3.7.13}$$

where

$$\omega_1 = f'(x), \quad \omega_2 = f'(x + \alpha f(x)/\omega_1(x)),$$

$$\omega_3 = f'(x + \beta f(x)/\omega_1(x) + \gamma f(x)/\omega_2(x)).$$

Among the methods of this group, a fifth order method of the form

$$x_{k+1} = x_k - \frac{6f(x_k)}{D}, \tag{3.7.14}$$

where $D = f'(x_i) + f'(x_k - f(x_k)/f'(x_k))$
$$+ 4f'(x_k - \tfrac{1}{8}f(x_i)/f'(x_i) - \tfrac{3}{8}f(x_k)/f'(x_k - f(x_k)/f'(x_k))),$$
requires one function and three derivative evaluations per iteration.

Now suppose that the equation $f(x) = 0$ is such that the derivative f' is very difficult to evaluate.

Kung and Traub (1974) have developed a family of multipoint iterative methods without memory and of order 2^{n-1} that require just n evaluations of f per iteration. The family is based on the inverse interpolating polynomials, and n is the number of points over which the interpolation is carried out.

An example of such a method for $n = 4$ is:

$$z_1 = x_k + \beta f(x_k),$$

$$z_2 = z_1 - \beta \frac{f(x_k)f(z_1)}{f(z_1) - f(x_k)},$$

$$z_3 = z_2 - \frac{f(x_i)f(z_1)}{f(z_2) - f(x_k)} \left\{ \frac{1}{f[z_1, x_k]} - \frac{1}{f[z_2, z_1]} \right\},$$

$$x_{k+1} = z_3 - \frac{f(x_k)f(z_1)f(z_2)}{f(z_3) - f(x_k)} \left\{ \frac{1}{f[z_1, x_k]} - \frac{1}{f[z_2, z_1]} - \frac{1}{f[z_3, z_2]} \right\}, \quad (3.7.15)$$

where β is a suitably chosen non-zero constant and $f[x_i, x_{i-1}]$ denotes, as usual, the first divided difference.

Optimality of Multipoint Iterative Methods

We can make the following general observation in relation to the multipoint iterative methods we have just discussed. If a multipoint method is composed of two or more one-point iterations directly, then such a method will retain the limitations on e and on p. For example, a multipoint method composed of n Newton–Raphson iterations is of order 2^n and requires n evaluations of f and n evaluations of f', that is, $2n$ evaluations per iteration.

On the other hand, Kung and Traub (1974) have shown that a multipoint iterative process without memory of order 2^{n-1} can be constructed from

either just n evaluations of f (e.g. formula (3.7.15))
or $n - 1$ evaluations of f and one evaluation of f'.

Their work confirms the view that the concept of a multipoint iterative method as a tool for constructing more efficient iterative methods as compared with presently existing methods is a potentially fruitful field of study.

Kung and Traub have conjectured that a multipoint iterative method without memory based on n evaluations has optimal order 2^{n-1}. Wozniakowski (1975a) has proved their conjecture for the so-called Hermitian situation, that is under the assumption that if $f^{(j)}(x_k)$, j-th derivative of f at point x_i, is known, then $f^{(0)}(x_k), \ldots, f^{(j-1)}(x_k)$ are also known.

3.8 Conditioning of the Root Finding Problem

Our concern in this section is the conditioning of the root finding problem and how it effects the computation of an accurate approximation to a simple root α of $f(x) = 0$. In what follows we shall use the approach of Dahlquist.

Let x_n be an approximation to $\alpha, f(\alpha) \equiv 0, f'(\alpha) \neq 0$. By the mean value theorem we have

$$f(x_n) = (x_n - \alpha)f'(\xi), \quad \xi \in \text{int}(x_n, \alpha) = J \tag{3.8.1}$$

from which the method-independent estimate of the error is obtained as

$$| x_n - \alpha | \leqslant f(x_n) | /M, \quad | f'(x) | \geqslant M, \quad x \in J. \tag{3.8.2}$$

Denote by $\bar{f}(x_n)$ the computed value of the function and let

$$\bar{f}(x_n) = f(x_n) + \delta(x_n) \tag{3.8.3}$$

where $| \delta(x) | \leqslant \delta$ independently of x. We now show that the accuracy with which the root α can be determined is restricted by the size of δ. To do this, we first observe that the best one can hope for is to find x_n such that $\bar{f}(x_n) = 0$. The exact value of the function, then, will satisfy the condition

$$| f(x) | \leqslant \delta. \tag{3.8.4}$$

Assuming that $f'(x)$ does not change very rapidly in the vicinity of $x = \alpha$, from (3.8.2) we obtain

$$| x_n - \alpha | \leqslant \frac{\delta}{M} \approx E_\alpha \quad \text{where } E_\alpha = \frac{\delta}{| f'(\alpha) |}. \tag{3.8.5}$$

This means that E_α is the best method-independent error estimate available. We shall call it the limiting error of the root α. We further observe that if $| f'(\alpha) |$ is small, then $E\alpha$ is large. In this case the problem of finding α is ill-conditioned. Two possible situations are illustrated in Fig. 3.8.1. Assume that root α is required to be computed with some prescribed tolerance ε. The error estimate, E_α, will not usually be known. If the prescribed tolerance happens to be such that $\varepsilon < E_\alpha$, then it is unlikely that the iterations will ever stop. Therefore, instead of specifying the tolerance ε it is often more convenient to determine the root with its limiting error E_α and then to estimate E. For

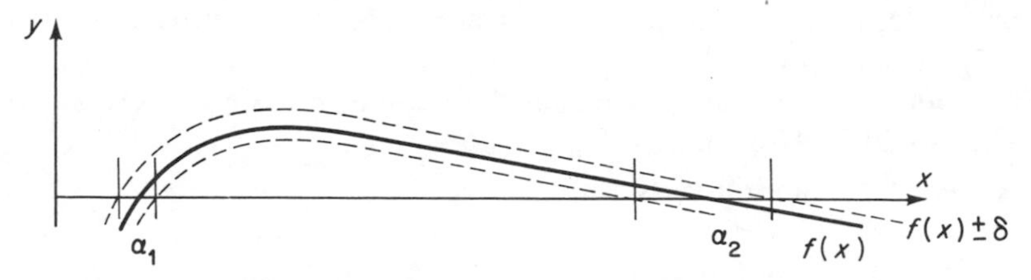

Figure 3.8.1 Examples of a well-conditioned root (α_1) and an ill-conditioned root (α_2)

iterative methods which converge faster than the first order, only a few extra iterations are normally needed to achieve an accuracy within the limiting error even when $\varepsilon \gg E_\alpha$.

We can therefore deduce a rule for a stopping criterion which is based on the following simple observation: Provided the sequence $\{x_n\}$ converges to the root α, the differences $|x_n - x_{n-1}|$ will decrease until $|x_n - \alpha| \approx E_\alpha$, at least when n is greater than some fixed value, n_0. Thereafter, the error will be dominated by round-off effects and the differences will vary randomly. The iterations can therefore be stopped and the values of x_n accepted as a satisfactory value of the root as soon as the following two conditions are satisfied simultaneously

$$|x_{n+1} - x_n| \geqslant |x_n - x_{n-1}| \quad \text{and} \quad |x_n - x_{n-1}| < \delta. \tag{3.8.6}$$

Here δ is some coarse tolerance which is introduced with the purpose of preventing the iterations being stopped before x_n even gets into the neighbourhood of α.

The error in x_n is then adequately estimated by $|x_{n+1} - x_n|$. Using this stopping criterion considerably reduces the risk of the iterations continuing forever in the case of an ill-conditioned problem, e.g. for large E_α. It also takes care of the situation when the tolerance prescribed happens to be below the computational accuracy that can be achieved for specific well-conditioned roots.

3.9 Iterative Methods for Solving Systems of Non-linear Equations

Consider the following set of non-linear algebraic equations:

$$\begin{aligned} &f_1(x_1, \ldots, x_n) = 0, \\ &\ldots \\ &f_n(x_1, \ldots, x_n) = 0, \end{aligned} \tag{3.9.1}$$

where functions f_j are assumed to be continuously differentiable. The Newton–Raphson method for determining a solution to equations (3.9.1) is given in a vector form by

$$\mathbf{x}_{k+1} = \mathbf{x}_k - \mathbf{J}^{-1}(\mathbf{x}_k)\mathbf{F}(\mathbf{x}_k), \quad x_0 \text{ arbitrary}, \tag{3.9.2}$$

where $\mathbf{J}^{-1}(\mathbf{x})$ is the inverse of the Jacobian matrix of $\mathbf{F}(\mathbf{x})$. (The Jacobian matrix contains the $n \times n$ partial derivatives of F in terms of $\mathbf{x}$, where $J_{ij} = \delta F_i / \delta x_j$.)

The order of convergence for this method is shown to be equal to 2 (Ortega and Rheinboldt, 1970), that is if

$$\mathbf{x}_k \to \mathbf{x}, \text{ the solution vector, as } k \to \infty,$$

then

$$\|\mathbf{x}_k - \mathbf{x}\| \leqslant K\theta^{2^k} \quad \text{for all } k,$$
$$\text{for some } K > 0, \text{ and for } \theta \in (0, 1). \tag{3.9.3}$$

Direct evaluation of the $n \times n$ Jacobian matrix $\mathbf{J}(\mathbf{x}_k)$ may pose a numerically tricky problem, instead the following finite difference approximation, $\mathbf{J}_k$, to $\mathbf{J}(\mathbf{x}_k)$ may be used. It then yields the method known as the discrete Newton–Raphson method and due to Shamanskii (1967).

Let $\mathbf{c}_j$ be the jth column of the $n \times n$ identity matrix $\mathbf{I}$, and let the jth column $\mathbf{J}_k\mathbf{c}_j$ of the matrix $\mathbf{J}_k$ be defined as

$$\mathbf{J}_k\mathbf{c}_j = [\mathbf{F}(\mathbf{x}_k + \varepsilon_k\mathbf{c}_j) - \mathbf{F}(\mathbf{x}_k)]/\varepsilon_k, \quad 1 \leqslant j \leqslant n, \tag{3.9.4}$$

where ε_k satisfies $0 < \varepsilon_k < \| \mathbf{F}(\mathbf{x}_k) \|$.

Given a solution estimate $\mathbf{x}_k$, the discrete Newton method generates the next estimate $\mathbf{x}_{k+1}$ as

$$\mathbf{x}_{k+1} = \mathbf{x}_k - \mathbf{J}_k{}^{-1}\mathbf{F}(\mathbf{x}_i) \tag{3.9.5}$$

The method converges with the order 2, the same as Newton's method (Traub, 1964; Brent, 1973). The following algorithm is based on this method.

Algorithm newton (root)

> *initialize root*
> *initialize stop*
> **while not** *stop* **do**
>   *compute J through* (3.9.4)
>   *compute* J^{-1}
> $root := root - J^{-1}f(root)$
>   *Compute new stop*
> **enddo**

The method (3.9.2) requires the Jacobian $\mathbf{J}(\mathbf{x}_k)$ and the method (3.9.5), $(n + 1)$ gradients $\mathbf{F}(\mathbf{x}_k)$, $\mathbf{F}(\mathbf{x}_k + \varepsilon_k c_j)$, $j = 1, \ldots, n$, to be evaluated at each iteration. This means that much computation time is devoted to evaluating the current Jacobian or its finite difference approximation. As a means of speeding up the process one can think of spreading out the computational work of the Jacobian evaluation over several iterations. Two approaches have emerged following these lines of thought.

Partial Updating Approach

One approach, following Mukai (1979), may be called partial updating. It has been elaborated in the works of Wolfe (1959), Barnes (1965), Polak (1974) and Mukai (1979), and involves the updating of only one or at most r, $r < n$, columns of the Jacobian at a time and retaining the remaining columns from the previous iteration.

Let $S(k)$ denote the indices of the columns which are updated at iteration k. In the partial updating approach the columns of the new Jacobian estimate

$\mathbf{J}_k$ are defined by

$$\mathbf{J}_k\mathbf{c}_j = [\mathbf{F}(\mathbf{x}_k + \varepsilon_k\mathbf{c}_j) - \mathbf{F}(\mathbf{x}_k)]/\varepsilon_k, \quad j \in S(k) \tag{3.9.6}$$

and

$$\mathbf{J}_k\mathbf{c}_j = \mathbf{J}_{k-1}\mathbf{c}_j, \quad j \notin S(k). \tag{3.9.7}$$

Then the new estimate $\mathbf{x}_{k+1}$ is determined by (3.9.5). Selection of the r columns to be updated is arbitrary so long as each column is updated at least once in $\lceil n/r \rceil$ iterations.

Since $\mathbf{J}_k$ differs from $\mathbf{J}_{k-1}$ only in r columns, the inverse $\mathbf{J}_k^{-1}$ in (3.9.5) can be computed using the following updating formula

$$\mathbf{J}_k^{-1} = \mathbf{J}_{k-1}^{-1} - (\mathbf{I}^r - \mathbf{J}_{k-1}^{-1}\mathbf{J}_k^r)[(\mathbf{J}_{k-1}^r)^r\mathbf{J}_k^r]^{-1}(\mathbf{J}_{k-1}^r)^r, \tag{3.9.8}$$

where $\mathbf{I}^r$, $\mathbf{J}_k^r$ and $\mathbf{J}_{k-1}^r$ are $n \times r$ submatrices respectively made of the r columns of $\mathbf{I}$, $\mathbf{J}_k$ and $\mathbf{J}_{k-1}$ corresponding to the most recently updated r columns of $\mathbf{J}_k$. The updating formula (3.9.8) only requires the inversion of a $r \times r$ matrix, $(\mathbf{J}_{k-1}^r)^r\mathbf{J}_k^r$. Therefore in general it requires less time than the straightforward inversion of an $n \times n$ matrix $\mathbf{J}_k$. At its disadvantage, however, is a higher susceptibility to the accumulation of round-off errors.

Deferred Updating Approach

The other approach may be called deferred updating and can be seen in methods proposed by Traub (1964), Shamanskii (1967), Brent (1973) and Mukai (1979). In the deferred updating strategy the Jacobian is only occasionally re-evaluated. Given a solution estimate x_k, the next estimate x_{k+1} is computed as a result of s Newton–Raphson iterations with the same Jacobian or its finite difference approximation. The following algorithm is based on the Shamanskii method.

Algorithm shamanskii (root, s)

> *initialize root*
> *initialize stop*
> **while not** *stop* **do**
>   *compute J through* (3.9.4)
>   *compute* J^{-1}
>   **while (not** *stop* **and** *iter* $< s$**) do**
>   *iter* := *iter* + 1
>   *root* := *root* − $J^{-1} \times F(root)$
>   *compute new stop*
>   **enddo**
> **enddo**

Shamanskii (1967) has shown that for this algorithm the order of convergence is at least $s+1$. Earlier Traub (1964) has obtained the same result for an algorithm based on the exact Jacobian.

3.10 Numerical Stability of Iterative Algorithms

Until recently it was generally believed that there was less danger of serious error accumulation in iterative algorithms than in direct algorithms, i.e. algorithms that correspond to computing methods involving a finite number of operations. The argument was that at each iteration one works with the iteration function, for which the values of the parameters are preserved unchanged throughout the sequence of iterations. In consequence, numerical stability problem for iterative algorithms did not seem important. In fact, for some specific groups of iterative methods this belief can be supported by readily carried out analysis. For example, consider iterative methods of the general form

$$x_{n+1} = \phi(x_n) \tag{3.10.1}$$

where the iteration function depends on an immediately preceding approximation, x_n, only.

Denote by $\bar{x}_1, \bar{x}_2, \ldots$ the computed sequence of approximations to a root α and by δ_n the error in computing $\phi(\bar{x}_n)$.

We have

$$\bar{x}_{n+1} = \phi(\bar{x}_n) + \delta_n, \quad n = 0, 1, 2, \ldots, \tag{3.10.2}$$

whence, subtracting the equation $\alpha = \phi(\alpha)$ and using the mean value theorem on the result, we obtain

$$\bar{x}_{n+1} - \alpha = \phi'(\xi_n)(\bar{x}_n - \alpha) + \delta_n, \quad \xi_n \in \operatorname{int}(\bar{x}_n, \alpha). \tag{3.10.3}$$

Rewrite the expression (3.10.3) in the form

$$[1 - \phi'(\xi_n)]\,(\bar{x}_{n+1} - \alpha) = \phi'(\xi_n)(\bar{x}_n - \bar{x}_{n+1}) + \delta_n. \tag{3.10.4}$$

Assuming further that

$$|\phi'(\xi_n)| \leqslant m < 1 \quad \text{and} \quad |\delta_n| < \delta$$

we obtain

$$|\bar{x}_{n+1} - \alpha| < \frac{m}{1-m}|\bar{x}_{n+1} - \bar{x}_n| + \frac{1}{1-m}\delta, \tag{3.10.5}$$

which gives a strict estimate of the error in $\bar{x}_{n+1}$ in terms of known quantities. The first term on the right hand side of (3.10.5) estimates the truncation error and the second term the round-off error.

We note that the final computing error depends only on the error, δ_n, occurring in the immediately preceding iteration, and $\bar{x}_n$ must be considered as an exact quantity when estimating δ_n. It follows that it is not necessary to work to full accuracy in the earlier iterations since round-off errors made at this stage do not effect the final accuracy. In this sense we can say that iterative methods are self-correcting.

In (3.10.5), for sufficiently large n, the round-off error becomes dominant. In this case further iterations do not give any improvement in accuracy while the error varies randomly with a constant size of order $\delta/(1-m)$. However, the research carried out by Wozniakowski (1975b, 1975c) has shown that for wider groups of iterative methods the problem of numerical stability is of importance. In his study, Wozniakowski defines the concepts of condition number, numerical stability and problem conditioning for solving single and sets of non-linear equations. He shows that only if a problem is well-conditioned and if the iterative algorithm used to solve the problem is numerically stable, can one expect to obtain a good approximation to the solution. He also demonstrates that under a natural assumption about the computed evaluation of f, the Newton–Raphson and the secant methods are numerically stable.

Exercises

1. Give an algorithm for the secant method.
2. Give an algorithm for Krautstengel's method.
3. Show that the multipoint iteration function

$$x_{i+1} = x_i - \frac{f(x_i)}{f'(x_i)} \frac{f(z_i) - f(x_i)}{2f(z_i) - f(x_i)}, \qquad \text{where} \qquad z_i = x_i - \frac{f(x_i)}{f'(x_i)}$$

   is of order 4.
4. A simple real zero α of a non-linear function $f(x)$ is computed using an iterative method of the general form

$$x_{i+1} = \phi(x_i, x_{i-1}, \ldots, x_{i-m}), \quad x_0 \text{ arbitrary.}$$

   For $m = 0$ and $m = 1$, derive the conditions under which the sequence of values $x_0, x_1, \ldots$ converges to the root α.
5. Prove that the equation $x^{m+1} - e \sum_{j=0}^{m} x^j = 0$ has a unique largest real root x which for any $m \geqslant 0$, lies in the interval $e \leqslant x \leqslant e+1$.
6. Prove that the largest attainable order of convergence of a one-point iterative method which uses no derivatives is 2.
7. Give algorithms for computing the square root of a number using the iterative formulae of (a) Newton–Raphson's method, (b) the secant method, (c) Traub's method, (d) Jarratt's method and (e) the fifth order method.
8. Obtain the square root of number 5 using Dekker's algorithm.
9. In exercise 7 obtain the algorithms for computing nth root of a number.
10. In the equation $x = a - bx^2$, where a and b are positive constants, the positive root is computed using the method $x_{i+1} = a - bx_i^2$. Under what conditions the method is convergent?

11. Derive a multipoint iterative method by combining the Newton–Raphson and the method of false position. Given the order of convergence of this new method.
12. Using the methods of Exercise 7 determine to an accuracy of 10^{-5} all roots of the equations:

$$2^x - 2x^2 - 1 = 0,$$
$$2 \log x - x/2 + 1 = 0,$$
$$e^{-x} - (x-1)^2 = 0,$$
$$2 - x\,e^x = 0.$$

13. The order of convergence of an iterative method $x_{i+1} = \phi(x_i)$, $i = 0, 1, \ldots$ which solves $f(x) = 0$ for a simple root α, can be also defined as the number p such that $\phi'(\alpha) = \phi''(\alpha) = \ldots = \phi^{(p-1)}(\alpha) = 0$ but $\phi^{(p)}(\alpha) \neq 0$. Using this definition show that both the Traub and Jarratt methods are of order 4.
14. Using the inverse polynomial interpolation and setting $e = 1$, $m = 3$, derive a one-point iterative method of the form $x_{i+1} = \phi_{e,m}(x_i, \ldots, x_{i-m})$ which finds a simple root α of the equation $f(x) = 0$.
15. Show that the multipoint method (3.7.2) is of order 3.
16. Given a general multipoint formula

$$x_{i+1} = \phi(x_i) = x_i - \alpha_1 w_1(x_i) - \alpha_2 w_2(x_i),$$

where $w_1(x) = f(x)/f'(x)$, $\quad w_2(x) = f(x)/f'(x + \beta w_1(x))$,

determine values for the parameters α_1, α_2, β so as to obtain a third order multipoint method for finding a simple zero of $f(x)$, that requires one function and two derivative evaluations per iteration.
17. Give a Newton-like algorithm for solving a set of non-linear equations which uses the exact Jacobian.
18. In partial updating approach at each iteration only one or at most r, $r < n$, columns of the Jacobian are updated and this is as given by formulae (3.9.6) and (3.9.7). In this context suggest how the function $S(k)$ can be defined so as to ascertain that though the selection of the r columns to be updated is arbitrary nevertheless each column of the $n \times n$ Jacobian is updated at least once in $\lceil n/r \rceil$ iterations.

[*Hint*: try the following function

$$S(k) = \{(kr+1) \bmod n,\ (k+2) \bmod n,\ \ldots,\ (kr+r) \bmod n\}.]$$

19. Prove relation (3.9.8).

4

Direct Methods for Solving a Set of Linear Equations

In this chapter computational complexity and numerical accuracy is studied of some of the better-known algorithms for solving a set of linear algebraic equations. The principal component of these algorithms is the procedure of matrix multiplication. Matrix multiplication is principally used in many other successful algorithms of linear algebra, for example, matrix inversion, the evaluation of the determinant, boolean matrix multiplication and the transitive closure of a graph. The computational time required for matrix multiplication dominates the total computational time required to solve any of these problems. We shall discuss some recently developed matrix multiplication algorithms which compute the product of two matrices using significantly fewer arithmetic operations compared with the standard techniques. Some of these 'fast' multiplication algorithms are yet of theoretical interest only. They can actually supersede the standard algorithms only when applied to solve problems of truly large size. Therefore the 'fast' algorithms are referred to as the 'asymptotically faster' multiplication algorithms. Furthermore, the problem of numerical error control for these algorithms is not yet understood sufficiently well. However the new algorithms draw attention to the fact that the obvious algorithms are not necessarily the best. The new algorithms also form a basis for the development of genuinely efficient algorithms for this important class of problems.

We shall restrict ourselves to the so called direct methods for solving a set of linear equations as opposed to iterative methods, which are not the subject of this chapter.

A direct method of finding $\mathbf{x}$ to satisfy $\mathbf{Ax} = \mathbf{b}$ is one that defines $\mathbf{x}$ in terms of a finite number of operations on scalars. One such method is to invert $\mathbf{A}$ and then to calculate $\mathbf{x}$ from $\mathbf{x} = \mathbf{A}^{-1}\mathbf{b}$. In practice the determination of the inverse, $\mathbf{A}^{-1}$, is a computationally difficult process in itself. Therefore, the method based on matrix inversion is never used for solving sets of linear equations. The inverse $\mathbf{A}^{-1}$ may however be important in contexts other than the solution of linear equations and hence techniques for the efficient computation of $\mathbf{A}^{-1}$ are sought along with the general methods for the solution of simultaneous linear equations. Majority of the direct methods are the methods of factorisation in which the matrix **A** is transformed into the product of two easily invertible matrices. For example, in triangular decomposition **A** is transformed into the product of a lower triangular matrix, **L**, and an upper triangular matrix, **U**.

Thus

$$\mathbf{LUx} = \mathbf{b},$$

giving

$$\mathbf{x} = \mathbf{U}^{-1}\mathbf{L}^{-1}\mathbf{b}.$$

We intend to study direct methods and to assess their efficiency basing our judgement upon two fundamental criteria, susceptibility to round-off errors and the number of arithmetic operations a particular method requires.

The analysis of round-off errors in direct methods has an interesting and educational history. The first analyses of computer round-off errors were attempted even before modern computers have been built. Then, in 1947 von Neumann and Goldstine published a paper concerning the inversion of a positive definite matrix on a digital computer using an elimination method. They noted that computer arithmetic would not possess the fundamental properties of associativity and distributivity. However, because of the way they handled pivoting and because they looked at the error rather than the residual they were very pessimistic about the reliability of solutions to systems of order greater than about 20. This pessimism discouraged the use of direct methods in favour of iterative methods because of the iterative methods numerical stability. The backward error analysis of Wilkinson has however led to a realization that the more stable direct methods are generally very accurate and the idea that more accurate results can usually be obtained only by iterative methods has been abandoned.

4.1 Gaussian Elimination

The method first developed by Gauss early in the nineteenth century, has been studied by various authors and with various conclusions on its reliability but its complete analysis became possible only with the introduction of Wilkinson's backward error analysis. It has been established that the method is generally reliable and efficient.

We shall first introduce the method, examining in particular its most commonly used variant, known as triangular decomposition, and shall then proceed to analyse its computational complexity and numerial stability.

Gaussian Elimination

Gaussian elimination is the process of subtracting suitable multiples of each row of the system of equations from all succeeding rows to produce a matrix with zeros below and to the left of the main diagonal. If the right-hand side, **b**, is carried along as an extra column these operations are known as forward elimination. The resulting augmented matrix represents a system of equations with the same solution as the original. This solution is easily found by a back substitution process, because the last equation involves only the last unknown, the next to last equation only the last two unknowns, and so on.

Consider the following system of equations:

$$
\begin{aligned}
&(1)\ a_{11}x_1 + a_{12}x_2 + \ldots\ldots + a_{1n}x_n = b_1 \\
&(2)\ a_{21}x_1 + a_{22}x_2 + \ldots\ldots + a_{2n}x_n = b_2 \\
&\quad \ldots \\
&(n)\ a_{n1}x_1 + a_{n2}x_2 + \ldots\ldots + a_{nn}x_n = b_n
\end{aligned}
\tag{4.1.1}
$$

The first step of the Gaussian elimination involves in (4.1.1) the elimination of the variable x_1 from the equations (2) to (n) as follows. For each $i = 2$ to n multiply the first equation by a_{i1}/a_{11} (assuming $a_{11} \neq 0$) and subtract it from the ith equation thereby obtaining a new set of equations (2) to (n). This transformation is also applied to **b**. We now have $(n-1)$ equations in the $(n-1)$ unknowns x_2 to x_n since the coefficients of x_1 have been made zero in the equations (2) to (n). The process is then repeated on the new equations (2) to (n), assuming that the leading coefficient of the equation (2) is non-zero. This second step produces $(n-2)$ equations in $(n-2)$ unknowns. Successive eliminations are made in this way until the process concludes with the nth equation involving only x_n. The matrix of the coefficients for the new system of equations is upper triangular. If at any point in the above process the leading coefficient (which is also called a pivot) of the first of the remaining equations is zero, then before the next elimination is performed the equations or the variables must be permuted so that one obtains a non-zero pivot. Schemes for doing this are known as pivoting. The 'partial pivoting', which is now used almost exclusively, involves searching the corresponding rth column for its largest element. Multiples of the row containing this element are then used to introduce the zeros in the other rows. The row containing the pivot is interchanged with the rth row. The other scheme, known as 'complete pivoting', uses the largest element in the entire unreduced matrix. With either form of pivoting none of the multipliers is greater than unity in magnitude. Complete pivoting can produce more accurate results for some matrices but it is so much more expensive that its use may be justified only in very special cases. The transformation of **A** into upper triangular form is known as *forward elimination*. The remainder of the solution is obtained by *back substitution* in which one starts by substituting the value for x_n into the previous equation which involves x_n and x_{n-1}, to obtain the value of x_{n-1}. This process of substitution is continued until the value of x_1 is obtained. In ordinary Gaussian elimination the transformation of the vector **b** is done simultaneously with the processing of **A**. The multipliers created during forward elimination are normally saved along with the new coefficients.

There are two reasons for this. First, if there is more than one right-hand side then the forward elimination process need only be carried once. Second, if the process of iterative improvement of the solution is used then starting with the first solution the multipliers along with the new coefficients are used to compute new solutions.

Evaluation of the Determinant

Gaussian elimination may also be applied to compute the determinant. Evaluation of the determinant is of importance in many applications involving matrices. One obvious example is the solution of a linear system $\mathbf{Ax} = \mathbf{b}$. Here the solution vector is given as $\mathbf{x} = \mathbf{A}^{-1}\mathbf{b}$ and hence the system has a unique solution if and only if its matrix is non-singular, that is the inverse $\mathbf{A}^{-1}$ exists. This in other words means that the determinant is non-zero. So, in general, in order to solve $\mathbf{Ax} = \mathbf{b}$ one first needs to evaluate det $\mathbf{A}$.

We shall now outline a computational scheme which evaluates the determinant using Gaussian elimination. Let the determinant of $\mathbf{A}$ be

$$\det \mathbf{A} = |\mathbf{A}| = \begin{vmatrix} a_{11} & a_{12} & \cdots & a_{1n} \\ a_{21} & a_{22} & \cdots & a_{2n} \\ \vdots & & & \\ a_{n1} & a_{n2} & \cdots & a_{nn} \end{vmatrix} \tag{4.1.2}$$

and let $a_{11} \neq 0$.

Taking the element a_{11} outside the first row we shall have

$$\det \mathbf{A} = |\mathbf{A}| = a_{11} \begin{vmatrix} 1 & b_{12} & \cdots & b_{1n} \\ a_{21} & a_{22} & \cdots & a_{2n} \\ \cdots & & & \\ a_{n1} & a_{n2} & \cdots & a_{nn} \end{vmatrix}, \text{ where } b_{1j} = \frac{a_{1j}}{a_{11}}, \; j = 2, \ldots, n.$$

Next, subtracting from each of the remaining rows the first row multiplied by the first element of the relevant row we obtain

$$|\mathbf{A}| = a_{11} \begin{vmatrix} 1 & b_{12} & \cdots & b_{1n} \\ 0 & a_{22}^{(1)} & \cdots & a_{2n}^{(1)} \\ & \cdots & & \\ 0 & a_{n2}^{(1)} & \cdots & a_{nn}^{(1)} \end{vmatrix} = a_{11} \begin{vmatrix} a_{22}^{(1)} & \cdots & a_{2n}^{(1)} \\ & \cdots & \\ a_{n2}^{(1)} & \cdots & a_{nn}^{(1)} \end{vmatrix},$$

where $a_{ij}^{(1)} = a_{ij} - a_{i,1}b_{1j}, \qquad i, j = 2, \ldots, n.$

With the determinant of the $(n-1)$th order that remains, we shall deal as before, provided $a_{22}^{(1)} \neq 0$. Carrying the process through, we find that the determinant is equal to the product of the leading elements:

$$\det\mathbf{A} = |\mathbf{A}| = a_{11}a_{22}^{(1)} \ldots a_{nn}^{(n-1)}.$$

If at any step it should turn out that $a_{ii}^{i-1} = 0$, pivoting can be performed so that a non-vanishing element appears in the upper left corner.

Inspection of the computation process shows that with the exception of the final multiplication the process is the same as the forward stage of the Gaussian elimination. It also shows that if one or more columns of **A** cannot offer a non-zero element at all, the determinant is equal to zero and hence **A** is singular.

We are now ready to present a Gaussian elimination algorithm to solve a linear system $\mathbf{Ax} = \mathbf{b}$ using partial pivoting. In the algorithm **A** is the system's matrix $\mathbf{A}[i,j]$, $i,j = 1, \ldots, n$, b is the system's right-hand vector $b[i]$, $i = 1, \ldots, n$, the vector $x[j]$, $j = 1, \ldots, n$ is the solution vector, and error is a boolean variable which assumes the value true when **A** is singular.

Algorithm gauss (A, b)

```
    error := false
    for i := 1 to n - 1 do
        max := ABS(A[i, j])
        rowmax := i
        jlow := i + 1
        for j := jlow to n do
        //search for largest in absolute value element//
            entry := ABS(A[j, i])
            if entry > max then
                max := entry
                rowmax := j
            endif
        enddo
        if max = 0.0 then error := true
        else
            if rowmax <> i then
            // interchange rows to put largest element on diagonal//
                for j := 1 to n do
                    atemp := A[rowmax, j]
                    A[rowmax, j] := A[i, j]
                    A[i, j] := atemp
                enddo
                btemp := b[rowmax]
                b[rowmax] := b[i]
                b[i] := btemp
            endif
            for j := jlow to n do
```

```
            pivot := A[j,i]/A[i,i]
            for k := i to n do
                A[j,k] := A[j,k] - pivot × A[i,k]
            enddo
            b[j] := b[j] - pivot × b[i]
        enddo
    endif
    if A[n,n] = 0.0 then error := true
    else
        x[n] := b[n]/A[n,n]
        i := n - 1
        //back substitution//
        repeat
            sum := 0.0
            for j := i + 1 to n do
                sum := sum + A[i, j] × x[n]
            enddo
            x[i] := (b[i] - sum)/A[i,i]
            i := i - 1
        until i = 0
    endif
    if error then matrix is singular
enddo
```

Triangular Decomposition

In the triangular decomposition the matrix $\mathbf{A}$ is first decomposed into a lower triangular matrix with unit diagonal, $\mathbf{L}$, and an upper triangular matrix, $\mathbf{U}$,

$$\mathbf{A} = \mathbf{LU} \tag{4.1.3}$$

The solution of equation $\mathbf{LUx} = \mathbf{b}$, is then obtained by solving two equations:

$$\begin{aligned} \mathbf{Ly} &= \mathbf{b} \\ \mathbf{Ux} &= \mathbf{y} \end{aligned} \tag{4.1.4}$$

If we denote the original set of equations by:

$$\mathbf{A}^{(1)}\mathbf{x} = \mathbf{b}^{(1)} \tag{4.1.5}$$

then each of the $(n-1)$ steps of Gaussian elimination leads to a new set of equations which are equivalent to the original set. We denote the rth equivalent set by

$$\mathbf{A}^{(r)}\mathbf{x} = \mathbf{b}^{(r)} \tag{4.1.6}$$

$\mathbf{A}^{(r)}$ is already upper triangular as far as its first r rows are concerned and it has a square matrix of order $(n - r + 1)$ in the bottom right-hand corner. As an example take $\mathbf{A}^{(3)}$ when $n=5$:

$$\mathbf{A}^{(3)} = \begin{bmatrix} a & a & a & a & a \\ 0 & a & a & a & a \\ 0 & 0 & a & a & a \\ 0 & 0 & a & a & a \\ 0 & 0 & a & a & a \end{bmatrix} \text{ where the } a\text{'s indicate non-zero elements}$$

The matrix $\mathbf{A}^{(r+1)}$ is derived from $\mathbf{A}^{(r)}$ by subtracting a multiplier m_{ir} of the rth row from the ith row for values of i from $r+1$ to n. The multipliers m_{ir} are defined by:

$$m_{ir} = a_{ir}^{(r)}/a_{rr}^{(r)} \qquad i = r+1, \ldots, n. \tag{4.1.7}$$

The rth row of $\mathbf{A}^{(r)}$ is called the rth pivotal row and $a_{rr}^{(r)}$ is called the rth pivot.

(a) Consider each element for which $i \leqslant j$.
The element is modified in each transformation until $\mathbf{A}^{(i)}$ is obtained, after which the element remains unchanged. We can write

$$\begin{aligned} a_{ij}^{(i)} &= a_{ij}^{(i-1)} - m_{i,i-1} a_{i-1,j}^{(i-1)} \\ &= a_{ij}^{(1)} - \sum_{l=1}^{i-1} m_{il} a_{lj}^{(1)} \end{aligned} \tag{4.1.8}$$

where the $a_{ij}^{(i)}$ are the elements of $\mathbf{A}^{(1)}$.

(b) Consider the elements for which $i > j$.
The element is modified as in (a) until $\mathbf{A}^{(j)}$ is obtained. $a_{jj}^{(j)}$ is then used to compute m_{ij}, and $a_{ij}^{(j+1)}$ to $a_{ij}^{(n)}$ are each taken to be exactly equal to zero, m_{ij} satisfies

$$m_{ij} = \frac{a_{ij}^{(j)}}{a_{jj}^{(j)}} \tag{4.1.9}$$

$$\text{and } a_{ij}^{(j)} = a_{ij}^{(j-1)} - m_{i,j-1} a_{j-1,j}^{(j-1)} = a_{ij}^{(1)} - \sum_{l=1}^{j-1} m_{il} a_{lj}^{(l)} \tag{4.1.10}$$

The set of n^2 equations given by (4.1.8) and (4.1.10) can be written in the matrix form as $\mathbf{A} = \mathbf{LU}$, where for $n=4$, say, we have

$$\mathbf{L} = \begin{bmatrix} 1 & 0 & 0 & 0 \\ m_{21} & 1 & 0 & 0 \\ m_{31} & m_{32} & 1 & 0 \\ m_{41} & m_{42} & m_{43} & 1 \end{bmatrix} \text{ and } \mathbf{U} = \begin{bmatrix} a_{11}^{(1)} & a_{12}^{(1)} & a_{13}^{(1)} & a_{14}^{(1)} \\ 0 & a_{22}^{(2)} & a_{23}^{(2)} & a_{24}^{(2)} \\ 0 & 0 & a_{33}^{(3)} & a_{34}^{(3)} \\ 0 & 0 & 0 & a_{44}^{(4)} \end{bmatrix} \tag{4.1.11}$$

It may happen that in (4.1.7) for some r the $a_{rr}^{(r)}$ will be either zero or very small. In such case division by this particular $a_{rr}^{(r)}$ must be prevented.

Storage Requirements for Triangular Decomposition

The memory requirements may be substantial in the problems involving processing of matrices. For example, when solving a set of n linear equations $\mathbf{Ax} = \mathbf{b}$ a two-dimensional input array of size n is used to store the matrix. In addition the intermediate results of the matrix manipulation are needed to be temporarily stored. The total volume of memory required may become an important factor which limits the size of the problem that can be solved on a particular computer.

The speed of the solution is obviously also influenced by the amount of auxiliary memory used. Thus whenever an algorithm solves $\mathbf{Ax} = \mathbf{b}$ in place, that is without any auxiliary memory required, it adds to the attractiveness of the algorithm. Triangular decomposition algorithm does not use any extra storage as the coefficients of the reduced equations do not need to be stored separately. The current equation is simply overwritten by the new coefficients derived from the elimination process and the multipliers are recorded in the spaces left by the coefficients that have been eliminated.

In place of the pivots their reciprocals are stored for further use in the formation of the multipliers and in the subsequent back substitution. For example, the intermediate matrix $\mathbf{A}^{(3)}$ for $n = 5$ will look like this

$$\mathbf{A}^{(3)} = \begin{bmatrix} [a_{11}^{(1)}]^{-1} & a_{12}^{(1)} & a_{13}^{(1)} & a_{14}^{(1)} & a_{15}^{(1)} \\ m_{21} & [a_{22}^{(2)}]^{-1} & a_{23}^{(2)} & a_{24}^{(2)} & a_{25}^{(2)} \\ m_{31} & m_{32} & [a_{33}^{(3)}]^{-1} & a_{34}^{(3)} & a_{35}^{(3)} \\ m_{41} & m_{42} & a_{43}^{(3)} & a_{44}^{(3)} & a_{45}^{(3)} \\ m_{51} & m_{52} & a_{53}^{(3)} & a_{54(3)} & a_{55}^{(3)} \end{bmatrix} \tag{4.1.12}$$

When the solution computed is not sufficiently accurate, an iterative technique, known as the iterative refirement method, can be used to improve the solution. In this case the original matrix $\mathbf{A}$ needs to be retained along with the triangular decomposition solution. This implies the use of an auxiliary array of the same size as the input array. An **LU** decomposition algorithm is implemented below. In the algorithm decomposition given is a non-singular square matrix $\mathbf{A}$ of size n. The algorithm decomposes the matrix into the product of two matrices $\mathbf{L}$ and $\mathbf{U}$, where $\mathbf{L}$ is a lower triangular matrix with unity entries on the main diagonal and $\mathbf{U}$ is an upper triangular matrix. The original matrix $\mathbf{A}$ is retained at the end of the process while the results are stored in the auxiliary array $\mathbf{B}$, where the strictly lower triangle contains the elements of the lower triangular matrix $\mathbf{L}$ and the upper triangle including the main diagonal contains the elements of the upper triangular matrix $\mathbf{U}$.

Algorithm decompostion (*A*)

```
for i := 1 to n do
    for j := 1 to n do
        B[i, j] := A[i, j]
    enddo
enddo
for i := 1 to n - 1 do
    max := ABS(B[i, i])
    rowmax := i
    jlow := i + 1
    for j := jlow to n do
    //search for largest in absolute value element//
        entry := ABS(B[j, i])
        if entry > max then
            max := entry
            rowmax := j
        endif
    enddo
    if rowmax < > i then
    // interchange rows to put largest element on diagonal//
        for j := 1 to n do
            temp := B[rowmax, j]
            B[i, j] := temp
        enddo
    endif
    for j := jlow to n do
        pivot := B[j, i]/B[i, i]
        for k := i to n do
            B[j, k] := B[j, k] - pivot × B[i, k]
        enddo
        B[j, i] := pivot
    enddo
enddo
```

Algebraic Complexity of Gaussian Elimination

We shall first consider the triangular decomposition phase. Formula (4.1.8) shows that if $i <= j$ then the ijth element of $\mathbf{A}$ is modified as

$$a_{ij}^{(i)} = a_{ij}^{(1)} - \sum_{l=1}^{i-1} m_{il} a_{ij}^{(l)}$$

To compute $a_{ij}^{(i)}$ calls for

$$(i - 1) \text{ multiplications and } (i - 1) \text{ additions.} \tag{4.1.13}$$

(Note that as usual 'additions' are taken to mean both additions and subtractions.)

Similarly, formula (4.1.10) shows that if $i > j$ then the ijth element of **A** is modified as

$$a_{ij}^{(j)} = a_{ij}^{(1)} - \sum_{l=1}^{j-1} m_{il} a_{ij}^{(l)}$$

To compute $a_{ij}^{(j)}$ calls for

$(j-1)$ multiplications and $(j-1)$ additions. (4.1.14)

Furthermore the computation of the $n(n-1)/2$ multipliers, m_{ij}, requires

n reciprocals and $n(n-1)/2$ multiplications. (4.1.15)

The total number of multiplications required to decompose **A** into **L** and **U** is given by

$$\sum_{j=1}^{n} \left[\sum_{i=1}^{j} (i-1) + \sum_{i=j+1}^{n} (j-1) \right] + \frac{n(n-1)}{2} = \frac{n^3}{3} - \frac{n}{3} \tag{4.1.16}$$

Similarly, the number of additions is given by

$$\sum_{j=1}^{n} \left[\sum_{i=1}^{j} (i-1) + \sum_{i=j+1}^{n} (j-1) \right] = \frac{n^3}{3} - \frac{n^2}{2} + \frac{n}{6}. \tag{4.1.17}$$

In total, the decompositions of **A** into **L** and **U** requires

$\frac{n^3}{3} - \frac{n}{3}$ multiplications,

$\frac{n^3}{3} - \frac{n^2}{2} + \frac{n}{6}$ additions, (4.1.18)

and n reciprocals.

Back Substitution

To complete the solution of $\mathbf{Ax} = \mathbf{b}$, the following steps are performed:

(i) y is evaluated from $\mathbf{Ly} = \mathbf{b}$.
(ii) x is evaluated from $\mathbf{Ux} = \mathbf{y}$.

(i) can be written in the form:

$$\begin{aligned}
&l_{11}y_1 &&= b_1 \\
&l_{21}y_1 + l_{22}y_2 &&= b_2 \\
&\dots \\
&l_{n1}y_1 + l_{n2}y_2 + \dots + l_{nn}y_n &&= b_n
\end{aligned}$$

The variables $y_1, y_2, \ldots, y_n$ are determined in succession from the first, second, ..., nth equations, respectively,

i.e. $y_1 = b_1/l_{11}$, $y_2 = (b_2 - l_{22}y_1)/l_{22}$, etc.

In a general step the evaluation of y_r requires one reciprocal (which has already been calculated), r multiplications and $r - 1$ additions. Hence to solve the system of equations $\mathbf{Ly} = \mathbf{b}$ requires $n(n-1)/2$ additions and $n(n+1)/2$ multiplications. The solution of $\mathbf{Ux} = \mathbf{y}$ requires the same amount of arithmetic. However in our case all $l_{ii} = 1$ and so the number of multiplications required is reduced by n.

To summarize, the amount of arithmetics required in the back substitution phase is

$$\begin{aligned} &n^2 \text{ multiplications} \\ \text{and } &n^2 - n \text{ additions,} \end{aligned} \tag{4.1.19}$$

and the grand total number of operations required to solve $\mathbf{Ax} = \mathbf{b}$ is

$$\begin{aligned} &\frac{n^3}{3} + n^2 - \frac{n}{3} \quad \text{multiplications,} \\ &\frac{n^3}{3} + \frac{n^2}{2} - \frac{5n}{6} \quad \text{additions,} \\ &n \text{ reciprocals,} \end{aligned} \tag{4.1.20}$$

which shows that the Gaussian elimination is of $O(n^3)$.

Matrix Inversion

An allied problem to solution of a set of linear equations is the inversion of a matrix. We shall give two methods to find the inverse of a matrix $\mathbf{A}$.

Method One

(a) $\mathbf{A}$ is decomposed into $\mathbf{L}$ and $\mathbf{U}$
(b) $\mathbf{L}$ and $\mathbf{U}$ are inverted
(c) $\mathbf{A}^{-1}$ is formed by premultiplying $\mathbf{L}^{-1}$ by $\mathbf{U}^{-1}$

Step (a) has already been discussed, so we shall turn to step (b). If a_{ij} are the elements of $\mathbf{U}$ and u_{ij} are the elements of $\mathbf{U}^{-1}$ then by the definition we have

$$\sum_{k=1}^{n} a_{ik}u_{kj} = \delta_{ij} = \begin{cases} 1 & \text{if } i = j, \\ 0 & \text{otherwise,} \end{cases}$$

for example, for $n = 4$ we have to find the u_{ij} such that

$$\begin{bmatrix} a_{11} & a_{12} & a_{13} & a_{14} \\ 0 & a_{22} & a_{23} & a_{24} \\ 0 & 0 & a_{33} & a_{34} \\ 0 & 0 & 0 & a_{44} \end{bmatrix} \begin{bmatrix} u_{11} & u_{12} & u_{13} & u_{14} \\ 0 & u_{22} & u_{23} & u_{24} \\ 0 & 0 & u_{33} & u_{34} \\ 0 & 0 & 0 & u_{44} \end{bmatrix} = \begin{bmatrix} 1 & 0 & 0 & 0 \\ 0 & 1 & 0 & 0 \\ 0 & 0 & 1 & 0 \\ 0 & 0 & 0 & 1 \end{bmatrix}$$

Here the u_{ij} are computed like this.

Compute $u_{ii} = 1/a_{ii}$ for $i = 1, \ldots, n$.
Compute u_{in} for $i = n - 1, \ldots, 1$.
Compute $u_{i,n-1}$ for $i = n - 2, \ldots, 1$.
etc.

The elements of $\mathbf{U}^{-1}$ are thus given by

$$u_{ii} = \frac{1}{a_{ii}}, \tag{4.1.21}$$

$$u_{ij} = -u_{ii} \sum_{p=i+1}^{j} a_{ip}u_{pj}, \quad i \neq j. \tag{4.1.22}$$

The computation of n recipocals in (4.1.21) is carried out as part of the decomposition of $\mathbf{A}$ into $\mathbf{L}$ and $\mathbf{U}$. To compute u_{ij}, $i \neq j$, *in* (4.1.22) calls for $(j - i + 1)$ multiplications and $(j - i - 1)$ additions.

Thus the total number of arithmetics required to compute the inverse of $\mathbf{U}$ is

$$\sum_{j=1}^{n} \sum_{i=1}^{j-1} (1 + j - i) = \frac{n(n-1)(n+4)}{6} \quad \text{multiplications}$$

and (4.1.23)

$$\sum_{j=1}^{n} \sum_{i=1}^{j-1} (j - i - 1) = \frac{n(n-1)(n-2)}{6} \quad \text{additions.}$$

It is easy to see that the same number of arithmetics is required to compute $\mathbf{L}^{-1}$, with exception that since the diagonal elements of $\mathbf{L}$ are unity the number of multiplications to compute $\mathbf{L}^{-1}$ is reduced by $n(n-1)$. Hence to invert $\mathbf{L}$ requires

$$\frac{n(n-1)(n-2)}{6} \quad \text{multiplications,}$$

$$\frac{n(n-1)(n-2)}{6} \quad \text{additions.} \tag{4.1.24}$$

The total number of arithmetics required to invert $\mathbf{L}$ and $\mathbf{U}$ is

$$\frac{n(n^2-1)}{3} \quad \text{multiplications,}$$

$$\frac{n(n-1)(n-2)}{3} \quad \text{additions.} \tag{4.1.25}$$

In a similar manner it can be shown that in step (c) of method one formation of $\mathbf{A}^{-1} = \mathbf{U}^{-1}\mathbf{L}^{-1}$ requires the number of multiplications given as

$$\sum_{j=1}^{n}\left[\sum_{i=1}^{j}(n-j)+\sum_{i=j+1}^{n}(n-i+1)\right]=\frac{n(n^2-1)}{3},$$

and the number of additions as

$$\sum_{j=1}^{n}\left[\sum_{i=1}^{j}(n-j)+\sum_{i=j+1}^{n}(n-i)\right]=\frac{n^3}{3}-\frac{n^2}{2}+\frac{n}{6}$$

The total number of arithmetic operations to invert a matrix by this method is

$$\begin{array}{ll} n^3-n & \text{multiplications} \\ n^3-2n^2+n & \text{additions} \end{array} \tag{4.1.26}$$

Method Two

(a) Decompose $\mathbf{A}$ into $\mathbf{L}$ and $\mathbf{U}$,
(b) Invert $\mathbf{L}$,
(c) Evaluate $\mathbf{A}^{-1}$ from $\mathbf{L}^{-1} = \mathbf{U}\mathbf{A}^{-1}$

Here only step (c) has not been previously considered. In this step to compute $\mathbf{A}^{-1}$ calls for

$$\frac{n^3}{2}+\frac{n^2}{2}-1 \quad \text{multiplications}$$

and (4.1.27)

$$\frac{n^3}{2}-n^2+\frac{n}{2} \quad \text{additions.}$$

The total number of arithmetics required by method two is

$$n^3-1 \quad \text{multiplications,}$$

and (4.1.28)

$$n^3-2n^2+n \quad \text{additions.}$$

It appears that an explicit evaluation of $\mathbf{U}^{-1}$ in method one calls for fewer multiplications. Yet the reduction of $n(n-1)$ multiplications in method one as compared with method two may be offset by the extra rounding error incurred during the process of computing $\mathbf{U}^{-1}$.

4.2 Error Analysis

We shall analyse numerical stability of the Gaussian elimination following Wilkinson's backward error analysis approach. For this a summary of basic results on the floating-point matrix multiplication and solution of a set of linear equations is in order. The summary is presented in Appendix A, and the reader is advised to consult Appendix A on all the necessary details.

Conditioning of the Problem

Consider the set

$$\mathbf{Ax} = \mathbf{b}. \tag{4.2.1}$$

Let $\mathbf{x} = \mathbf{A}^{-1}\mathbf{b}$ be the exact solution of the set and $\mathbf{x}^{(c)}$ be its computed solution.

We shall distinguish between the difference

$$\mathbf{A}^{-1}\mathbf{b} - \mathbf{x}^{(c)} = \mathbf{x} - \mathbf{x}^{(c)} \tag{4.2.2}$$

which is called the error and the difference

$$\mathbf{b} - \mathbf{Ax}^{(c)} \tag{4.2.3}$$

which is called the residual.

Studies on the conditioning of problem (4.2.1) show that if both the error and the residual vectors are small then the problem is well-conditioned. If the residual is small while the error is very large then the problem is called ill-conditioned. The matrix of an ill-conditioned problem is usually 'nearly singular' or even exactly singular.

If an instance of problem (4.2.1) is ill-conditioned then no computational algorithm will solve it accurately. For a well-conditioned problem one expects that a reasonable approximation to the exact solution can be computed if a stable computational algorithm is used. Numerical stability of a particular algorithm is studied using the concept of the error as defined by (4.2.2), or of the residual as defined by (4.2.3). If the solution, $\mathbf{x}^{(c)}$, is computed using a particular algorithm and the error $\mathbf{x} - \mathbf{x}^{(c)}$ is small or the residual $\mathbf{b} - \mathbf{Ax}^{(c)}$ is small then the algorithm is numerically stable and vice versa.

In the forward error analysis (of pre-Wilkinson era) one attempts to bound the difference between the exact and computed solutions at every step of the computation. As the computation progresses this becomes more and more difficult, the bounds on the difference between the solutions become 'loose', and this leads to far too pessimistic conclusions on the method's reliability. The Wilkinson's backward error analysis, on the other hand, studies the residual and not the error. The approach is simpler and gives reasonably sharp bound estimates on the residual which in turn leads to more realistic conclusions on the reliability of the method.

Since the introduction of backward error analysis many important results have been established, and the general reliability of the Gaussian elimination is one of them. Indeed the technique of backward error analysis is considered to be one of the most important successes in mathematical error analysis.

Numerical Stability of Triangular Decomposition

In the error analysis which is carried out below we follow closely the work of Wilkinson (1963, 1965) and Forsythe and Moler (1967).

Wilkinson in his backward error analysis has proved the following theorem.

> Consider $\mathbf{Ax} = \mathbf{b}$ where $\mathbf{A}$ and $\mathbf{b}$ are a given matrix and vector with elements which are floating-point numbers. Let $\mathbf{x}^{(c)}$ be the vector of floating-point numbers computed by Gaussian elimination. Then there is a matrix $\mathbf{E}$ (with real number elements) such that $\mathbf{x}^{(c)}$ is the exact solution of
>
> $$(\mathbf{A} + \mathbf{E})\mathbf{x}^{(c)} = \mathbf{b}. \tag{4.2.4}$$
>
> The matrix $\mathbf{E}$ is called the *perturbation matrix.*
> For a well-conditioned problem the perturbation matrix is always small in comparison with $\mathbf{A}$.

Equation (4.2.1) can be written as

$$\mathbf{b} - \mathbf{Ax}^{(c)} = \mathbf{Ex}^{(c)} \tag{4.2.5}$$

which shows that the bounds on the residual $\mathbf{b} - \mathbf{Ax}^{(c)}$ are known as soon as the bounds on the perturbation $\mathbf{E}$ are known.

We shall now study the bounds on the perturbation $\mathbf{E}$ under various conditions on the computed solution, $\mathbf{x}^{(c)}$. Recall that the process of computing the $\mathbf{x}^{(c)}$ by triangular decomposition consists of two stages.

First, the factors $\mathbf{L}$ and $\mathbf{U}$ are computed such that $\mathbf{A} = \mathbf{LU}$. In fact, the computed factors $\mathbf{L}$ and $\mathbf{U}$ will satisfy the relation

$$\mathbf{LU} = \mathbf{A} + \partial\mathbf{A}, \tag{4.2.6}$$

where $\delta\mathbf{A}$ is the error matrix due to the round-off errors.

Second, two sets of equations,

$$\mathbf{Ly} = \mathbf{b} \tag{4.2.7}$$

and $\mathbf{Ux} = \mathbf{y}$, (4.2.8)

are solved. The computed solution $\mathbf{y}^{(c)}$ of set (4.2.7) is the exact solution of the set

$$(\mathbf{L} + \partial\mathbf{L})\mathbf{y}^{(c)} = \mathbf{b} \tag{4.2.9}$$

and similarly the computed solution $\mathbf{x}^{(c)}$ of set (4.2.8) is the exact solution of the set

$$(\mathbf{U} + \partial\mathbf{U})\mathbf{x}^{(c)} = \mathbf{y}, \tag{4.2.10}$$

where each of the matrices, $\partial\mathbf{L}$ and $\partial\mathbf{U}$, is what we may call the local perturbation matrix.

Hence, the computed solution, $\mathbf{x}^{(c)}$, of the set (4.2.4) is the exact solution of the set

$$(\mathbf{L}+\partial\mathbf{L})(\mathbf{U}+\partial\mathbf{U})\mathbf{x}^{(c)}=\mathbf{b} \quad \text{where } \mathbf{LU}=\mathbf{A}+\partial\mathbf{A},$$

which can be written as

$$(\mathbf{LU}+\partial\mathbf{A}+\mathbf{U}\partial\mathbf{L}+\mathbf{L}\partial\mathbf{U})\mathbf{x}^{(c)}=\mathbf{b}.$$

Denoting the sum $\partial\mathbf{A}+\mathbf{U}\partial\mathbf{L}+\mathbf{L}\partial\mathbf{U}$ by $\mathbf{E}$, we can write

$$(\mathbf{LU}+\mathbf{E})\mathbf{x}^{(c)}=\mathbf{b}. \tag{4.2.11}$$

In order to derive a bound on the perturbation matrix **E** consider the perturbations $\partial\mathbf{A}$, $\partial\mathbf{L}$ and $\partial\mathbf{U}$.

Bound Estimate on $\partial\mathbf{A}$

To estimate the bound on $\partial\mathbf{A}$ consider again the process of triangular decomposition defined by equations (4.1.6) to (4.1.11).

(a) Consider the elements a_{ij} where $i \leqslant j$. The element is modified in each transformation until $\mathbf{A}^{(i)}$ is obtained after which it remains fixed. We can write

$$\begin{aligned}
a_{ij}^{(2)} &\equiv a_{ij}^{(1)} - m_{i1}a_{1j}^{(1)} + d_{ij}^{(2)},\\
a_{ij}^{(3)} &= a_{ij}^{(2)} - m_{i2}a_{2j}^{(2)} + d_{ij}^{(3)},\\
&\dots\\
a_{ij}^{(i)} &\equiv a_{ij}^{(i-1)} - m_{i,i-1}a_{i-1,j}^{(i-1)} + d_{ij}^{(i)},
\end{aligned} \tag{4.2.12}$$

$d_{ij}^{(k)}$ is the difference between the accepted $a_{ij}^{(k)}$ and the exact value which would be obtained using the computed $a_{ij}^{(k-1)}$, $m_{i,k-1}$ and $a_{k-1,j}^{(k-1)}$.

Summing up equations (4.2.12) we obtain

$$a_{ij}^{(i)} \equiv a_{ij}^{(1)} - \sum_{l=1}^{i-1} m_{il}a_{lj}^{(l)} + e_{ij}, \tag{4.2.13}$$

where

$$e_{ij} = \sum_{k=2}^{i} d_{ij}^{(k)}. \tag{4.2.14}$$

(b) Consider the elements where $i > j$. The element is modified as in (a) until $\mathbf{A}^{(j)}$ is obtained; $a_{ij}^{(j)}$ is then used to compute m_{ij}, and $a_{ij}^{(j+1)}$ to $a_{ij}^{(n)}$ are taken to be exactly equal to zero. The computed multiplier satisfies

$$m_{ij} \equiv a_{ij}^{(j)}/a_{jj}^{(j)} + q_{ij}, \tag{4.2.15}$$

where q_{ij} is the rounding error involved in division. (4.2.15) can be written as

$$0 \equiv a_{ij}^{(i)} - m_{ij}a_{jj}^{(j)} + d_{ij}^{(j+1)}, \tag{4.2.16}$$

where

$$d_{ij}^{(j+1)} = a_{jj}^{(j)}q_{ij}. \tag{4.2.17}$$

Thus we can write

$$\begin{aligned} a_{ij}^{(2)} &\equiv a_{ij}^{(1)} - m_{i1}a_{1j}^{(1)} + d_{ij}^{(2)}, \\ a_{ij}^{(3)} &\equiv a_{ij}^{(2)} - m_{i2}a_{2j}^{(2)} + d_{ij}^{(3)}, \\ &\dots \\ a_{ij}^{(j)} &\equiv a_{ij}^{(j-1)} - m_{i,j-1}a_{j-1,j}^{(j-1)} + d_{ij}^{(j)}, \\ 0 &\equiv a_{ij}^{(j)} - m_{ij}a_{jj}^{(j)} + d_{ij}^{(j+1)}. \end{aligned} \tag{4.2.18}$$

Summing up (4.2.18) we obtain

$$0 \equiv a_{ij}^{(1)} - \sum_{l=1}^{j} m_{il}a_{lj} + e_{ij}, \tag{4.2.19}$$

where

$$e_{ij} = \sum_{k=2}^{j+1} d_{ij}^{(k)}. \tag{4.2.20}$$

Note that the set of n^2 equations defined by (4.2.13) and (4.2.19) is equivalent to the matrix equation

$$\mathbf{LU} \equiv \mathbf{A}^{(1)} + \partial\mathbf{A}, \tag{4.2.21}$$

where the elements ∂a_{ij} of $\partial\mathbf{A}$ are defined by (4.2.14) and (4.2.20).

The Bounds on $\partial\mathbf{A}$*: Pivoting*

The computed $a_{ij}^{(k)}$ is defined by

$$\begin{aligned} a_{ij}^{(k)} &\equiv \mathrm{fl}(a_{ij}^{(k-1)} - m_{i,k-1}a_{k-1,j}^{(k-1)}) \\ &\equiv [a_{ij}^{(k-1)} - m_{i,k-1}a_{k-1,j}^{(k-1)}(1+\delta_1)](1+\delta_2), \end{aligned} \tag{4.2.22}$$

and $d_{ij}^{(k)}$ is defined by

$$\begin{aligned} d_{ij}^{(k)} &= a_{ij}^{(k)} - [a_{ij}^{(k-1)} - m_{i,k-1}a_{k-1,j}^{(k-1)}] \\ &= a_{ij}^{(k)} - \left[\frac{a_{ij}^{(k)}}{1+\delta_2} + m_{i,k-1}a_{k-1,j}^{(k-1)}\,\delta_1\right] \\ &= \frac{a_{ij}^{(k)}\,\delta_2}{1+\delta_2} + m_{i,k-1}a_{k-1,j}^{(k-1)}\,\delta_1. \end{aligned} \tag{4.2.23}$$

The bounds on $d_{ij}^{(k)}$ obviously depend on the bounds on $a_{ij}^{(k)}$ and $m_{i,k-1}$. The expression for the multipliers,

$$m_{ij} = a_{ij}^{(j)}/a_{jj}^{(j)},$$

shows that if $a_{jj}^{(j)} = 0$ then the method of triangular decomposition will break down. However, when the method works the pivotal rows, i, and the pivots, $a_{ij}^{(j)}$, can at every stage be selected so as to ensure the smallest possible value for the m_{ij}.

Indeed, this value can always be made to satisfy

$$|m_{ij}| \leqslant 1. \tag{4.2.24}$$

To satisfy (4.2.24) by partial pivoting:
Eliminate the columns in natural order. At the rth stage take the pivotal row be one of the remaining $n-r+1$ rows and such that its element of the largest modulus is in column r. Exchange this new pivotal row with the rth row.

To satisfy (4.2.24) by complete pivoting:
Select as a pivot at the rth stage the element of the largest modulus in the whole of the remaining $n-r+1$ square array.

Of the two kinds of pivoting, the partial pivoting has some advantages over the complete pivoting. In particular, for sparse matrices with a specific pattern the pattern may be preserved if partial pivoting is used while complete pivoting is most likely to destroy the pattern.

If partial pivoting is used and if all elements of $A^{(1)}$ satisfy $|a_{ij}^{(1)}| \leqslant a$, then all elements of $\mathbf{A}^{(r)}$ satisfy

$$|a_{ij}^{(r)}| \leqslant 2^{r-1}a. \tag{4.2.25}$$

This bound on the computed elements $a_{ij}^{(r)}$ is not very sharp, and in practice it is rare that any element attains a value even as large as $8a$.

If complete pivoting is used then as shown by Wilkinson

$$|a_{rr}^{(r)}| < r^{1/2}(2^1 3^{1/2} 4^{1/3} \ldots r_{rr}^{1/r-1})^{1/2} a. \tag{4.2.26}$$

However Wilkinson again suggests that this is a severe overestimate. Assume that some form of pivoting has been used and that $|m_{ij}| \leqslant 1$ and $|a_{ij}^{(i)}| \leqslant 1$.

Denote the maximum element in any $A^{(r)}$ by g, then from (4.2.23)

$$\begin{aligned} |d_{ij}^{(k)}| &\leqslant \frac{g2^{-t}}{1-2^{-t}} + g2^{-t} \\ &< (2.01)g^{2-t} \quad \text{(say).} \end{aligned} \tag{4.2.27}$$

This applies to all $d_{ij}^{(k)}$, except $d_{ij}^{(j+1)}$ when $i > j$. These differences are estimated from (4.2.15) and (4.2.17) as

$$m_{ij} \equiv \text{fl}(a_{ij}^{(j)}/a_{jj}^{(j)}) = [a_{ij}^{(j)}/a_{jj}^{(j)}](1+\delta), \tag{4.2.28}$$

where

$$\delta < 2^{-1}, \quad \text{and} \quad q_{ij} = a_{ij}^{(j)}\,\delta/a_{jj}^{(j)}. \tag{4.2.29}$$

The estimate is given by

$$|d_{ij}^{(j+1)}| = |a_{jj}^{(j)} a_{ij}^{(j)} \delta / a_{jj}^{(j)}| = |a_{ij}^{(j)}\sigma| \leqslant g2^{-1}. \tag{4.2.30}$$

Noting that $g2^{-1} < (2.01)g2^{-1}$ the bound (4.2.27) is then used to cover all the d_{ij}. Using the bounds (4.2.14), (4.2.20) and (4.2.27) the bound on $\mathbf{A}$ is obtained as this:

$$|\partial\mathbf{A}| \leqslant (2.01)g2^{-1}\begin{bmatrix} 0 & 0 & 0 & \cdots & 0 & 0 \\ 1 & 1 & 1 & \cdots & 1 & 1 \\ 1 & 2 & 2 & \cdots & 2 & 2 \\ 1 & 2 & 3 & \cdots & 3 & 3 \\ \cdots & & & & & \\ \cdots & & & & & \\ 1 & 2 & 3 & & (n-1) & (n-1) \end{bmatrix} \tag{4.2.31}$$

Bound Estimates on $\partial\mathbf{L}$ *and* $\partial\mathbf{U}$

Assume now that the set of equations $\mathbf{Ax} = \mathbf{b}$ is expressed in the form

$$\mathbf{LUx} = \mathbf{b} \tag{4.2.32}$$

and to obtain its solution, we proceed by solving the two sets of equations,

$$\mathbf{Ly} = \mathbf{b}$$

and

$$\mathbf{Ux} = \mathbf{y}$$

Consider the solution of the lower triangular set, $\mathbf{Ly} = \mathbf{b}$,

$$\begin{aligned} &l_{11}y_1 &&= b_1, \\ &l_{21}y_1 + l_{22}y_2 &&= b_2, \\ &\cdots \\ &l_{n1}y_1 + l_{n2}y_2 + \ldots + l_{nn}y_n &&= n_n \end{aligned} \tag{4.2.33}$$

At the time when y_r is computed from the rth equation, $y_1, y_2, \ldots, y_{r-1}$ have already been computed. In floating-point form we have

$$Y_r \equiv fl\left(\frac{-l_{r1}y_1 - l_{r2}y_2 - \ldots - l_{r,r-1}y_{r-1} + b_r}{l_{rr}}\right)$$

$$= \frac{\left(\sum_{j=1}^{r-1} - l_{rj}y_j(1+\varepsilon_{rj})\right) + b_r(1+\delta_r)}{l_{rr}(1+\beta_r)}, \tag{4.2.34}$$

where

$$\begin{aligned} &|\beta_r|, |\delta_r| \leqslant 2^{-t_1}, \\ &|\varepsilon_{rj}| \leqslant (r+2-j)2^{-t_1} \end{aligned} \tag{4.2.35}$$

and

$$2^{-t_1} = (1.06)2^{-t}.$$

In (4.2.34), dividing the numerator and the denominator of the right-hand side expression by $(1+\delta_r)$ and denoting $(1+\varepsilon_{rj})/(1+\delta_r)$ and $(1+\beta_r)/(1+\delta_r)$ by $1+\alpha_{ri}$ and $1+\alpha_{rr}$, respectively, we obtain

$$y_r \equiv \frac{\sum_{j=1}^{r-1} (-l_{rj}y_j)(1+\alpha_{rj}) + b_r}{l_{rr}(1+\alpha_{rr})}, \tag{4.2.36}$$

where

$$|\alpha_{rj}| \leqslant (r+1-i)2^{-t}$$

and

$$|\alpha_{rr}| \leqslant 2^{-t+1}. \tag{4.2.37}$$

Expression (4.2.36) can be written in the form

$$\sum_{j=1}^{r} l_{rj}y_j(1+\alpha_{rj}) \equiv b_r, \tag{4.2.38}$$

which shows that the computed vector solution is the exact solution of

$$(\mathbf{L}+\partial\mathbf{L})\mathbf{y} = \mathbf{b}, \tag{4.2.39}$$

where, say, for $n=6$, we have

$$|\partial\mathbf{L}| \leqslant 2^{-t} \begin{bmatrix} 2|l_{11}| & 0 & 0 & 0 & 0 & 0 \\ 2|l_{21}| & 2|l_{22}| & 0 & 0 & 0 & 0 \\ 3|l_{31}| & 2|l_{32}| & 2|l_{33}| & 0 & 0 & 0 \\ 4|l_{41}| & 3|l_{42}| & 2|l_{43}| & 2|l_{44}| & 0 & 0 \\ 5|l_{51}| & 4|l_{52}| & 3|l_{53}| & 2|l_{54}| & 2|l_{55}| & \\ 6|l_{61}| & 5|l_{62}| & 4|l_{63}| & 3|l_{64}| & 2|l_{65}| & 2|l_{66}| \end{bmatrix} \tag{4.2.40}$$

If the maximum element in any $|\mathbf{A}^{(r)}|$ is greater than or equal to g, i.e. $|l_{ij}| \leqslant g$, then

$$\|\partial\mathbf{L}\|_1 = \|\partial\mathbf{L}\|_\infty \leqslant \tfrac{1}{2}(n^2+n+2)g2^{-t}. \tag{4.2.41}$$

Note that for these matrix norms, we have

$$\|\partial\mathbf{L}\| \leqslant n\|\mathbf{L}\|2^{-t}. \tag{4.2.42}$$

Now, rewrite (4.2.39) as

$$\mathbf{b} - \mathbf{L}\mathbf{y} = \partial\mathbf{L}\mathbf{y}. \tag{4.2.43}$$

If we assume that **L** and **b** are scaled so that all elements are of modulus <1, then we can take $g=1$ and so from (4.2.41) the following bound for the residual follows:

$$\| \mathbf{b} - \mathbf{Ly} \|_\infty \leqslant \tfrac{1}{2}(n^2 + n + 2) \| \mathbf{y} \|_\infty 2^{-t}. \tag{4.2.44}$$

Note that the bound for the residual is directly dependent on $\| \mathbf{y} \|_\infty$.

If $\| \mathbf{y} \|_\infty$ is of order unity, then since in practice the factor $(n^2 + n + 2)$ is very unlikely to be attained, the residual (4.2.44) is of necessity very small whether or not $\mathbf{y}$ is an accurate solution.

This completes the analysis on the local perturbation matrix, $\partial\mathbf{L}$. A similar analysis is readily carried out for the local perturbation matrix of the second set, $\partial\mathbf{U}$.

We are now in a position to give a bound estimate for the total perturbation matrix, $\mathbf{E}$, given by

$$\mathbf{E} = \partial\mathbf{A} + \mathbf{U}\partial\mathbf{L} + \mathbf{L}\partial\mathbf{U}.$$

Indeed, we have shown that assuming pivoting has been used and all $| l_{ij} | \leqslant 1$, the following bounds hold

$$\| \partial\mathbf{A} \|_\infty \leqslant (2.01)g\left(\frac{n}{2} + 1\right)(n-1)2^{-t}$$

$$\| \delta\mathbf{L} \|_\infty \leqslant \tfrac{1}{2}(n^2 + n + 2)2^{-t},$$

$$\| \partial\mathbf{U} \|_\infty \leqslant \tfrac{1}{2}g(n^2 + n + 2)2^{-t},$$

$$\| \mathbf{L} \|_\infty \leqslant n,$$

$$\| \mathbf{U} \|_\infty \leqslant gn,$$

where g is the maximum element of any $| \mathbf{A}^{(r)} |$, $r = 1, \ldots, n$. A bound for $\| E \|_\infty$, for sufficiently large n, is then given as

$$\| \mathbf{E} \|_\infty \leqslant g(n^3 + 2.005n^2)2^{-t}, \tag{4.2.45}$$

and the bound on the residual as

$$\| \mathbf{b} - \mathbf{Ax} \|_\infty \leqslant g(n^3 + 2.005n^2) \| \mathbf{x} \|_\infty 2^{-t}. \tag{4.2.46}$$

In (4.2.46) the dominant term involves n^3. This term comes from the errors in the solution of the triangular sets of equations, but as was pointed out by Wilkinson, the bounds for $\| \partial\mathbf{L} \|$ and $\| \partial\mathbf{U} \|$ may only be attained in exceptional circumstances, and, in fact, a more realistic bound for the residual is

$$\| \mathbf{b} - \mathbf{Ax} \|_\infty \leqslant gn \| \mathbf{x} \|_\infty 2^{-t}. \tag{4.2.47}$$

4.3 Iterative Refinement of the Solution

If matrix $\mathbf{A}$ is mildly ill-conditioned then the computed solution $x^{(1)}$ of $\mathbf{Ax} = \mathbf{b}$ may not be sufficiently accurate. This solution can be improved by a process of iterative refinement in which a sequence of vectors $\mathbf{x}^{(1)}, \mathbf{x}^{(2)}, \ldots, \mathbf{x}^{(r)}$ is computed which in certain circumstances converges to the true solution $\mathbf{x}$.

At the start of the iterative procedure we already have **L** and **U** such that

$$\mathbf{LU} = \mathbf{A} + \mathbf{E}.$$

Now define the residual vectors $\mathbf{r}^{(s)}$ as

$$\mathbf{r}^{(s)} = \mathbf{b} - \mathbf{A}\mathbf{x}^{(s)}, \tag{4.3.2}$$

where

$$\mathbf{x}^{(s+1)} = \mathbf{x}^{(s)} + (\mathbf{LU})^{-1}\mathbf{r}^{(s)}. \tag{4.3.3}$$

The sth iteration of the iterative refinement process involves the following three steps.

Compute the residuals $\mathbf{r}^{(s)} = \mathbf{b} - \mathbf{A}\mathbf{x}^{(s)}$.
Solve the system $(\mathbf{LU})^{-1}\mathbf{r}^{(s)} = \mathbf{d}^{(s)}$.
Compute the new solution $\mathbf{x}^{(s+1)} = \mathbf{x}^{(s)} + \mathbf{d}^{(s)}$.

Each system $(\mathbf{LU})\mathbf{d}^{(s)} = \mathbf{r}^{(s)}$ has the same matrix **LU** and so the use of multipliers is made retained from the first solution. This is why iterative improvement of the first solution adds only a moderate amount, about 25%, to the computational tie of the algorithm. It is essential that the residuals $\mathbf{r}^{(s)}$ are computed with higher precision than the other components.

If the sequence of residuals and vectors can be solved without further rounding error then from (4.3.2) and (4.3.3) we have

$$\mathbf{x}^{(s+1)} = \mathbf{x}^{(s)} + (\mathbf{LU})^{-1}(\mathbf{Ax} - \mathbf{A}\mathbf{x}^{(s)})$$

$$\text{or} \quad \mathbf{x}^{(s+1)} - \mathbf{x} = [\mathbf{I} - (\mathbf{LU})^{-1}\mathbf{A}]^{s}(\mathbf{x}^{(1)} - \mathbf{x}) \tag{4.3.4}$$

and

$$\mathbf{r}^{(s+1)} = \mathbf{A}(\mathbf{x} - \mathbf{x}^{(s+1)} = [\mathbf{I} - \mathbf{A}(\mathbf{LU})^{-1}]\mathbf{r}^{(s)} \tag{4.3.5}$$

From (4.3.4) and (4.3.5) it follows that the iterations converge if

$$[\mathbf{I} - (\mathbf{A} + \mathbf{E})^{-1}\mathbf{A}]^{s} \to 0 \quad \text{as} \quad s \to \infty, \tag{4.3.6}$$

i.e. if $\| \mathbf{I} - (\mathbf{A} + \mathbf{E})^{-1}\mathbf{A} \| < 1,$ (4.3.7)

which is satisfied if

$$\| \mathbf{A}^{-1} \| \, \| \mathbf{E} \| < \tfrac{1}{2}. \tag{4.3.8}$$

If **A** has been triangulated using Gaussian elimination with the partial pivoting and floating-point with accumulation then in general

$$\| \mathbf{E} \| \leqslant n2^{-t}, \tag{4.3.9}$$

and the iterative process is convergent if

$$\| \mathbf{A}^{-1} \| \leqslant n2^{t-1}. \tag{4.3.10}$$

Now if

$$n2^{-t} \| \mathbf{A}^{-1} \| < 2^{-p}, \quad \text{for } p < 1 \tag{4.3.11}$$

then from (4.3.34) and (4.3.5)

$$\| \mathbf{x}^{(s+1)} - \mathbf{x} \| \leqslant [2^{-p}/(1-2^{-p})] \, \| \mathbf{x}^{(s)} - \mathbf{x} \| \tag{4.3.12}$$

and

$$\| \mathbf{r}^{(s+1)} \| \leqslant [2^{-p}/(1-2^{-p})] \, \| \mathbf{r}^{(s)} \| \, . \tag{4.3.13}$$

Both the error and the residual decrease by at least the factor $2^{-p}/(1-2^{-p})$ with each iteration so that if p is appreciably greater than 2 we effectively gain at least p binary digits per iteration.

In practice the iterative process cannot be performed exactly and thus the effect of rounding errors could nullify the whole process. If the iteration is performed using t-digit arithmetic, the accuracy of $\mathbf{x}^{(s)}$ cannot increase indefinitely with s since only t digits are used to represent the components. In the floating-point arithmetic with accumulating the residual of the first solution is almost certain to be of the same order of magnitude as that of the correctly rounded solution. Therefore it can scarcely be expected that the residual would diminish by the factor 2^{-p} with each iteration, and it appears improbable that the accuracy of the successive solutions could steadily improve if the residuals remain roughly constant. Nevertheless this is precisely what does happen and the performance of the practical process using any mode of arithmetic does not fall short of that corresponding to exact iteration provided only that inner products are accumulated in the computation of the residuals.

For further details of iterative improvement see Forsythe and Moler (1967), Moler (1967), and Wilkinson (1963).

4.4 Cholesky Decomposition of Symmetric Matrices

A special case important in practice is given by the set $\mathbf{Ax} = \mathbf{b}$, where $\mathbf{A}$ is the so-called positive definite. A real symmetric matrix is called positive definite if the quadratic form $\mathbf{x}^T\mathbf{Ax} > 0$ for all real $\mathbf{x} \neq 0$. (A necessary and sufficient condition for the matrix to be positive definite is that all its eigenvalues are positive.)

When matrix $\mathbf{A}$ is symmetric and positive definite, the diagonal elements of the lower triangular matrix $\mathbf{L}$ can be chosen so that $\mathbf{L}$ is real and the upper triangular matrix $\mathbf{U}$ is the transpose of $\mathbf{L}$. In this case

$$\mathbf{A} = \mathbf{LL}^T \tag{4.4.1}$$

and the process is known as the cholesky decomposition.

Cholesky's Method

Computation of $\mathbf{L}$ by Cholesky's method is illustrated in the following example. Let $\mathbf{A}$ be a 3×3 matrix. We wish to find the elements l_{ij} for which the following matrix equation holds:

$$\begin{bmatrix} l_{11} & 0 & 0 \\ l_{21} & l_{22} & 0 \\ l_{31} & l_{32} & l_{33} \end{bmatrix} \begin{bmatrix} l_{11} & l_{21} & l_{31} \\ 0 & l_{22} & l_{32} \\ 0 & 0 & l_{33} \end{bmatrix} = \begin{bmatrix} a_{11} & a_{12} & a_{13} \\ a_{12} & a_{22} & a_{23} \\ a_{13} & a_{23} & a_{33} \end{bmatrix} \tag{4.4.2}$$

(4.4.42) can be written as a set of six equations

$$\begin{aligned} l_{11}^2 &= a_{11}, \\ l_{11}l_{21} = l_{21}l_{11} &= a_{12}, \\ l_{11}l_{31} = l_{31}l_{11} &= a_{13}, \\ l_{21}^2 + l_{22}^2 &= a_{22}, \\ l_{21}l_{31} + l_{22}l_{32} = l_{31}l_{21} + l_{32}l_{22} &= a_{23}, \\ l_{31}^2 + l_{32}^2 + l_{33} &= a_{33} \end{aligned} \tag{4.4.3}$$

To solve the set, solve the first three equations to find l_{11}, l_{21} and l_{31}. Then solve the fourth and fifth equations to find l_{22} and l_{32}. Finally find l_{33} from the sixth equation. For an $n \times n$ matrix the expressions to compute the elements l_{ij} are given as follows:

$$l_{ii} = \left(a_{ii} - \sum_{j=1}^{i-1} l_{ij}^2 \right)^{1/2}, \qquad i = 1, \ldots, n, \tag{4.4.4}$$

and

$$l_{ij} = \frac{a_{ij} - \sum_{k=1}^{j-1} l_{ik} l_{jk}}{l_{jj}} \qquad \begin{array}{l} i = 1, \ldots, n. \\ j - 1, \ldots, i - 1. \end{array} \tag{4.4.5}$$

Division in (4.4.5) is usually replaced by computing instead the reciprocal of the square root in (4.4.4) and then multiplying the numerator in (4.4.5) by the reciprocal. In fact, the computed reciprocals are stored in the diagonal position of **L** and are used directly in the subsequent steps of computation. Only matrix **L** needs to be stored and not $\mathbf{L}^T$. There is no need to compute the inverse of **L** in order to solve the set $\mathbf{Ax} = \mathbf{b}$. By (4.4.1) we have

$$\mathbf{LL}^T\mathbf{x} = \mathbf{b}. \tag{4.4.6}$$

Writing $\mathbf{L}^T\mathbf{x} = \mathbf{y}$ we obtain two sets of equations with a triangular matrix:

$$\mathbf{Ly} = \mathbf{b} \tag{4.4.7}$$

$$\mathbf{L}^T = \mathbf{y} \tag{4.4.8}$$

and two processes of back substitution give the solution **x**.

A useful property of Cholesky's decomposition is that if the matrix **A** is originally scaled so that all the elements of **A** are bounded by unity then so too are all the elements of **L**. This fact can be easily observed when equating the diagonal elements in the equation (4.4.1), i.e.

$$l_{i1}^2 + l_{i2}^2 + \ldots + l_{ii}^2 = a_{ii}, \qquad i = 1, \ldots, n. \tag{4.4.9}$$

No pivoting is required in the Cholesky decomposition.

If **A** is symmetric but not positive definite then Cholesky's method is not possible without complex arithmetic as in general $\mathbf{A} = \mathbf{L}\mathbf{L}^T$ does not hold. In fact the symmetric decomposition of a matrix which is not positive definite enjoys none of the stability of the positive definite case. To be sure of stability we must use interchanges and this will destroy the symmetry.

The total number of multiplications required by the process is readily given by

$$\sum_{j=1}^{n} \left[(j-1) + \sum_{i=j+1}^{n} j \right] = \frac{n^3}{6} + \frac{n^2}{2} - \frac{2}{3}\, n \tag{4.4.10}$$

and the total number of additions by

$$\sum_{j=1}^{n} \left[(j-1) + \sum_{i=j+1}^{n} (j-1) \right] = \frac{n(n^2-1)}{6}. \tag{4.4.11}$$

In order to solve simultaneous equations $\mathbf{L}\mathbf{y} = \mathbf{b}$ and $\mathbf{L}^T\mathbf{x} = \mathbf{y}$ a further $n(n-1)$ additions and $n(n+1)$ multiplications are required. The n divisions needed for each triangular system do not have to be carried out as the reciprocals of the diagonal elements l_{ii} are already stored in the diagonal positions of **L**. Thus the total arithmetics required by Cholesky's method to solve a set of n linear equations is

$$\frac{n^3}{6} + \frac{3}{2}\, n^2 + \frac{n}{3} \qquad \text{multiplications,}$$

$$\frac{n^3}{6} + n^2 - \frac{7}{6}\, n \qquad \text{additions,} \tag{4.4.12}$$

$$n \qquad \text{square root reciprocal evaluations.}$$

A variant of Cholesky decompositions which avoids calculation of square roots altogether is suggested in Martin, Peters and Wilkinson (1965).

Matrix Inversion

To invert a symmetric positive definite matrix **A** write

$$\mathbf{A} = \mathbf{L}\mathbf{L}^T,$$

$$\mathbf{A}^{-1} = (\mathbf{L}^T)^{-1}\mathbf{L}^{-1}. \tag{4.4.13}$$

In (4.4.13) an explicit inversion of **L** is not required since by writing $\mathbf{L}^T\mathbf{A}^{-1} = \mathbf{L}^{-1}$ and then letting

l_{ij} be the elements of $\mathbf{L}^T$,

d_{ij} be the elements of $\mathbf{L}^{-1}$,

and a_{ij} be the elements of $\mathbf{A}^{-1}$ (which is symmetric) we obtain a matrix expression of the form:

$$\begin{bmatrix} l_{11} & l_{21} & l_{31} \\ 0 & l_{22} & l_{23} \\ 0 & 0 & l_{33} \end{bmatrix} \begin{bmatrix} a_{11} & a_{12} & a_{13} \\ a_{12} & a_{22} & a_{23} \\ a_{13} & a_{23} & a_{33} \end{bmatrix} = \begin{bmatrix} d_{11} & 0 & 0 \\ d_{21} & d_{22} & 0 \\ d_{31} & d_{32} & d_{33} \end{bmatrix},$$

where

$$\begin{aligned} a_{33} &= d_{33}/l_{33}, \\ a_{23} &= -l_{23}a_{33}/l_{22}, \\ a_{22} &= (d_{22} - l_{23}a_{23})/l_{22}, \\ a_{13} &= -(l_{21}a_{13} + l_{31}a_{33})/l_{11}, \\ a_{12} &= -(l_{21}a_{22} + l_{31}a_{23})/l_{11}, \\ a_{11} &= (d_{11} - l_{21}a_{12} - l_{31}a_{13})/l_{11}. \end{aligned} \tag{4.4.14}$$

In (4.4.14) the only required elements of $\mathbf{L}^{-1}$ are the diagonal elements but these are simply the reciprocals of the diagonal elements of $\mathbf{L}$ which are already stored in $\mathbf{L}$.

The general expressions for the elements of $\mathbf{A}^{-1}$ are given by

$$a_{ii} = \frac{1}{l_{ii}}\left[\frac{1}{l_{ii}} - \sum_{k=i+1}^{n} l_{ik}a_{ik}\right]$$

and (4.4.15)

$$a_{ij} = -\frac{1}{l_{ii}}\sum_{k=i+1}^{n} l_{ki}a_{kj}, \qquad i < j.$$

It is not difficult to deduce the total number of arithmetics required to invert a symmetric positive definite matrix:

$$\begin{aligned} &\frac{n^3}{2} + n^2 - \frac{n}{2} && \text{multiplications,} \\ &\frac{n^3}{2} - \frac{n^2}{2} && \text{additions,} \\ &n && \text{square root reciprocals.} \end{aligned} \tag{4.4.16}$$

Numerical Stability

The error analysis of the Cholesky decomposition may be carried out in a way similar to the analysis of the general triangulation method. Wilkinson (1961) has shown that the Cholesky decomposition has a guaranteed stability even without pivoting, provided the original matrix $\mathbf{A}$ is scaled so that all its elements are bounded by unity. For the computation in floating-point with accumulation, the bounds on the error matrix $\mathbf{E}$, where $\mathbf{LL}^T = \mathbf{A} + \mathbf{E}$, are given as

$$e_{ij} = \begin{cases} l_{ij}l_{jj}2^{-t}, & i < j. \\ l_{ij}l_{ii}2^{-t}, & i > j, \\ l_{jj}^2 2^{-t}, & i = j. \end{cases} \tag{4.4.17}$$

4.5 The Orthogonal Reduction Methods

Another group of methods for solving a set of linear equations is the orthogonal reduction methods. In an orthogonal reduction method the original matrix is transformed into a triangular form using the operation of permutation by orthogonal matrices.

The original matrix is transformed successively to $\mathbf{A}_1, \mathbf{A}_2, \ldots, \mathbf{A}_s$, where each member of the sequence has at least one more zero below the diagonal than the previous matrix. These zeroes may be produced one at a time or a whole subcolumn at a time.

In Gaussian elimination the main obstacle to finding error bounds is the possibility of a progressive increase in size of the elements of the reduced matrices, since in general the current rounding error is proportional to the current size of the elements. With orthogonal or nearly orthogonal matrices this problem is no longer present. The Euclidean norm of each column of $\mathbf{A}_1$ is preserved by exact orthogonal transformations. Hence, if originally the sum of the squares in each column is less than unity, it will remain true for all subsequent matrices.

The methods of Givens (1958) and of Householder (1958b) are the best known and most efficient methods of the group.

We shall only briefly mention the Givens method and concentrate our attention on the Householder method. Strictly speaking the methods achieve the triangulation of the matrix of the set $\mathbf{Ax} = \mathbf{b}$, after which any standard technique may be employed to complete the solution of the set. In the Givens method the triangular form is brought about by means of a sequence of $n(n-1)/2$ plane rotations, whose product is an orthogonal matrix.

Each rotation requires the extraction of a square root. An important advantage of the Givens method as compared with Gaussian elimination lies in the fact that an orthogonal matrix is perfectly conditioned and so the conditioning of the matrix under transformation does not deteriorate through the successive transformations.

The Householder Reduction

The method developed by Householder (1958b) may be considered as a variant of the Givens method in the sense that it preserves all the advantages of the former method but requires fewer arithmetic operations. In terms of the computational complexity Householder's method is the most efficient of the orthogonal reduction methods.

To illustrate how the method works let us consider an intermediate step in the Householder reduction. Let $\mathbf{A}_r$ be upper triangular in its first r columns so that

$$\mathbf{A}_r = \begin{bmatrix} \mathbf{B}_r & \mathbf{C}_r \\ \mathbf{0} & \mathbf{W}_{n-r} \end{bmatrix}, \tag{4.5.1}$$

where $\mathbf{B}_r$ is an upper triangular matrix of order r. In the $(r+1)$th step matrix $\mathbf{P}_r$ is used of the form

$$\mathbf{P}_r = \begin{bmatrix} \mathbf{I} & \mathbf{0} \\ \mathbf{0} & \mathbf{I} - 2\mathbf{u}\mathbf{u}^* \end{bmatrix}, \qquad \| \mathbf{u} \|_2 = 1. \tag{4.5.2}$$

We then compute

$$\mathbf{A}_{r+1} = \mathbf{P}_r\mathbf{A}_r = \begin{bmatrix} \mathbf{B}_r & \mathbf{C}_r \\ \mathbf{0} & (\mathbf{I} - 2\mathbf{u}\mathbf{u}^*)\mathbf{W}_{n-r} \end{bmatrix} \tag{4.5.3}$$

and can see that only $\mathbf{W}_{n-r}$ is modified. Vector $\mathbf{u}$ must be chosen so that the first column of

$(\mathbf{I} - 2\mathbf{u}\mathbf{u}^*)\mathbf{W}_{n-r}$ is null apart from its first element.

This choice is based on the following *Lemma* (Householder, 1958b).
For any vector $\mathbf{a} \neq \mathbf{0}$ and any unit vector $\mathbf{v}$, a unit vector $\mathbf{u}$ exists such that

$$(\mathbf{I} - 2\mathbf{u}\mathbf{u}^*)\mathbf{a} = \| \mathbf{a} \|_2\mathbf{v}, \tag{4.5.4}$$

where

$$\| \mathbf{a} \|_2 = (\mathbf{a}^*\mathbf{a})^{1/2} = \left[\sum_{i=1}^{n} | a_i |^2 \right]^{1/2}.$$

(A unit vector is a vector with unity norm.)

The computation of the vector $\mathbf{u}$ requires two square root and one reciprocal evaluations. The calculation of $\| \mathbf{a} \|_2$ requires one of the necessary square root evaluations. Then, by (4.5.4) it is required that

$$\mathbf{a} - 2\mathbf{u}\mathbf{u}^*\mathbf{a} = \| \mathbf{a} \|_2\mathbf{v},$$

which, for convenience we rewrite as

$$\mathbf{a} - 2(\mathbf{u}^*\mathbf{a})\mathbf{u} = \| \mathbf{a} \|_2\mathbf{v}. \tag{4.5.5}$$

Now, let

$$q = 2\mathbf{u}^*\mathbf{a}, \tag{4.5.6}$$

then

$$q\mathbf{u} = \mathbf{a} - \| \mathbf{a} \|_2\mathbf{v} \tag{4.5.7}$$

and

$$(q\mathbf{u})^2 = (\mathbf{a} - \| \mathbf{a} \|_2\mathbf{v})^2$$

gives

$$q^2 = 2\| \mathbf{a} \|_2(\| \mathbf{a} \|_2 - \mathbf{v}^*\mathbf{a}), \tag{4.5.8}$$

since

$$\| \mathbf{u} \|_2 = \| \mathbf{v} \|_2 = 1.$$

The evaluation of q from (4.5.8) accounts for the other square root evaluation. Further from (4.5.8) it follows that q is real since

$$\| \mathbf{a} \|_2 - \mathbf{v}^*\mathbf{a} \geqslant 0, \tag{4.5.9}$$

as $\| \mathbf{a} \|_2$ is the length of vector $\mathbf{a}$ and $\mathbf{v}^*\mathbf{a}$ is the projection of $\mathbf{a}$ upon the unit vector $\mathbf{v}$. Hence, q can be taken to be non-negative. Now, if $\| \mathbf{a} \|_2 = \mathbf{v}^*\mathbf{a}$ then (4.5.4) is verified with $\mathbf{u} = 0$. Otherwise we can take $q > 0$ defined by (4.5.8) and then the vector $\mathbf{u}$ can always be computed from (4.5.7). This computing process would require evaluation of a reciprocal, q^{-1}.

Now let $\mathbf{a}$ be the first column of $\mathbf{A}$ and take $\mathbf{v} = \mathbf{e}_1$, the first column of the identity matrix. Application of the lemma provides a unitary matrix

$$\mathbf{P}_1 = (\mathbf{I} - 2\mathbf{u}_1\mathbf{u}_1^*) \tag{4.5.10}$$

such that the first column of $\mathbf{A}_1 = \mathbf{P}_1\mathbf{A}$ is null except for the first element (A matrix $\mathbf{B}$ is called unitary if $\mathbf{B}^{-1} = \mathbf{B}^*$.)

Matrix $\mathbf{A}_2$ is computed in a similar manner after suppressing the first row and the first column of the transformed matrix $\mathbf{A}_1$. The process continues for at most $(n - 2)$ steps after which the original matrix $\mathbf{A}$ is brought to a triangular form. We then have

$$\mathbf{PA} = \mathbf{A}_{n-1}, \tag{4.5.11}$$

where $\mathbf{A}_{n-1}$ is an upper triangular matrix and

$$\mathbf{P} = \mathbf{P}_{n-1}\mathbf{P}_{n-2} \ldots \mathbf{P}_2\mathbf{P}_1.$$

We note that the computation process requires $2(n - 2)$ square root evaluations. However, half of the square root evaluations can be evaded using an elegant device developed by Wilkinson (1965).

Consider the following problem. Given a vector $\mathbf{x}$ of length n, construct an elementary Hermitian matrix $\mathbf{P} = \mathbf{I} - 2\mathbf{ww}^*$, such that

$$\mathbf{Px} = k\mathbf{e}_1. \tag{4.5.12}$$

Since the l_2-norm is invariant, if we write

$$S^2 = x_1^2 + x_2^2 + \ldots + x_n^2$$

then

$$k = \pm S;$$

From (4.5.12) we have

$$x_1 - 2w_1(\mathbf{w}^T\mathbf{x}) = \pm S,$$
$$x_i - 2w_i(\mathbf{w}^T\mathbf{x}) = 0, \qquad i = 2, \ldots, n.$$

and, hence,

$$2Kw_1 = x_1 \mp S,$$
$$2Kw_i = x_i, \qquad i = 2, \ldots, n, \tag{4.5.13}$$

where

$$K = \mathbf{w}^T\mathbf{x}.$$

Squaring the equations (4.5.13), then adding them together, and dividing the result by factor 2 we obtain

$$2K^2 = S^2 \mp x_1 S \tag{4.5.14}$$

(In the computation the sign in (4.5.14) is decided so as to avoid K becoming very small, that is

if $x_1 > 0$ then $2K^2 = S^2 + x_1 S$ is used, and
if $x_1 < 0$ then $2K^2 = S^2 - x_1 S$ is used.)

If we write

$$\mathbf{u}^* = (x_1 \mp S, x_2, x_3, \ldots, x_n), \tag{4.5.15}$$

then

$$\mathbf{w} = \mathbf{u}/2K \tag{4.5.16}$$

and

$$\mathbf{P} = \mathbf{I} - \mathbf{u}\mathbf{u}^*/2K^2. \tag{4.5.17}$$

Hence, only one square root evaluation is required in the process, namely that of S^2 to give K.

We shall now sketch the Householder triangulation process for a 5×5 matrix.

Given

$$\mathbf{A} = \mathbf{A}_1 = \left[\begin{array}{c|cccc} a_{11} & a_{12} & a_{13} & a_{14} & a_{15} \\ \hline a_{21} & a_{22} & a_{23} & a_{24} & a_{25} \\ a_{31} & a_{32} & a_{33} & a_{34} & a_{35} \\ a_{41} & a_{42} & a_{43} & a_{44} & a_{45} \\ a_{51} & a_{52} & a_{53} & a_{54} & a_{55} \end{array}\right] = \left[\begin{array}{c|c} \mathbf{B}_1 & \mathbf{C}_1 \\ \hline a_{21} & \\ a_{31} & \mathbf{W}_{n-1} \\ a_{41} & \\ a51 & \end{array}\right].$$

(a) Form a symmetric orthogonal matrix

$$\mathbf{P}_1 = \left[\begin{array}{c|cccc} 1 & 0 & 0 & 0 & 0 \\ \hline 0 & & & & \\ 0 & & \mathbf{I} - 2\mathbf{u}_1\mathbf{u}_1^* & & \\ 0 & & & & \\ 0 & & & & \end{array}\right],$$

where

$$\| \mathbf{u}_1 \|_2 = 1$$

and

$$\mathbf{u}_1\mathbf{u}_1^* = (\mathbf{v}_1\mathbf{v}_1^*)/4K^2$$

with

$$\begin{aligned} S^2 &= a_{21}^2 + a_{31}^2 + a_{41}^2 + a_{51}^2, \\ \mathbf{v}_1^T &= (a_{21} \mp S, a_{31}, a_{41}, a_{51}), \\ 2K^2 &= S^2 \mp a_{21}S. \end{aligned}$$

Then obtain

$$\mathbf{A}_2 = \mathbf{P}_1\mathbf{A}_1 = \left[\begin{array}{c|cccc} x & x & x & x & x \\ \hline 0 & x & x & x & x \\ 0 & 0 & x & x & x \\ 0 & 0 & x & x & x \\ 0 & 0 & x & x & x \end{array}\right] = \left[\begin{array}{cc|ccc} x & x & x & x & x \\ 0 & x & x & x & x \\ \hline 0 & 0 & & & \\ 0 & 0 & & \mathbf{W}_{n-2} & \\ 0 & 0 & & & \end{array}\right] = \left[\begin{array}{c|c} \mathbf{B}_2 & \mathbf{C}_2 \\ \hline \mathbf{0} & \mathbf{W}_{n-2} \end{array}\right]$$

(the lower right 4×4 block of the first matrix: this is matrix $(\mathbf{I} - 2\mathbf{u}_1\mathbf{u}_1^*)\mathbf{W}_{n-1}$)

(b) Form a symmetric orthogonal matrix

$$\mathbf{P}_2 = \left[\begin{array}{cc|ccc} 1 & 0 & 0 & 0 & 0 \\ 0 & 1 & 0 & 0 & 0 \\ \hline 0 & 0 & & & \\ 0 & 0 & & \mathbf{I} - 2\mathbf{u}_2\mathbf{u}_2^* & \\ 0 & 0 & & & \end{array}\right],$$

where

$$\| \mathbf{u}_2 \|_2 = 1$$

and

$$\mathbf{u}_2\mathbf{u}_2^* = (\mathbf{v}_2\mathbf{v}_2^*)/4K^2$$

with

$$\begin{aligned} S^2 &= a_{32}'^2 + a_{42}'^2 + a_{52}'^2 \\ \mathbf{v}_2^T &= (a_{32}' \mp S, a_{42}', a_{52}') \\ 2K^2 &= S^2 \mp a_{32}'S. \end{aligned}$$

Then obtain

$$\mathbf{A}_3 = \mathbf{P}_2\mathbf{A}_2 = \left[\begin{array}{cc|ccc} x & x & x & x & x \\ 0 & x & x & x & x \\ \hline 0 & 0 & x & x & x \\ 0 & 0 & 0 & x & x \\ 0 & 0 & 0 & x & x \end{array}\right] = \left[\begin{array}{ccc|cc} x & x & x & x & x \\ 0 & x & x & x & x \\ 0 & x & x & x & x \\ \hline 0 & 0 & 0 & x & x \\ 0 & 0 & 0 & x & x \end{array}\right] = \left[\begin{array}{c|c} \mathbf{B}_3 & \mathbf{C}_3 \\ \hline \mathbf{0} & \mathbf{W}_{n-3} \end{array}\right]$$

this is matrix $(\mathbf{I} - 2\mathbf{u}_2\mathbf{u}_2^*)\mathbf{W}_{n-2}$

(c) Form a symmetric orthogonal matrix

$$\mathbf{P}_3 = \left[\begin{array}{ccc|cc} 1 & 0 & 0 & 0 & 0 \\ 0 & 1 & 0 & 0 & 0 \\ 0 & 0 & 1 & 0 & 0 \\ \hline 0 & 0 & 0 & \multicolumn{2}{c}{\mathbf{I} - 2\mathbf{u}_3\mathbf{u}_3^*} \\ 0 & 0 & 0 & & \end{array}\right]$$

where

$$\| \mathbf{u}_3 \|_2 = 1$$

and

$$\mathbf{u}_3\mathbf{u}_3^* = (\mathbf{v}_3\mathbf{v}_3^*)/4\mathrm{K}^2$$

with

$$\begin{aligned} S^2 &= a_{43}''^2 + a_{53}''^2 \\ \mathbf{v}_2^T &= (a_{43}'' \mp S, a_{53}'') \\ 2K^2 &= S^2 \mp a_{43}'S. \end{aligned}$$

Then obtain

$$\mathbf{A}_4 = \mathbf{P}_3\mathbf{A}_3 = \left[\begin{array}{ccc|cc} x & x & x & x & x \\ 0 & x & x & x & x \\ 0 & 0 & x & x & x \\ \hline 0 & 0 & 0 & x & x \\ 0 & 0 & 0 & 0 & x \end{array}\right] = \left[\begin{array}{cccc|c} x & x & x & x & x \\ 0 & x & x & x & x \\ 0 & 0 & x & x & x \\ 0 & 0 & 0 & x & x \\ \hline 0 & 0 & 0 & 0 & x \end{array}\right] = \left[\begin{array}{c|c} \mathbf{B}_4 & \mathbf{C}_4 \\ \hline \mathbf{0} & \mathbf{W}_{n-4} \end{array}\right]$$

this is matrix
$(\mathbf{I} - 2\mathbf{u}_3\mathbf{u}_3^*)\mathbf{W}_{n-3}$

The matrix $\mathbf{A}_4$ is upper triangular. The Householder process is complete.

Now consider a general $n \times n$ matrix $\mathbf{A}$.

Let $\mathbf{A} = \mathbf{A}_1 = \{a_{ij}^{(1)}\}$ and let $\mathbf{A}_2, \mathbf{A}_3, \ldots, \mathbf{A}_{n+1}$ be defined as follows:

$$\mathbf{A}_{k+1} = \mathbf{P}_k\mathbf{A}_k, \qquad k = 1, 2, \ldots, n. \tag{4.5.18}$$

$\mathbf{P}_k$ is a unitary (or a symmetric orthogonal in the case of a real $\mathbf{A}$) matrix of the form

$$\mathbf{P}_k = I - \beta_k \mathbf{u}_k \mathbf{u}_k, \tag{4.5.19}$$

where the elements of $\mathbf{P}_k$ are derived so that

$$a_{i,k+1}^{(k+1)} = 0, \qquad i = k+2, \ldots, n.$$

The components of the matrix $\mathbf{P}_k$ are computed as this:

$$\beta_k = [S_k(S_k + |a_{k,k}^{(k)}|)]^{-1}, \quad \text{where} \quad S_k = \left[\sum_{i=k}^{n} (a_{i,k}^{(k)})^2\right]^{1/2}, \quad \text{and}$$

$$u_i^{(k)} = \begin{cases} 0, & i < k, \\ sgn(a_{kk}^{(k)})(S_k + a_{kk}^{(k)}), & i = k, \\ a_{ik}^{(k)}, & i > k. \end{cases}$$

However, the matrix $\mathbf{P}_k$ itself does not need to be computed explicitly. Instead the following is noted:

$$\mathbf{A}_{k+1} = (I - \beta_k \mathbf{u}_k \mathbf{u}_k^*)\mathbf{A}_k = \mathbf{A}_k - \mathbf{u}_k \mathbf{w}^{\mathrm{T}}{}_k, \tag{4.5.20}$$

where

$$\mathbf{w}^T{}_k = \beta_k \mathbf{u}_k^* \mathbf{A}_k. \tag{4.5.21}$$

When computing the vector $\mathbf{w}_k$ and subsequently the matrix $\mathbf{A}_{k+1}$ the advantage is taken of the fact that the first $(k-1)$ components of u_k are equal to zero.

In order to ensure the best accuracy of the computation as soon as the matrix $\mathbf{A}_k$ is computed and before the start of the computation of $\mathbf{A}_{k+1}$, among the last $(n-k+1)$ columns of $\mathbf{A}_k$ the column is found which contains the elements with the maximum sum

$$\max_{k \leqslant j \leqslant n} \left\{ s_j^{(k)} = \sum_{i=k}^{n} (a_{ij}^{(k)})^2 \right\}. \tag{4.5.22}$$

This column is then exchanged with the current kth column of $\mathbf{A}_k$. The new $\mathbf{A}_k$ is ready to be made use of in computing matrix $\mathbf{A}_{k+1}$. After $\mathbf{A}_{k+1}$ has been computed the new sums $s_j^{(k+1)}$ are expressed as

$$s_j^{(k+1)} = s_j^{(k)} - (a_{kj}^{(k+1)})^2, \quad j = k+1, \ldots, n, \tag{4.5.23}$$

and thus can be easily computed.

The Householder reduction method is implemented in the algorithm householder.

Algorithm householder ($\mathbf{A}_1$)

for $k = 1$ **to** n **do**

//prepare matrix $\mathbf{A}_k$//

compute $\max_{k \leqslant j \leqslant n} \left\{ s_j^{(k)} = \sum_{i=k}^{n} (a_{ij}^{(k)})^2 \right\}$

exchange kth *column of* $\mathbf{A}_k$ *with column for which* $s_j^{(k)} = \max_j$

//compute matrix $\mathbf{A}_{k+1}$*//*

compute $S_k = SQRT\left(\sum_{i=k}^{n} (a_{ik}^{(k)})^2 \right)$

compute $\beta_k = [S_k(S_k + ABS(a_{kk}^{(k)}))]^{-1}$

compute $u_k = \{u_i^{(k)}, \quad i = 1, \ldots, n\}$

$$u_i^{(k)} = \begin{cases} 0, & i < k, \\ sgn(a_{kk}^{(k)})(S_k + ABS(a_{kk}^{(k)})), & i = k, \\ a_{ij}^{(k)}, & i > k \end{cases}$$

compute $w_k^T = \beta_k \mathbf{u}_k^* \mathbf{A}_k$

compute $\mathbf{A}_{k+1} = \mathbf{A}_k - \mathbf{u}_k \mathbf{w}_k^T$

enddo

Matrix Inversion

The Householder triangulation gives

$$\mathbf{PA} = \mathbf{R} \tag{4.5.24}$$

where $\mathbf{R}$ is upper triangular matrix. To invert $\mathbf{A}$ we use

$$\mathbf{A}^{-1}\mathbf{P}^{-1} = \mathbf{R}^{-1},$$

which gives

$$\mathbf{A}^{-1} = \mathbf{R}^{-1}\mathbf{P}. \tag{4.5.25}$$

(1) The Householder reduction requires

$$\frac{2}{3}n^3 + n^2 + \frac{4}{3}n - 3 \text{ multiplications.} \tag{4.5.26}$$

(2) The inversion of $\mathbf{R}$ requires

$$\frac{n(n-1)(n+4)}{6} \text{ multiplications.} \tag{4.5.27}$$

(3) $\mathbf{P} = \mathbf{P}_{n-1}\mathbf{P}_{n-2} \ldots \mathbf{P}_2\mathbf{P}_1$, where $\mathbf{P}_r = \mathbf{I} - 2\mathbf{u}_r\mathbf{u}_r^*$.

To evaluate all the $\mathbf{P}_r$'s requires

$$\sum_{r=2}^{n} r^2 = \frac{(2n^3 + 3n^2 + n)}{6} - 1 \text{ multiplications.}$$

To evaluate $\mathbf{P}$ by multiplying the $\mathbf{P}_r$'s requires a further

$$\sum_{r=2}^{n-1} r^2 = \frac{(2n^3 - 3n^2 + n)}{6} - 1 \text{ multiplications.}$$

Thus, to calculate $\mathbf{P}$ requires

$$\frac{2n^3}{3} + \frac{1}{3}n - 2 \text{ multiplications.} \tag{4.5.28}$$

(4) To multiply $\mathbf{R}^{-1}$ by $\mathbf{P}$ requires

$$\frac{n(n+1)}{2} \text{ multiplications.} \tag{4.5.29}$$

From (4.5.26) to (4.5.29) the number of multiplications required to invert $\mathbf{A}$ is given as

$$\frac{3n^3}{2} + 2n^2 + \frac{3}{2}n - 5. \tag{4.5.30}$$

Error Analysis of the Householder Reduction

We shall now study numerical stability of the Wilkinson variant of Householder's reduction. In this method the unitary matrices $\mathbf{P}_r$, $r = 1, \ldots, n-1$, which are used in the process of triangulation of matrix $\mathbf{A}$, are determined using the algorithms given by (4.5.12) to (4.5.17). The error analysis of the method is due to Wilkinson (1965) and we shall highlight its main points.

The first basic step in the triangulation is the computation of the current unitary matrix $\mathbf{P}_r$. To obtain some bounds on the errors accumulated in the computed matrix $\mathbf{P}_r^{(c)}$ we estimate the difference $\| \mathbf{P}_r^{(c)} - P_r \|$.

To do this consider accumulation of the errors in the sequence of computational steps to compute $\mathbf{P}_r^{(c)}$. Suppose we have a vector $\mathbf{x}$ of length n, and we wish to construct an elementary Hermitian matrix $\mathbf{P}$, such that

$$\mathbf{Px} = k\mathbf{e}_1$$

where $\mathbf{e}_1$ is the first column of the identity matrix. Suppose that $\mathbf{x}$ has floating-point components and we use fl_2 computation, i.e. floating-point with accumulation, to derive the computed matrix $\mathbf{P}^{(c)}$. As usual the double-length floating-point representation of a number, y, will be denoted by either $\bar{y}$ or $\text{fl}_2(y)$.

The steps in the computation are

(1) Calculate $\bar{a}$, where

$$\bar{a} \equiv \text{fl}_2(\bar{x}_1^2 + \bar{x}_1^2 + \ldots + \bar{x}_n^2). \tag{4.5.31}$$

Assuming that the 2t-digit mantissa is retained as it will be used in later steps, from (2.5.16) we have

$$\bar{a} \equiv (x_1^2 + x_2^2 + \ldots + x_n^2)(1 + \varepsilon) = S^2(1 + \varepsilon), \tag{4.5.32}$$

where

$$|\varepsilon| < \tfrac{3}{2}\, n 2^{-2t_2}, \qquad 2^{-t_2} = (1.06)2^{-2t}.$$

(2) Calculate $\bar{S}$.

$$\bar{S} \equiv \text{fl}_2((\bar{a})^{1/2}) = [S^2(1 + \varepsilon)]^{1/2}(1 + \alpha), \tag{4.5.33}$$

where it may be assumed that

$$|\alpha| < (1.00001)2^{-t}.$$

From (4.5.33) we obtain

$$\bar{S} = S(1 + \beta), \tag{4.5.34}$$

where

$$|\beta| < (1.00002)2^{-t}.$$

(3) Calculate $2\bar{K}^2$

$$\begin{aligned} 2\bar{K}^2 &= \text{fl}_2(\bar{x}_1^2 + \bar{x}_2^2 + \ldots + \bar{x}_n^2 + \bar{x}_1\bar{S}) \\ &\equiv [x_1^2(1 + \varepsilon_1) + x_2^2(1 + \varepsilon_2) + \ldots + x_n^2(1 + \varepsilon_n) + x_1\bar{S}(1 + \varepsilon_{n+1})](1 + e) \end{aligned} \tag{4.5.35}$$

where again from (2.5.16)

$$|\varepsilon_i| < \frac{3}{2}(n + 1)2^{-2t_2}, \quad |e| \leqslant 2^{-t}.$$

In the formula (4.5.14) there is a choice of the sign. To ensure numerical stability of the method the correctly chosen sign should always give the larger of the two possible values for $2K^2$ because the quantity is used as a denominator in the operation of division, and division by small numbers, as a rule, leads to disastrous loss of significant digits.

Rewrite (4.5.35) as

$$2\bar{K}^2 = (x_1{}^2 + x_2{}^2 + \ldots + x_n{}^2 + x_1\bar{S})(1+\varepsilon)(1+e), \tag{4.5.36}$$

where

$$\varepsilon = \max\{\,|\varepsilon_i|\,\},\quad \varepsilon \leqslant \frac{3}{2}(n+1)2^{-2t_2},\quad |\varepsilon| \leqslant 2^{-t}.$$

Assuming a positive x_1, set

$$2K^2 = S^2 + x_1 S \tag{4.5.37}$$

(No generality is lost in taking x_1 to be positive, provided we take the stable choice).

We also have

$$\begin{aligned} 2\bar{K}^2 &\equiv (x_1{}^2 + x_2{}^2 + \ldots + x_n{}^2 + x_1\bar{S})(1+\varepsilon)(1+e) \\ &\equiv [S^2 + x_1 S(1+\gamma)]\,(1+\varepsilon)(1+e) \end{aligned} \tag{4.5.38}$$

where

$$|\varepsilon| < \frac{3}{2}(n+1)2^{-2t_2}.$$

Now, if we write

$$S^2 + x_1 S(1+\gamma) \equiv (S^2 + x_1 S)(1+\varkappa), \tag{4.5.39}$$

then

$$\varkappa = \left(\frac{x_1 S}{S^2 + x_1 S}\right)\gamma. \tag{4.5.40}$$

Since $x_1 S$ and S^2 are positive and $x_1 \leqslant S$ this gives

$$|\varkappa| \leqslant |\gamma|/2. \tag{4.5.41}$$

Substituting (4.5.39) into (4.5.38) we obtain

$$2\bar{K}^2 \equiv [S^2 + x_1 S]\,(1+\varkappa)(1+\varepsilon)(1+e), \tag{4.5.42}$$

giving

$$2\bar{K}^2 \equiv 2K^2(1+\theta),\quad \text{where}\quad |\theta| < (1.501)2^{-t} \tag{4.5.43}$$

(4) Finally compute $\mathbf{u}$ where $\mathbf{u}^* = (x_1 + S, x_2, \ldots, x_n)$. As can be seen only the first element of $\mathbf{u}$ requires evaluation and we have

$$\begin{aligned} u_1 &= \mathrm{fl}(x_1 + \bar{S}) = (x_1 + \bar{S})(1+\phi) \\ &= [x_1 + S(1+\gamma)]\,(1+\phi),\quad |\phi| = < 2^{-t}. \end{aligned} \tag{4.5.44}$$

Here again the choice of the sign ensures low relative error and we have

$$\bar{u}_1 \equiv (x_1 + S)(1+\psi) \equiv u_1(1+\psi),\quad |\psi| < (1.501)2^{-t} \tag{4.5.45}$$

We can write

$$\bar{\mathbf{u}}^* \equiv (\bar{u}_1, u_2, u_3, \ldots, u_n) \equiv \mathbf{u}^* + \partial\mathbf{u}^*, \tag{4.5.46}$$

where

$$\|\partial\mathbf{u}\|_2 = |u_1|\,|\psi| < (1.501)2^{-t}\|\mathbf{u}\|_2 \tag{4.5.47}$$

and thus

$$\|\bar{\mathbf{u}}\|_2 = \|\mathbf{u} + \partial\mathbf{u}\|_2 < [1 + (1.501)2^{-t}]\,\|\mathbf{u}\|_2. \tag{4.5.48}$$

Now,

$$\mathbf{P}^{(c)} - \mathbf{P} = \frac{\bar{\mathbf{u}}\bar{\mathbf{u}}^*}{2\bar{K}^2} - \frac{\mathbf{u}\mathbf{u}^*}{2K}.$$

In order to estimate the norm $\|\mathbf{P}^{(c)} - \mathbf{P}\|_2$, consider first the norm

$$\left\|\frac{\mathbf{u}\mathbf{u}^*}{2K^2}\right\|_2.$$

We have

$$\|\mathbf{u}\|_2 = \|\mathbf{u}^*\|_2 = ((x_1 + S)^2 + x_2^2 + \ldots + x_n^2)^{1/2} = (2S(S + x_1))^{1/2}, \tag{4.5.49}$$

where $S = \|\mathbf{x}\|_2$.

Using (4.5.49) and (4.5.37) we find

$$\left\|\frac{\mathbf{u}\mathbf{u}^*}{2K^2}\right\| = \frac{\|\mathbf{u}\|_2\,\|\mathbf{u}^*\|_2}{2K^2} = \frac{2S(S + x_1)}{S(S + x_1)} = 2. \tag{4.5.50}$$

Now consider

$$\|\mathbf{P}^{(c)} - \mathbf{P}\|_2 = \left\|\frac{\bar{\mathbf{u}}\bar{\mathbf{u}}^*}{2\bar{K}^2} - \frac{\mathbf{u}\mathbf{u}^*}{2K^2}\right\|_2.$$

From (4.5.42), (4.5.46), (4.5.47) and (4.5.49) we have

$$\frac{\bar{\mathbf{u}}\bar{\mathbf{u}}^*}{2\bar{K}^2} = \frac{(\mathbf{u} + \mathbf{du})(\mathbf{u}^* + \mathbf{du}^*)}{2K^2(1 + \theta)} = \frac{\mathbf{u}\mathbf{u}^* + (\mathbf{u} + \mathbf{u}^*)\mathbf{du} + \mathbf{du}\,\mathbf{du}^*}{2K^2(1 + \theta)},$$

giving for the norm

$$\begin{aligned}
\left\|\frac{\bar{\mathbf{u}}\bar{\mathbf{u}}^*}{2\bar{K}^2} - \frac{\mathbf{u}\mathbf{u}^*}{2K^2}\right\|_2 &= \left\|\frac{\mathbf{u}\mathbf{u}^*}{2K^2}\left(1 - \frac{1}{1 + \theta}\right) + \frac{(\mathbf{u} + \mathbf{u}^*)\mathbf{du} + \mathbf{du}\,\mathbf{du}^*}{2K^2(1 + \theta)}\right\|_2 \\
&\leqslant \left\|\frac{\mathbf{u}\mathbf{u}^*}{2K^2}\right\|_2\left(1 - \frac{1}{1 + \theta}\right) + 2\left\|\frac{\mathbf{u}\,\mathbf{du}}{2K^2}\right\|_2\frac{1}{1 + \theta} + \left\|\frac{\mathbf{du}\,\mathbf{du}^*}{2K^2}\right\|_2\frac{1}{1 + \theta} \\
&\leqslant 2\left(1 - \frac{1}{1 + \theta}\right) + 2\,\frac{2\theta}{1 + \theta} + \frac{\theta^2}{1 + \theta} \\
&= \frac{2\theta + 4\theta + \theta^2}{1 + \theta} < 6\theta < (9.01)2^{-t}.
\end{aligned}$$

Thus,

$$\| \mathbf{P}^{(c)} - \mathbf{P} \|_2 < (9.01)2^{-t}. \tag{4.5.51}$$

In (4.5.51) matrix $\mathbf{P}$ may be any one of the matrices $\mathbf{P}_r$, $r = 1, \ldots, n-1$, which are computed for the purpose of triangulation of $\mathbf{A}$.

The second basic step in the triangulation of $\mathbf{A}$ is premultiplication of $\mathbf{A}$ by the elementary Hermitian matrix $\bar{\mathbf{P}}_r$, $r = 1, \ldots, n-1$.

We again consider a general case for the matrix $\bar{\mathbf{P}}$, where $\bar{\mathbf{P}}$ stands for any of the matrices $\bar{\mathbf{P}}_r$, $r = 1, \ldots, n-1$. The matrix $\bar{\mathbf{P}}$ has, in general, the following structure

$$\bar{\mathbf{P}} = \begin{bmatrix} \mathbf{I} & \mathbf{0} \\ \mathbf{0} & \mathbf{R}^{(c)} \end{bmatrix} \tag{4.5.52}$$

with the appropriate dimension $(n-r) \times (n-r)$ for the matrix $\mathbf{R}^{(c)}$.

It follows that the premultiplications, $\bar{\mathbf{P}}\mathbf{A}$, leave the first r rows of $\mathbf{A}$ unaltered and the last $(n-r)$ rows are premultiplied by $\mathbf{R}^{(c)}$, and so, the error matrix

$$\mathrm{fl}_2(\bar{\mathbf{P}}\mathbf{A}) - \bar{\mathbf{P}}\mathbf{A} \tag{4.5.53}$$

has nulls in its first r rows.

The bounds for the error matrix (4.5.53) can be obtained in terms of the Euclidean norm of $\mathbf{A}$ to give

$$\| \mathrm{fl}_2(\bar{\mathbf{P}}\mathbf{A}) - \bar{\mathbf{P}}\mathbf{A} \|_E \leqslant (3.55) \| \mathbf{A} \|_E 2^{-t}. \tag{4.5.54}$$

Further, if $\bar{\mathbf{A}}_n$ is the computed triangular matrix, $\mathbf{A}_1$ the original matrix and $\mathbf{G}$ the error matrix, then for

$$\bar{\mathbf{A}}_n = \bar{\mathbf{P}}_{n=1}\bar{\mathbf{P}}_{n-2} \ldots \bar{\mathbf{P}}_1(\mathbf{A}_1 + \mathbf{G}) \tag{4.5.55}$$

the bounds on $\mathbf{G}$ can readily be verified to yield

$$\| \mathbf{G} \|_E \leqslant 3.55(n-1)[1 + (9.01)2^{-t}]^{n-2} 2^{-t} \| \mathbf{A}_1 \|_E. \tag{4.5.56}$$

Similarly, if

$$\bar{\mathbf{b}}_n = \bar{\mathbf{P}}_{n-1}\bar{\mathbf{P}}_{n-3} \ldots \bar{\mathbf{P}}_1(\mathbf{b}_1 + \mathbf{k}) \tag{4.5.57}$$

where $\bar{\mathbf{b}}_n$ is the final computed right-hand side vector, then the bounds on k are given as

$$\| \mathbf{k} \|_2 < 3.55(n-1)[1 + (9.01)2^{-t}]^{n-1} 2^{-t} \| \mathbf{b}_1 \|_2. \tag{4.5.58}$$

Comparing formulae (4.5.56) and (4.5.58) with the bounds on the relative error in the solution x of the set $\mathbf{Ax} = \mathbf{b}$ (see Appendix A) it can be seen that the Householder reduction is a stable method.

4.6 How to Speed up Matrix Methods

We have seen that various direct methods for solving a set of linear equations all require the number of arithmetics of $O(n^3)$. The difference may be found in the values of the coefficient in the leading term of the time complexity function. For instance, Gaussian elimination requires in total fewer arithmetics than the other methods, except for the Cholesky decomposition, though Cholesky's method is of course not as general as the other methods.

With regard to the number of arithmetic operations we can distinguish a 'fast' or a 'slow' method. What is then a minimum number of arithmetics required to solve a set of linear equations in n unknowns? The answer depends to an extent on what type of operations are allowed for an algorithm to use. Klyuev and Kokovkin-Shcherbak (1965) have proved that $n^3/3 + n^2 - n/3$ multiplications and $n^3/3 + n^2/2 - 5n/6$ additions (the same as the number of operations in Gaussian elimination) is the minimum number of these operations required to solve a system of n linear equations, and this is for the methods which employ linear combinations of rows and columns only. However, Winograd (1967, 1968, 1970a), Strassen (1969) and Pan (1978) have put forward methods based on an approach different from manipulation of rows and columns. Winograd's method takes only $n^3/3$ multiplications by $n^3/2$ additions. Assuming, for example, that multiplication takes longer than addition, Winograd's method should be 'faster' than Gaussian elimination. Strassen's method requires less than $6n^{\log_2 7}$ multiplications and additions together. This method would be even 'faster' than Winograd's for large n.

Several years passed since these discoveries without much progress on further speed-up methods. Then in 1978 Pan has proposed a new way of reducting even further the order of computational complexity in the problems involving matrices. The current record stands at $O(n^{2.496})$ and it undoubtedly will be improved sooner than later (Pan, 1984). We shall consider some of these new methods.

The new methods are typically shown to be more 'efficient' or 'faster' compared with the traditional or 'classical' methods for problem instances of very large size. The key point about the new methods is their importance in the theory of algorithms as they demonstrate conceptually new ways of the algorithm design where the 'tradeoff' of more 'expensive' computational operations for 'cheaper' operations can bring about an overall improvement in the algorithm performance. The concept of the 'tradeoff' of one kind of operations for another commands at present wide support as one of the ways to improve efficiency of computations.

4.7 The Winograd Method

The Winograd Identity

Consider the identity

$$x_1y_1 + x_2y_2 = (x_1 + y_2)(x_2 + y_1) - x_1x_2 - y_1y_2. \qquad (4.7.1)$$

The Winograd identity is an expansion of (4.7.1) for the even number, $n = 2k$, of pairwise products

$$\sum_{i=1}^{2k} x_i y_i = \sum_{u=1}^{k} (x_{2u-1} + y_{2u})(x_{2u} + y_{2u-1}) - \sum_{u=1}^{k} x_{2u-1}x_{2u} - \sum_{u=1}^{k} y_{2u-1}y_{2u}. \quad (4.7.2)$$

Suppose that we need to compute $\mathbf{c} = \mathbf{Ab}$, where $\mathbf{A}$ is an $m \times n$ matrix and $\mathbf{b}$ is an n-vector. Assume that n is even, i.e. $n = 2k$. (If n is odd, simply apply the algorithm to the matrix $\mathbf{A}'$ and vector $\mathbf{b}'$ of dimensions $m \times (n+1)$ and $(n+1)$, respectively, where $a'_{i,n+1} = 0$ for all i and $b'_{n+1} = 0$.)

Using the Winograd identity, we can write

$$\begin{aligned} c_i &= \sum_{j=1}^{n} a_{ij} b_j \\ &= \sum_{u=1}^{k} (a_{i,2u-1} + b_{2u})(a_{i,2u} + b_{2u-1}) \\ &\quad - \sum_{u=1}^{k} a_{i,2u-1} a_{i,2u} - \sum_{u=1}^{k} b_{2u-1} b_{2u}, \qquad i = 1, \ldots, m. \end{aligned} \quad (4.7.3)$$

In (4.7.3) to compute all c_i requires

$$\begin{aligned} &\frac{3}{2} nm && \text{multiplications} \\ \text{and } &\left(\frac{5}{2} n - 1\right) m && \text{additions.} \end{aligned} \quad (4.7.4)$$

In comparison the standard method requires only nm multiplications and $(n-1)m$ additions. However, the method (4.7.3) becomes advantageous over the standard method, when the same matrix has to be multiplied by many vectors as is, for example, the case in computing the product of two matrices. The savings in the number of arithmetic operations comes from the fact that the product $f_i = \sum_{u=1}^{k} a_{i,2u-1} a_{i,2u}$ has to be calculated only once.

The Matrix Multiplication Algorithm

Let $\mathbf{A}$ and $\mathbf{B}$ be two matrices of dimension $m \times n$ and $n \times p$, respectively. The algorithm to compute $\mathbf{C} = \mathbf{AB}$ using the Winograd identity is given as

(assuming $n = 2k$)

Compute $$f_i = \sum_{u=1}^{k} a_{i,2u-1} a_{i,2u}, \qquad i = 1, \ldots, m. \quad (4.7.5)$$

Compute $$g_j = \sum_{u=1}^{k} b_{2u-1,j} b_{2u,j}, \qquad j = 1, \ldots, p. \quad (4.7.6)$$

Compute $$c_{ij} = \sum_{u=1}^{k} (a_{i,2u-1} + b_{2u,j})(a_{i,2u} + b_{2u-1,j}) - f_i - g_j, \quad (4.7.7)$$

$$i = 1, \ldots, m,$$
$$j = 1, \ldots, p.$$

The total number of operations required by the process is

$$\frac{nmp}{2} + \frac{n}{2}(m+p) \qquad \text{multiplications}$$

$$\text{and } \frac{3}{2}nmp + mp + \left(\frac{n}{2} - 1\right)(m+p) \qquad \text{additions.} \tag{4.7.8}$$

This number may be compared with nmp multiplications and $(n-1)mp$ additions required by the standard method. We note that with the use of Winograd's identity, the number of multiplications is about halved while some price for this reduction is paid in terms of an increased number of additions.

Optimality of Winograd's Formula

We have seen that the Winograd method is based on the identity involving the sum of two pairwise products. Perhaps it is then possible to derive a matrix multiplication algorithm similar to Winograd's based on, say, the identity involving three pairwise products, $a_1b_1 + a_2b_2 + a_3b_3$, or even higher than three? Would these further algorithms allow even greater savings in the number of arithmetics? In fact, it has been shown that the answer to this question is no. The Winograd formula is the best of its kind (Harter, 1972).

Algebraic Complexity of Winograd's Method

We first consider the matrix inversion.

Matrix Inversion

Suppose that an $n \times n$ matrix $\mathbf{A}$ is to be inverted. We can view $\mathbf{A}$ as an $m \times m$ matrix whose entries are $k \times k$ matrices. Gaussian elimination is then performed on this $m \times m$ matrix and whenever a multiplication of two $k \times k$ matrices is required use of Winograd's identity is made.

By (4.1.32) the number of $k \times k$ matrix multiplications required to invert $\mathbf{A}$ is $m^3 - 1$. To multiply two $k \times k$ matrices using Winograd's identity requires, by (4.7.8), $k^3/2 + k^2$ multiplications. (We assume that the vectors f_i and g_i of (4.7.5) and (4.7.6) are computed when needed and are not stored even though some of the $k \times k$ entries are used more than once in the computation.)

Further, m reciprocals are required to invert $\mathbf{A}$. These reciprocals are treated as inversions of $k \times k$ matrices. The inversions are carried out by ordinary Gaussian elimination and thus will call for $k^3 - 1$ multiplications and k reciprocals. This gives altogether $k^3 + k - 1$ multiplicative operations. Thus the number of multiplications to invert A is

$$(m^3 - 1)\left(\frac{k^3}{2} + k^2\right) + m(k^3 + k - 1), \quad n = mk,$$

or

$$\frac{n^3}{2}+\frac{n^3}{2k}+nk^2+n-\frac{n}{k}-\frac{k^3}{2}-k^2. \tag{4.7.9}$$

In (4.7.9) the terms involving the powers of k only are insignificant since normally $k \ll n$. To minimize the function (4.7.9) in terms of k we write

$$F(k)=\frac{n^3}{2}+\frac{n^3}{2k}+nk^2+n-\frac{n}{k},$$

and the minimum of the function occurs when

$$\frac{dF}{dk}=-\frac{n^3}{2k^2}+2nk+\frac{n}{k^2}=0$$

i.e. when $k \approx (n^2/4)^{1/3}$. (4.7.10)

This gives the number of required multiplications as

$$\frac{n^3}{2}+1.19n^{7/3}+n-0.63n^{1/3}. \tag{4.7.11}$$

In a similar manner the number of additions is obtained as

$$\frac{3n^3}{2}+3.75n^{7/3}-2n^2+O(n^{5/3}). \tag{4.7.12}$$

Solution of a Set of Linear Equations

Next we consider computational complexity of the Winograd method when applied to solve a set $\mathbf{Ax}=\mathbf{b}$. The set is solved by first decomposing $\mathbf{A}$ into the product $\mathbf{LU}$ and then applying back substitution. All matrix operations are accomplished using Winograd's matrix multiplication method. Here we have the following arithmetic estimates.

The decomposition of $\mathbf{A}$ into $\mathbf{LU}$ requires $m(m^2-1)/3$ $k \times k$ matrix multiplications and m reciprocals of a $k \times k$ matrix. This gives

$$\frac{1}{3}m(m^2-1)\left(\frac{k^3}{2}+k^2\right)+m(k^3+k-1) \quad \text{multiplications} \tag{4.7.13}$$

The process of back substitution calls for m^2 $k \times k$ matrix multiplications by k-size vectors. Each of these multiplications involves $k^2+k/2$ matrix entry multiplications. This gives

$$m^2\left(k^2+\frac{k}{2}\right) \quad \text{multiplications.} \tag{4.7.14}$$

The total number of multiplications to solve a set of n linear equations is

$$F(k)=\frac{1}{3}\left(m^3-m\left(\frac{k^3}{2}+k^2\right)+m^2\left(k^2+\frac{k}{2}\right)+m(k^3+k-1)\right.$$

$$=\frac{n^3}{6}+\frac{n^3}{3k}+n^2+\frac{5nk^2}{6}+\frac{n^2}{2k}-\frac{nk}{3}+n-\frac{n}{k},\quad n=mk. \tag{4.7.15}$$

The function $F(k)$ assumes its minimum value when

$$\frac{dF}{dk}=-\frac{n^3}{3k^2}+\frac{5nk}{3}-\frac{n^2}{2k^2}-\frac{n}{3}-\frac{n}{k^2}=0, \tag{4.7.16}$$

which gives $k\approx(n^2/5)^{1/3}$. (4.7.17)

Substituting the value of k into (4.7.15) we obtain the number of multiplications as

$$\frac{n^3}{6}+0.86n^{7/3}+n^2+O(n^{5/3}). \tag{4.7.18}$$

In a similar fashion we obtain for the total number of additions:

$$\frac{n^3}{2}+\frac{n^3}{k}+\frac{nk^2}{2}+\frac{3n^2}{2}, \tag{4.7.19}$$

where an optimum value of $k\approx(n^2/5)^{1/3}$. This gives

$$\frac{n^3}{2}+1.88n^{7/3}+\frac{3}{2}n^2+O(n^{5/3}). \tag{4.7.20}$$

Thus the Winograd method reduces the number of required multiplications by one-half as compared with the standard methods, at the same time it requires the number of additions increased by a factor of $\frac{3}{2}$.

Error Analysis

In the analysis we follow closely the method of Brent (1970b). To find the inner product $\sum_{i=1}^{n}x_iy_i$ using Winograd's method for n even compute

$$\bar{w}=\mathrm{fl}(\bar{p}-\bar{r}), \tag{4.7.25}$$

where

$$\bar{p}=\mathrm{fl}\left(\sum_{i=1}^{n/2}(x_{2i-1}+y_{2i})(x_{2i}+y_{2i-1})\right), \tag{4.7.26}$$

$$\bar{r}=\mathrm{fl}(\bar{f}+\bar{g}), \tag{4.7.27}$$

$$\bar{f}=\mathrm{fl}\left(\sum_{i=1}^{n/2}x_{2i-1}x_{2i}\right) \tag{4.7.28}$$

and

$$\bar{g} = \text{fl}\left(\sum_{i=1}^{n/2} y_{2i-1}y_{2i}\right). \tag{4.7.29}$$

Define the error e_f in f so that

$$e_f = f - \text{fl}\left(\sum_{i=1}^{n/2} x_{2i-1}x_{2i}\right), \tag{4.7.30}$$

and define the errors e_g, e_r, e_p, e_w in a similar way. Next assume that the computations are carried out under the conditions of normalized t-digit rounded floating-point arithmetic, and further assue that n is such that $n2^{-t} \leqslant 1$, which for our purposes may be more conveniently written as

$$\frac{n2^{1-t}}{2} \leqslant 1. \tag{4.7.31}$$

Furthermore, let

$$t_1 = t - \frac{n2^{1-t}}{\log 2}. \tag{4.7.32}$$

This choice of t_1 ensures the inequality

$$2^{t-t^1} \geqslant 1 + n2^{1-t}. \tag{4.7.33}$$

In practice condition (4.7.27) is usually satisfied very easily and the distinction between t and t_1 may be ignored. The following inequalities can now be proved.

If $a = \max_i\{|a_i|\}$ and $b = \max_i\{|b_i|\}$

then the error in the computed inner product

$S = \sum_{i=1}^{n} a_i b_i$ for $n \geqslant 1$ is bounded by

$$|e_S| \leqslant \tfrac{1}{4}ab(n^2 + 3n)2^{1-t_1}, \tag{4.7.30}$$

where S stands for ‘standard method’ if the product S is computed in the usual way,

$$|e_W| \leqslant \tfrac{1}{8}(a+b)^2(n^2 + 16n)2^{1-t_1}, \tag{4.7.31}$$

where W stands for ‘Winograd’s method’ if the product S is computed by Winograd’s method,

$$|e_W| \leqslant \tfrac{3}{8}a^2(n^2 + 17n)2^{1-t_1}, \tag{4.7.32}$$

if S is computed by Winograd’s method and $a = b$.

These three bounds can be proved as follows. For the inner product by standard method due to Wilkinson we have

$$e_n = \sum_{i=1}^{n} a_i b_i - \text{fl}\left(\sum_{i=1}^{n} a_i b_i\right) = \sum_{i=1}^{n} a_i b_i e_i,$$

where

$$|e_1| \leqslant (1+\tfrac{1}{2}2^{1-t})^n - 1$$

and

$$|e_i| \leqslant (1+\tfrac{1}{2}2^{1-t})^{n+2-i} - 1, \qquad \text{for } 2 \leqslant i \leqslant n.$$

It follows that

$$|e_S| = |S - \bar{S}| \leqslant ab\left[\left(\sum_{k=2}^{n} (1+\tfrac{1}{2}2^{1-t})^k - 1\right) + (1+\tfrac{1}{2}2^{1-t})^n - 1\right]. \tag{4.7.33}$$

In order to estimate the term involving summation consider expression $(1+z)^k$ and expand it as a binomial series to obtain

$$(1+z)^k - 1 = kz + \frac{k(k-1)}{2} z^2\left(1 + \frac{(k-2)z}{3}\right) + O(z^4)$$

$$\leqslant kz + \frac{k(k-1)z^2}{2}\left[1 - \frac{(k-2)z}{3}\right]^{-1}. \tag{4.7.34}$$

The maximum value of $\left[1 - \frac{(k-2)z}{3}\right]^{-1}$ occurs when k and z assume their maximum values.

If we take bounds on k and z as

$$k \leqslant n+2 \quad \text{and} \quad z \leqslant \frac{1}{n} = \frac{1}{2}2^{1-t}, \tag{4.7.35}$$

then

$$\left[1 - \frac{(k-2)z}{3}\right]^{-1} \leqslant \frac{3}{2}. \tag{4.7.36}$$

Substituting (4.7.34) into (4.7.33) and using (4.7.35) we obtain

$$|e_S| \leqslant abz\,\frac{n^2+3n-2}{2}\left(1 + \frac{nz}{2}\right)$$

$$\leqslant \tfrac{1}{4}ab(n^2+3n-2)2^{1-t}2^{t-t_1}$$

$$\leqslant \tfrac{1}{4}ab(n^2+3n)2^{1-t_1},$$

since $\left(1 + \frac{nz}{2}\right) \leqslant \left(1 + \frac{n_2^{1-t}}{4}\right) \leqslant 2^{t-t_1}$.

This proves the bound (4.7.30).

In order to prove the bound (4.7.31) for the Winograd method we first assume that $n = 2m$ is even, so that by (4.7.27)

$$\tfrac{1}{2}m2^{1-t} \leqslant \tfrac{1}{2}.$$

Now we have

$$|e_f| \leqslant a^2\left[\left(\sum_{k=2}^{m} (1+z)^k - 1\right) + (1+z)^m - 1\right], \tag{4.7.37}$$

where the same bound as (4.7.35) is valid for z. Using the arguments analogous to those of (4.7.34)–(4.7.36) we obtain

$$\begin{aligned} |e_f| &\leqslant \tfrac{1}{2}a^2 z(m^2+3m-2)(1+\tfrac{2}{5}mz) \\ &\leqslant \tfrac{6}{5}a^2 m^2 z. \end{aligned} \tag{4.7.38}$$

Similarly we obtain

$$|e_g| \leqslant \tfrac{6}{5}b^2 m^2 z. \tag{4.7.39}$$

From (4.7.26) we have

$$|f| \leqslant a^2 m + |e_f|,$$

and thus

$$|f| \leqslant a^2 m(1+\tfrac{6}{5}mz). \tag{4.7.40}$$

Again, a similar estimate, with a replaced by b, is valid for the sum g. Hence, we can write

$$|f| + |g| \leqslant (a^2+b^2)m(1+\tfrac{6}{5}mz). \tag{4.7.41}$$

Further the error in term r of (4.7.23) is

$$\begin{aligned} |e_r| &\leqslant |e_f| + |e_g| + z(|f| + |g|) \\ &\leqslant \tfrac{1}{2}z(a^2+b^2)(m^2+5m-2)(1+2mz). \end{aligned} \tag{4.7.42}$$

Similarly it can be shown that the error in term p of (4.7.22) is given by

$$|e_p| \leqslant \tfrac{1}{2}z(a+b)^2(m^2+7m-2)(1+\tfrac{7}{6}mz) \tag{4.7.43}$$

The total error for (4.7.21) is

$$|e_w| \leqslant |e_p| + |e_r| + z(|p| + |r|), \qquad z \leqslant \tfrac{1}{2}2^{1-t}.$$

Using (4.7.42) and (4.7.43) we obtain

$$\begin{aligned} |e_W| &\leqslant z(a+b)^2(m^2+8m-2)(1+3mz) \\ &\leqslant \frac{1}{2}(a+b)^2\left(\frac{n^2}{4}+4n-2\right)\left(1+\frac{3}{4}n2^{1-t}\right)2^{1-t}, \end{aligned} \tag{4.7.44}$$

where we used $m = \dfrac{n}{2}$ and $z \leqslant \dfrac{1}{2}2^{1-t}$.

Finally, since by (4.7.29) we have

$$1+\tfrac{3}{4}n2^{1-t} \leqslant 2^{t-t_1},$$

it follows that

$$|e_W| = |e_w| \leqslant \tfrac{1}{8}(a+b)^2(n^2+16n)2^{1-t_1}.$$

This completes the proof of the bound (4.7.31).

For $a=b$ a stronger result can be derived. Equation (4.7.31) shows that Winograd's method is badly unstable when a/b is either very large or very small compared with unity as then $(a+b)^2 \gg ab$. By comparison the bound for the standard method is not influenced by the ratio a/b. For Winograd's method to be effective some sort of scaling must be employed. Ignoring the simple cases of $a=0$ and $b=0$, there exists an integer k such that

$$2^{-1/2} \leqslant \frac{a}{b}\,2^k < 2^{1/2}. \tag{4.7.45}$$

If we replace x by $2^k x$, computed without rounding errors, then using the Winograd identity and multiplying the final result by 2^{-k}, we obtain

$$|e_W| \leqslant \tfrac{1}{8}(2+2^{1/2}+2^{-1/2})ab(n^2+16n)2^{1-t_1} \tag{4.7.46}$$

and this expresion lacks the term $(a+b)^2$.

If we require the matrix product $\mathbf{C}=\mathbf{AB}$, then even a crude scaling, say, $\mathbf{A}$ into $2^k\mathbf{A}$ and $\mathbf{B}$ into $2^{-k}\mathbf{B}$, such that

$$2^{-1} \leqslant \frac{2^k a}{2^{-k} b} < 2,$$

followed by the application of the Winograd algorithm, gives the result $\mathbf{AB}+\mathbf{E}$ where

$$\max|e_{ij}| \leqslant \tfrac{9}{8}\,ab(n^2+16n)2^{-t_1}, \tag{4.7.47}$$

$$a=\max|a_{ij}|, \qquad b=\max|b_{ij}|.$$

We conclude that Winograd's method with scaling is nearly as accurate as the standard method (without accumulation of inner products) and the scaling takes the time of $O(n^2)$ compared with the total time of $O(n^3)$.

Winograd's algorithm can be used to solve set $\mathbf{Ax}=\mathbf{b}$ by partitioning the $n\times n$ matrix $\mathbf{A}$ into $m\times m$ submatrices of dimension $k\times k$ and then applying the matrix algorithm where appropriate. Brent (1970b) has shown that for the method to be stable some form of pivoting must be used. He has also proposed one such algorithm. Brent's algorithm uses scaling and partial pivoting to obtain an error bound which is satisfactory unless the so-called growth factor, G, is too large. The growth factor is defined as

$$G=\max_{i,j}\{|u_{ij}|\}, \tag{4.7.48}$$

where u_{ij} are the elements of matrix $\mathbf{U}$ in $\mathbf{A}=\mathbf{LU}$. In fact, G affects also the error bound in Gaussian elimination with partial pivoting and in the same way.

Winograd's algorithm can be used to obtain the Cholesky decomposition $\mathbf{A}=\mathbf{LL}^T$ with $n^3/12$ multiplications and $n^3/4$ additions instead of the usual

$n^3/6$ of each. There is no need for pivoting and, if $|a_{ij}| \leqslant 1$ and the square roots are computed with a relative error of 2^{1-t}, then

$$\mathbf{LL}^T = \mathbf{A} + \mathbf{E}, \qquad (4.7.49)$$

where

$$|e_{ij}| \leqslant \tfrac{3}{8}(n^2 + 15n)2^{1-t_1}. \qquad (4.7.50)$$

4.8 The Strassen Method

In 1969 Strassen suggested a recursive algorithm for computing the product of two square matrices of order n in less than $4.7n^{\log_2 7}$ arithmetics. the algorithm can be used to invert a matrix, to solve a set of linear equations, to compute the determinant, all requiring the total number of arithmetics of $O(n^{\log_2 7})$. We shall first consider the basic Strassen algorithm to find the product of two matrices.

Matrix Multiplication

Let $\mathbf{A}$ and $\mathbf{B}$ be two matrices of dimension $n = 2m$, an even number, and their product be denoted by $\mathbf{C}$.

$$\mathbf{A} = \begin{bmatrix} \mathbf{A}_{11} & \mathbf{A}_{12} \\ \mathbf{A}_{21} & \mathbf{A}_{22} \end{bmatrix}, \quad \mathbf{B} = \begin{bmatrix} \mathbf{B}_{11} & \mathbf{B}_{12} \\ \mathbf{B}_{21} & \mathbf{B}_{22} \end{bmatrix}, \quad \mathbf{C} = \begin{bmatrix} \mathbf{C}_{11} & \mathbf{C}_{12} \\ \mathbf{C}_{21} & \mathbf{C}_{22} \end{bmatrix}$$

The standard method to compute the product as we know calls for n^3 multiplications and $n^2(n-1)$ additions. To evaluate $\mathbf{C} = \mathbf{AB}$ using Strassen's method we compute

$$\begin{aligned}
\mathbf{Q}_1 &= (\mathbf{A}_{11} + \mathbf{A}_{22})(\mathbf{B}_{11} + \mathbf{B}_{22}), \\
\mathbf{Q}_2 &= (\mathbf{A}_{21} + \mathbf{A}_{22})\mathbf{B}_{11}, \\
\mathbf{Q}_3 &= \mathbf{A}_{11}(\mathbf{B}_{12} - \mathbf{B}_{22}), \\
\mathbf{Q}_4 &= \mathbf{A}_{22}(-\mathbf{B}_{11} + \mathbf{B}_{21}), \\
\mathbf{Q}_5 &= (\mathbf{A}_{11} + \mathbf{A}_{12})\mathbf{B}_{22}, \\
\mathbf{Q}_6 &= (-\mathbf{A}_{11} + \mathbf{A}_{21})\mathbf{B}_{11} + \mathbf{B}_{12}), \\
\mathbf{Q}_7 &= (\mathbf{A}_{12} - \mathbf{A}_{22})(\mathbf{B}_{21} + \mathbf{B}_{22}),
\end{aligned} \qquad (4.8.1)$$

then

$$\begin{aligned}
\mathbf{C}_{11} &= \mathbf{Q}_1 + \mathbf{Q}_4 - \mathbf{Q}_5 + \mathbf{Q}_7, \\
\mathbf{C}_{21} &= \mathbf{Q}_2 + \mathbf{Q}_4, \\
\mathbf{C}_{12} &= \mathbf{Q}_3 + \mathbf{Q}_5, \\
\mathbf{C}_{22} &= \mathbf{Q}_1 + \mathbf{Q}_3 - \mathbf{Q}_2 + \mathbf{Q}_6.
\end{aligned} \qquad (4.8.2)$$

Here are required seven multiplications of $m \times m$ matrices and 18 additions of $m \times m$ matrices. The new algorithm becomes advantageous in the number of arithmetics for large values of n. The key to the economy lies in the fact that the algorithm calls for only seven multiplications of $m \times m$ matrices rather than eight. Since the product of two $m \times m$ matrices uses m^3 multiplications and $m^2(m-1)$ additions and the sum of two $m \times m$ matrices uses m^2 additions, the algorithm utilizes

$$7m^3 = (7/8)n^3 \qquad \text{multiplications}$$

and

$$7(m^3 - m^2) + 18m^2 7m^3 + 11m^2 = (7/8)n^3 + (11/4)n^2 \qquad \text{additions.}$$

Whenever $n > 30$ we have both

$$(7/8)n^3 < n^3 \quad \text{and} \quad (7/8)n^3 + (11/4)n^2 \leqslant n^3 - n^2.$$

Whenever m itself is an even number we can reduce the number of arithmetic operations by using the same algorithm to compute each of the seven products of $m \times m$ matrices.

Taking $n = 2^s$ to be a power of 2 and denoting by $M(s)$ the number of multiplications needed to multiply two $n \times n$ matrices and by $A(s)$ the number of additions, we obtain

$$\begin{aligned} M(s+1) &= 7M(s), \\ A(s+1) &= 7A(s) + 18(4^s). \end{aligned} \tag{4.8.3}$$

Using the initial conditions $M(0) = 1$ and $A(0) = 0$ we obtain

$$\begin{aligned} M(s) &= 7^s = (2^s)^{\log_2 7} = n^{\log_2 7} = n^{2.807}, \\ A(s) &= 6(7^s - 4^s) = 6((2^s)^{\log_2 7} - (2^s))^2 = 6(n^{\log_2 7} - n^2). \end{aligned} \tag{4.8.4}$$

In the case where n is not the power of 2, appropriate 'padding' of the two matrices by zeros is possible so as to make their dimension a power of 2. We then obtain $M(n) \leqslant 7n^{\log_2 7} - 42n^2$, where $M(n)$ denotes the number of multiplications needed to multiply two $n \times n$ matrices, and a similar result for $A(n)$. The Strassen algorithm thus requires only $4.7n^{2.807}$ operations.

Memory Requirements

Due to the necessity to store the intermediate matrix components Strassen's algorithm requires substantial auxiliary memory space.

At each level, k, of recursion in the algorithm a declaration of seven new auxiliary matrices of order $n/2^k$ is required and the eighth matrix is needed for the result. Hence the straightforward application of Strassen's method needs at least

$$n^2 + 8(n^2/4 + n^2/16 + \ldots) = 11/3n^2$$

additional memory locations.

A version of the Strassen algorithm which performs *in situ* has been proposed by Kreczmar (1976). It reduces the extra storage requirements to $2/3n^2$ additional memory locations.

Winograd's Enhancement of the Strassen Algorithm

Winograd (1973) reduced the number of additions required by Strassen's algorithm to 15. This improved the constant factor in the order of algebraic complexity but not the order-of-magnitude.

The Winograd algorithm is as follows: Let matrices **A, B** and **C** be defined as in the Strassen matrix multiplication algorithm. To compute the elements of the product matrix **C** the following procedure is used.

$$\begin{array}{lll}
\mathbf{Q}_1 = \mathbf{A}_{21} - \mathbf{A}_{11}, & \mathbf{Q}_5 = \mathbf{B}_{22} - \mathbf{B}_{12} & \\
\mathbf{Q}_2 = \mathbf{A}_{11} + \mathbf{A}_{12}, & \mathbf{Q}_6 = \mathbf{B}_{12} - \mathbf{B}_{11}, & \\
\mathbf{Q}_3 = \mathbf{A}_{12} - \mathbf{Q}_1, & \mathbf{Q}_7 = \mathbf{B}_{11} + \mathbf{Q}_5, & \\
\mathbf{Q}_4 = \mathbf{A}_{22} - \mathbf{Q}_3, & \mathbf{Q}_8 = \mathbf{B}_{21} - \mathbf{Q}_7, & \\
\mathbf{P}_1 = \mathbf{A}_{21}\mathbf{B}_{11}, & \mathbf{P}_5 = \mathbf{Q}_4\mathbf{B}_{22}, & \mathbf{Q}_9 = \mathbf{P}_1 + \mathbf{P}_7, \\
\mathbf{P}_2 = \mathbf{A}_{22}\mathbf{B}_{21}, & \mathbf{P}_6 = \mathbf{A}_{12}\mathbf{Q}_8, & \mathbf{Q}_{10} = \mathbf{Q}_9 + \mathbf{P}_3, \\
\mathbf{P}_3 = \mathbf{Q}_1\mathbf{Q}_5, & \mathbf{P}_7 = \mathbf{Q}_3\mathbf{Q}_7, & \mathbf{Q}_{11} = \mathbf{P}_4 + \mathbf{P}_5, \\
\mathbf{P}_4 = \mathbf{Q}_2\mathbf{Q}_6, & & \\
\mathbf{C}_{11} = \mathbf{Q}_{10} + \mathbf{P}_6, & \mathbf{C}_{21} = \mathbf{P}_1 + \mathbf{P}_2, & \\
\mathbf{C}_{12} = \mathbf{Q}_{10} + \mathbf{P}_4, & \mathbf{C}_{22} = \mathbf{Q}_9 + \mathbf{Q}_{11}. &
\end{array} \tag{4.8.5}$$

The algorithm requires seven multiplications and 15 additions of $m \times m$ matrices, and this is three additions fewer than in the Strassen algorithm.

The Karatsuba–Makarov Method

Yet another algorithm of matrix multiplication was proposed by Karatsuba and Ofman (1962) and again discussed by Makarov (1975a and b). The algorithm uses fewer additions but the same number of multiplications as the standard algorithm.

To derive the algorithm consider again the matrices **A** and **B** of dimension $n = 2m$ and their product **C**. The elements $\mathbf{C}_{ij}$ are computed using the formula:

$$\begin{aligned}
\mathbf{C}_{11} &= \mathbf{A}_{11}(\mathbf{B}_{11} + \mathbf{B}_{21}) + (\mathbf{A}_{12} - \mathbf{A}_{11})\mathbf{B}_{21}, \\
\mathbf{C}_{12} &= \mathbf{A}_{12}(\mathbf{B}_{22} + \mathbf{B}_{12}) - (\mathbf{A}_{12} - \mathbf{A}_{11})\mathbf{B}_{12}, \\
\mathbf{C}_{21} &= \mathbf{A}_{21}(\mathbf{B}_{11} + \mathbf{B}_{21}) + (\mathbf{A}_{22} - \mathbf{A}_{21})\mathbf{B}_{21}, \\
\mathbf{C}_{22} &= \mathbf{A}_{22}(\mathbf{B}_{22} + \mathbf{B}_{12}) - (\mathbf{A}_{22} - \mathbf{A}_{21})\mathbf{B}_{12}.
\end{aligned} \tag{4.8.6}$$

Another possibility is to use

$$\begin{aligned} \mathbf{C}_{11} &= \mathbf{A}_{12}(\mathbf{B}_{11}+\mathbf{B}_{21}) - \mathbf{A}_{12} - \mathbf{A}_{11})\mathbf{B}_{11}, \\ \mathbf{C}_{12} &= \mathbf{A}_{11}(\mathbf{B}_{11}+\mathbf{B}_{21}) + (\mathbf{A}_{12} - \mathbf{A}_{11})\mathbf{B}_{11}, \\ \mathbf{C}_{22} &= \mathbf{A}_{12}(\mathbf{B}_{12}+\mathbf{B}_{22}) - (\mathbf{A}_{22} - \mathbf{A}_{21})\mathbf{B}_{12}, \\ \mathbf{C}_{22} &= \mathbf{A}_{21}(\mathbf{B}_{12}+\mathbf{B}_{22}) + (\mathbf{A}_{22} - \mathbf{A}_{21})\mathbf{B}_{12}. \end{aligned} \tag{4.8.7}$$

When used recursively, the algorithm calls for n^3 multiplications and $n^3 - 2n^2$ additions which compares favourably with n^3 multiplications and $n^3 - n^2$ additions required by the standard method.

The Pan Algorithm

Since the publication of the Strassen algorithm considerable efforts have been directed to speed up even further the matrix multiplication process. The Strassen algorithm became possible due to the successful design for the product of two 2×2 matrices in seven multiplications. It was subsequently shown that two 2×2 matrices cannot be multiplied in fewer than seven multiplications (Winograd, 1970a; Hopcroft and Kerr, 1969). The next step is then to seek a further asymptotic speed up by designing fast algorithms which would multiply two 3×3 matrices in no more than 21 multiplications, since for an algorithm to be faster than Strassen's it must use the number of multiplications (M) such that $\log_3 M < \log_2 7$, where M is the number of multiplications used in multiplying two 3×3 matrices.

However, the algorithm which would multiply two 3×3 matrices in 21 or fewer multiplications either is well hidden from the designers or does not exist at all since it has not been uncovered yet! The closest to the number 21 obtained so far is 23 multiplications (Laderman, 1976). Another algorithm to multiply two 5×5 matrices in 103 multiplications was proposed by Schachtel (1978). These results have suggested that perhaps still larger size matrices would yield better gains in speed up. In 1978 Pan has proved just that. He proposed a new algorithm which reduces the problem of the asymptotic acceleration of matrix multiplication to the problem of multiplying some matrices of a particular size, say 70×70, and is asymptotically faster than Strassen's.

To derive the new algorithm Pan has used the fact that the evaluation of the product of $n \times p$ by $p \times n$ matrices and decomposing the trace of product of three matrices of dimension $n \times p$ by $p \times m$ by $m \times n$ are two equivalent problems. Example 4.8.1 illustrates this point.

Example 4.8.1

Consider the product of 3×2 and 2×3 matrices:

$$\begin{bmatrix} a_{11} & a_{12} & a_{13} \\ a_{21} & a_{22} & a_{23} \end{bmatrix} \times \begin{bmatrix} b_{11} & b_{12} \\ b_{21} & b_{22} \\ b_{31} & b_{32} \end{bmatrix} = \begin{bmatrix} c_{11} & c_{12} \\ c_{21} & c_{22} \end{bmatrix}, \tag{4.8.8}$$

and the bilinear form of the traditional algorithm,

$$c_{ij} = \sum_{k=1}^{3} a_{ik}b_{kj}, \qquad i, j = 1, 2, \tag{4.8.9}$$

where

$$c_{11} = a_{11}b_{11} + a_{12}b_{21} + a_{13}b_{31}, \qquad c_{12} = a_{11}b_{12} + a_{12}b_{22} + a_{13}b_{32},$$
$$c_{21} = a_{21}b_{11} + a_{22}b_{21} + a_{23}b_{31}, \qquad c_{22} = a_{21}b_{12} + a_{22}b_{22} + a_{23}b_{32}.$$

The equivalent representation in trilinear form is

$$\begin{aligned}\sum_{i,j,k} a_{ij}b_{jk}c_{ki}
&= a_{11}b_{11}c_{11} + a_{11}b_{12}c_{21} + a_{12}b_{21}c_{11} + a_{12}b_{22}c_{21} + a_{13}b_{31}c_{11} \\
&\quad + a_{13}b_{32}c_{21} + a_{21}b_{11}c_{12} + a_{21}b_{12}c_{22} + a_{22}b_{21}c_{12} + a_{22}b_{22}c_{22} \\
&\quad + a_{23}b_{31}c_{12} + a_{23}b_{32}c_{22} \\
&= (a_{11}b_{11} + a_{12}b_{21} + a_{13}b_{31})c_{11} + (a_{21}b_{11} + a_{22}b_{21} + a_{23}b_{31})c_{12} \\
&\quad + (a_{11}b_{12} + a_{12}b_{22} + a_{13}b_{32})c_{21} + (a_{21}b_{12} + a_{22}b_{22} + a_{23}b_{32})c_{22}. \end{aligned} \tag{4.8.10}$$

In equation (4.8.10) the coefficient of c_{ij} is the (j, i) element of the product C in equation (4.8.9).

There is more than one way in which the trilinear form can be represented. The Pan algorithm represents the trilinear form for computational purposes in such a way as to achieve reduction in the total number of multiplications required.

Example 4.8.2

Given are two 4×4 matrices

$$\mathbf{A} = \begin{bmatrix} 5 & 3 & -1 & 4 \\ 2 & -5 & 7 & 9 \\ -7 & 4 & 0 & 5 \\ 8 & -3 & -4 & 1 \end{bmatrix} \quad \text{and} \quad \mathbf{B} = \begin{bmatrix} 2 & 1 & -4 & -6 \\ 4 & 3 & -9 & 1 \\ 6 & 7 & 7 & 5 \\ -8 & 0 & 2 & 1 \end{bmatrix}.$$

Find their product, **C**.

Solution

Consider the trilinear form as

$$\sum_{i,j,k} a_{ij}b_{jk}c_{ki}$$

$$= \sum_{\substack{i+j+k \\ \text{is even}}} (a_{ij} + a_{k+1,i+1})(b_{jk} + b_{i+1,j+1})(c_{ki} + c_{j+1,k+1})$$

$$-\sum_{i,k} a_{k+1,i+1} \sum_{\substack{j;i+j+k \\ \text{is even}}} (b_{jk} + b_{i+1,j+1})c_{ki}$$

$$-\sum_{i,j} a_{ij}b_{i+1,j+1} \sum_{\substack{k;i+j+k \\ \text{is even}}} (c_{ki} + c_{j+1,k+1})$$

$$-\sum_{j,k} \sum_{\substack{i;i+j+k \\ \text{is even}}} (a_{ij} + a_{k+1,j+1})b_{jk}c_{j+1,k+1}.$$

Substituting for a_{ij}'s and b_{ij}'s and collecting the coefficients of c_{ij} we get:

$$\begin{aligned}\sum_{i,j,k} a_{ij}b_{jk}c_{ki} = & -16c_{11} - 46c_{12} - 38c_{13} - 28c_{14} + 7c_{21} \\ & + 36c_{22} + 5c_{23} - 29c_{24} - 46c_{31} + 104c_{32} + 2c_{33} \\ & - 31c_{34} - 28c_{41} + 27c_{42} + 51c_{43} - 70c_{44}.\end{aligned}$$

Since the coefficient of c_{ij} is the (j, i)th entry in the matrix $\mathbf{C}$ we have

$$\mathbf{C} = \begin{bmatrix} -16 & 7 & -46 & -28 \\ -46 & 36 & 104 & 27 \\ -38 & 5 & 2 & 51 \\ -28 & -29 & -31 & -70 \end{bmatrix}$$

The number of multiplications required to calculate the product matrix $\mathbf{C}$ is 80, this is inferior to both Strassen's method (with 49 multiplications) and the traditional method (with 64 multiplications). However, when $n > 6$ the algorithm becomes faster than the traditional method.

Having studied various representations of the trilinear form, Pan has suggested an algorithm which yields results superior to Strassen's method and is as follows:

$$\sum_{i,j,k} a_{ij}b_{jk}c_{ki} = T_0 - T_1 - T_2 \quad - T_3, \tag{4.8.11}$$

where

$$\begin{aligned}T_0 = & \sum_{i,j,k \in S'(s)} (a_{ij} + a_{jk} + a_{ki})(b_{jk} + b_{ki} + b_{ij})(c_{ki} + c_{ij} + c_{jk}) \\ & - (a_{ij} - a_{j\bar{k}} + a_{k\bar{i}})(b_{j\bar{k}} + b_{ki} - b_{\bar{i}\bar{j}})(-c_{\bar{k}i} + c_{i\bar{j}} + c_{jk}) \\ & - (-a_{\bar{i}\bar{j}} + a_{j\bar{k}} + a_{ki})(b_{jk} - b_{ki} + b_{i\bar{j}})(c_{ki} + c_{ij} - c_{\bar{j}k}) \\ & - (a_{i\bar{j}} + a_{jk} - a_{\bar{k}i})(-b_{\bar{j}k} + b_{ki} + b_{ij})(c_{ki} - c_{\bar{i}\bar{j}} + c_{j\bar{k}}) \\ & - (a_{\bar{i}\bar{j}} + a_{\bar{j}\bar{k}} - a_{k\bar{i}})(-b_{j\bar{k}} + b_{ki} + b_{\bar{i}\bar{j}})(c_{\bar{k}\bar{i}} - c_{ij} + c_{\bar{j}k}) \\ & - (-a_{i\bar{j}} + a_{\bar{j}k} + a_{\bar{k}\bar{i}})(b_{\bar{j}\bar{k}} - b_{ki} + b_{\bar{i}})(c_{\bar{k}i} + c_{\bar{i}\bar{j}} - c_{j\bar{k}}) \\ & - (a_{\bar{i}\bar{j}} - a_{\bar{j}k} + a_{\bar{k}\bar{i}})(b_{\bar{j}\bar{k}} + b_{ki} - b_{\bar{i}\bar{j}})(-c_{\bar{k}i} + c_{\bar{i}\bar{j}} + c_{j\bar{k}}) \\ & + (a_{\bar{i}\bar{j}} + a_{\bar{j}\bar{k}} + a_{\bar{k}\bar{i}})(b_{\bar{j}\bar{k}} + b_{ki} + b_{\bar{i}\bar{j}})(c_{\bar{k}\bar{i}} + c_{\bar{i}\bar{j}} + c_{\bar{j}\bar{k}}),\end{aligned}$$

$$T_1 = \sum_{1 \leqslant i,j \leqslant s} a_{ij}b_{ij}[(s-2w_{ij})c_{ij} + \Sigma^*(c_{ki}+c_{jk})]$$
$$+ a_{ij}b_{\bar{i}\bar{j}}[(s-w_{ij})c_{ij} + \Sigma^*(-c_{\bar{k}i}+c_{jk})]$$
$$+ a_{\bar{i}\bar{j}}b_{i\bar{j}}[(s-w_{ij})c_{ij} + w_{ji}c_{j\bar{i}} + \Sigma^*(c_{k\bar{i}}-c_{\bar{j}k})]$$
$$+ a_{ij}\bar{b}_{ij}[(s-w_{ij})c_{ij} - \Sigma^*(c_{ki}+c_{j\bar{k}})]$$
$$+ a_{\bar{i}\bar{j}}b_{\bar{i}\bar{j}}[(s-w_{ij})c_{i\bar{j}} - \Sigma^*(c_{\bar{k}\bar{i}}+c_{\bar{j}k})]$$
$$+ a_{ij}\bar{b}_{\bar{i}\bar{j}}[(s-w_{ij})c_{\bar{i}\bar{j}} - w_{\bar{j}\bar{i}}c_{ji} + \Sigma^*(c_{\bar{k}i}-c_{j\bar{k}})]$$
$$+ a_{\bar{i}\bar{j}}\bar{b}_{i\bar{j}}[(s-w_{ij})c_{ij} + \Sigma^*(-c_{k\bar{i}}+c_{\bar{j}\bar{k}})]$$
$$+ a_{\bar{i}\bar{j}}\bar{b}_{i\bar{j}}[(s-2w_{ij})c_{\bar{i}\bar{j}} + \Sigma^*(c_{\bar{k}\bar{i}}+c_{\bar{j}\bar{k}})],$$

$$T_2 = \sum_{1 \leqslant i,j \leqslant s} \{a_{ij}\Sigma^*(b_{ki}+b_{jk})c_{ij} - a_{ij}\Sigma^*(b_{ki}+b_{j\bar{k}})c_{i\bar{j}}$$
$$+ a_{i\bar{j}}\Sigma^*(b_{jk}-b_{\bar{k}i})c_{ij} + a_{i\bar{j}}\Sigma^*[(b_{k\bar{i}}-b_{\bar{j}k}) - w_{ji}b_{\bar{j}\bar{i}})c_{\bar{i}\bar{j}}$$
$$+ a_{\bar{i}\bar{j}}[\Sigma^*(b_{\bar{k}i}- \ b_{j\bar{k}}) - w_{\bar{j}\bar{i}}\bar{b}_{ji}]c_{i\bar{j}} + a_{i\bar{j}}\Sigma^*(b_{\bar{j}\bar{k}}-b_{k\bar{i}})c_{\bar{i}\bar{j}}$$
$$- a_{\bar{i}\bar{j}}\Sigma^*(b_{\bar{k}\bar{i}}+b_{\bar{j}k})c_{\bar{i}\bar{j}} + a_{\bar{i}\bar{j}}\Sigma^*(b_{\bar{k}\bar{i}}+b_{\bar{j}\bar{k}})c_{\bar{i}\bar{j}}\}.$$

$$T_3 = \sum_{1 \leqslant i,j \leqslant s} \{\Sigma^*(a_{ki}+a_{jk})b_{ij}c_{ij} + \Sigma^*[(a_{k\bar{i}}-a_{\bar{j}k}) - w_{ji}a_{\bar{j}\bar{i}}]b_{\bar{i}\bar{j}}c_{i\bar{j}}$$
$$- \Sigma^*(a_{ki}+a_{j\bar{k}})b_{i\bar{j}}\bar{c}_{ij} + \Sigma^*(a_{jk}-a_{\bar{k}i})b_{ij}c_{\bar{i}\bar{j}}$$
$$+ \Sigma^*(a_{\bar{j}\bar{k}}-a_{k\bar{i}})b_{\bar{i}\bar{j}}c_{i\bar{j}} - \Sigma^*(a_{\bar{k}\bar{i}}+a_{\bar{j}k})b_{\bar{i}\bar{j}}c_{\bar{i}\bar{j}}$$
$$+ \Sigma^*[(a_{\bar{k}i}-a_{j\bar{k}}) - w_{\bar{j}\bar{i}}\bar{a}_{ji}]b_{i\bar{j}}c_{\bar{i}\bar{j}} + \Sigma^*(a_{\bar{k}\bar{i}}+a_{\bar{j}\bar{k}})b_{\bar{i}\bar{j}}c_{\bar{i}\bar{j}}.$$

Here $S'(s) = S_1'(s) \cup S_2'(s)$,

$$S_1'(s) = \{(i,\ j,\ k),\ 1 \leqslant i \leqslant j < k \leqslant s\}.$$
$$S_2'(s) = \{(i,\ j,\ k),\ 1 \leqslant k < j \leqslant i \leqslant s\},$$
$$n = 2s, \qquad \bar{i} = i+s, \qquad \bar{j} = j+s, \qquad \bar{k} = k+s,$$
$$w_{pq} = \begin{cases} 1 & \text{if } p = q, \\ 0 & \text{if } p \neq q, \end{cases}$$

and

$$\Sigma^* = \sum_{k=1}^{s} \text{ if } i = j \text{ then } k \neq i.$$

It can be seen that the number of terms in T_0 in $8(s^3 - s)/3$, and each of T_1, T_2, T_3 has $8s^2$ terms. Therefore the complexity of the algorithm is

$$8(s^3 - s)/3 + 24s^2 = (n^3 - 4n)/3 + 6n^2; \qquad \text{this is still of } O(n^3).$$

The overall reduction in the number of multiplications comes from the low coefficients of the dominant term in the complexity function. In Table 4.8.1 complexity of the Pan algorithm is shown for different matrix sizes, n. Table 4.8.2 gives comparative figures on the number of multiplications required by the different methods.

Table 4.8.1 Complexity function growth of the Pan algorithm for different matrix sizes

n	Multiplications (M)	$\log_n M$
16	2 880	2.873
32	17 024	2.811 1
64	111 872	2.795 25
66	121 880	2.795 17
68	132 464	2.795 13
70	143 640	2.795 12
72	155 424	2.795 15
74	167 832	2.795 2
76	180 880	2.795 3
78	194 584	2.795 4
80	208 960	2.795 5
128	797 184	2.801

Table 4.8.2 The number of multiplications required to calculate the product of two $n \times n$ matrices using different methods

n	Traditional	Strassen's	Pan's
2	8	7	24
4	64	49	112
6	216	189	280
8	512	343	544
16	4 096	2 401	2 880
32	32 768	16 807	17 024
64	262 144	117 649	111 872
128	2 097 152	823 723	797 184

4.9 Lower Bounds on Matrix Multiplication

By expressing the formulae for matrix multiplication of two 2×2 matrices with seven multiplications, Strassen was able to construct an $O(n^{2.81})$ matrix multiplication algorithm. If one could multiply 2×2 matrices in 6 multiplications, we would have an $O(n^{\log_2 6})$ or $O(n^{2.59})$ matrix multiplication algorithm. However, as was mentioned earlier seven multiplications is the minimum number required to find the product of two 2×2 matrices. More generally, multiplying two small $k \times k$ matrices in only m multiplications could give rise to a Strassen-like recursive construction of complexity $O(n^{\log_k m})$. Pan's algorithm in 1978 was the first real progress since Strassen's algorithm in an effort to reduce the algebraic complexity of matrix multiplication. Pan's algorithm of 1978 multiplies two 70×70 matrices in $O(n^{2.795})$ multiplications. Several new improvements followed Pan's algorithm. The asymptotically fastest known matrix multiplication procedure uses $O(n^{2.496})$ arithmetics (Pan, 1984).

So what is the minimum number of arithmetics needed to evaluate the product of two matrices? At present it can only be said that since the problem has $2n^2$ inputs and n^2 outputs, a lower bound on the number of arithmetics must be of $O(n^2)$, and we can write

$$O(n^2) \leqslant m \leqslant O(n^{2.496}),$$

where m is the number of arithmetic operations required to find the product of two $n \times n$ matrices.

Concerning numerical accuracy, we have seen that in the Gaussian elimination with complete pivoting, a bound for the maximum growth of the pivotal elements is quite small and there are strong reasons for believing (Wilkinson, 1965) that this bound cannot be approached. If partial pivoting is used, the final pivot can be 2^{n-1} times as large as the maximum element in the original matrix. However, although such a bound is attainable it is usually irrelevant for practical purposes, as it is quite unusual for any growth to take place at all, and if inner products can be accumulated than the equivalent perturbations in the original matrix are exceptionally small. Hence, Gaussian elimination is considered to be practically stable.

The numerical stability of the Householder method is guaranteed unconditionally in that most satisfactory *a priori* bounds for the equivalent perturbations of the original matrix are available. No dangerous growth in size of elements during the reduction is possible since apart from round-off errors the Euclidean norm of each column of the original matrix is preserved by exact orthogonal transformations.

We have also seen that an error analysis due to Brent of algorithms for matrix multiplications and triangular decomposition using the Winograd identity, shows that the Winograd method can be very bad numerically. However, by employing an appropriate technique of scaling, the method can be made to work with accuracy comparable to the Gaussian elimination when the latter is used without accumulation of inner products. However, scaling takes the computer time and this lowers the overall efficiency of the method.

Very little is yet understood about error control in the Strassen and Pan methods. Again, Brent (1970b) has established a satisfactory error bound for floating point matrix multiplication by Strassen's method. However, it seems that no such bounds may be contemplated for matrix inversion or the solution of sets of linear equations by the same method, because no pivoting is possible there.

Practical tests by Brent indicate that Winograd's algorithm, even with the necessary scaling is faster than Strassen's for $n > 250$, though the precise changeover point depends on the machine and compiler used (Brent, 1970a). The Winograd algorithm is also easier to program than Strassen's. And again, one has to bear in mind that in the newer methods the fewer multiplications become of some significance for only quite large n. Also the auxiliary computer memory needs for these methods have to be considered.

Finally, as a practical note, from the point of view of overall performance, which includes efficiency, accuracy, reliability, generality and ease of use, the conventional methods, like Gaussian elimination, still remain the better methods for most general matrices.

Exercises

1. Give the matrix produced by the forward elimination phase of Gaussian elimination gauss when used to solve the equations

$$\begin{aligned} x_{11} + x_{12} + x_{13} &= 6 \\ 2x_{21} + x_{22} + 3x_{23} &= 12 \\ 3x_{31} + x_{32} + 3x_{33} &= 14. \end{aligned}$$

2. Suppose that $\mathbf{Ax} = \mathbf{b}$ with an $n \times n$ matrix $\mathbf{A}$ which has only $3n$ non-zero elements is solved by Gaussian elimination. What is the storage requirement for the problem?
3. Explain what effect an interchange of columns in a matrix would have on the corresponding simultaneous equations.
4. Suppose that $\mathbf{Ax} = \mathbf{b}$ has an $m \times n$ matrix $\mathbf{A}$, where $m < n$. Can Gaussian elimination be of use to solve such a system?
5. The set $\mathbf{Ax} = \mathbf{b}$ is solved by (i) the standard Gaussian elimination and (ii) computing the inverse $\mathbf{A}^{-1}$ first, using Gaussian elimination, and then forming the product $\mathbf{A}^{-1}\mathbf{b}$. Show by an operation count that the method (ii) is more costly even if we have to solve $\mathbf{Ax} = \mathbf{b}$ for many different $\mathbf{b}$ vectors. Analyse how the situation will change if in (ii) the technique of complexity $O(n^{2.81})$ were used for computing $\mathbf{A}^{-1}$.
6. Show that the solution of $\mathbf{Ax} = \mathbf{b}$ where $\mathbf{A}$ is a tradiagonal matrix, by Gaussian elimination without pivoting can be done using $3(n-1)$ operations for the elimination procedure and $3(n-1)+1$ operations in the back-substitution process.
7. With reference to exercise 6, show that partial pivoting results in a triangular matrix with 3 non-zero diagonals instead of 2, so that the back substitution requires $5(n-2)+4$ operations.
8. Estimate the number of arithmetic operations required to compute the matrix determinant using Gaussian elimination.
9. Give an algorithm for Cholesky's decomposition.
10. Using Cholesky's method, find the inverse of the matrix

$$\mathbf{A} = \begin{bmatrix} 2 & 4 & 1 & 2 \\ 4 & 10 & 0 & -4 \\ 1 & 0 & 4.5 & 15 \\ 2 & -4 & 15 & 54 \end{bmatrix}$$

11. What happens if an $\mathbf{LU}$ decompostion is applied to a singular matrix?
12. Prove that the inverse of an upper (lower) triangular matrix is upper (lower) triangular.

13. Find (a) **LU** decomposition, (b) the inverse and (c) the determinant of the matrix

$$\begin{bmatrix} 2.12 & 0.42 & 1.34 & 0.88 & 0.75 & 1.21 & 0.39 & 2.11 \\ 1.34 & 0.95 & 1.87 & 0.43 & 0.54 & 1.25 & 0.75 & 0.91 \\ 0.88 & 1.87 & 0.98 & 0.46 & 1.71 & 0.41 & 1.87 & 0.19 \\ 0.75 & 0.43 & 0.46 & 2.44 & 1.14 & 0.36 & 1.13 & -2.54 \\ 1.21 & 0.54 & 1.71 & -2.14 & 1.49 & 1.71 & 0.39 & 0.87 \\ 0.39 & 1.25 & 0.41 & 0.36 & 1.71 & -1.51 & 0.42 & 1.14 \\ -2.11 & 0.75 & 1.87 & 1.13 & 0.39 & 0.42 & 1.91 & 0.45 \\ 1.52 & 0.91 & 0.19 & -2.54 & 0.87 & 1.14 & 0.45 & 2.35 \end{bmatrix}$$

using standard methods.

14. Give the total number of arithmetics required by the algorithm *householder.*

15. Show that the number of arithmetic operations required to solve a set of n linear algebraic equations by the Householder reduction requires

$$\frac{2}{3}n^3 + \frac{5}{2}n^2 + \frac{17}{6}n - 5 \qquad \text{multiplications,}$$

$$\frac{2n^3}{3} + \frac{3n^2}{2} + \frac{5n}{6} - 3 \qquad \text{additions}$$

$$2n - 1 \qquad \text{reciprocals, and}$$

$$n - 2 \text{ or } 2n - 4 \qquad \text{square roots.}$$

16. Derive an algorithm to compute the inverse of a matrix which is based on the Strassen method of matrix multiplication. [*Solution.* If **A** is a matrix of order $n = 2m$ to be inverted then write

$$\mathbf{A} = \begin{bmatrix} \mathbf{A}_{11} & \mathbf{A}_{12} \\ \mathbf{A}_{21} & \mathbf{A}_{22} \end{bmatrix} \qquad \mathbf{A}^{-1} = \begin{bmatrix} \mathbf{C}_{11} & \mathbf{C}_{12} \\ \mathbf{C}_{21} & \mathbf{C}_{22} \end{bmatrix}$$

To evaluate $\mathbf{A}^{-1}$ compute

$$\begin{aligned} &\mathbf{Q}_1 = \mathbf{A}_{11}^{-1}, \\ &\mathbf{Q}_2 = \mathbf{A}_{21}\mathbf{Q}_1, && \mathbf{C}_{12} = \mathbf{Q}_3\mathbf{Q}_6, \\ &\mathbf{Q}_3 = \mathbf{Q}_1\mathbf{A}_{12} && \mathbf{C}_{21} = \mathbf{Q}_6\mathbf{Q}_2, \\ &\mathbf{Q}_4 = \mathbf{A}_{21}\mathbf{Q}_3, && \mathbf{Q}_7 = \mathbf{Q}_3\mathbf{C}_{21}, \\ &\mathbf{Q}_5 = \mathbf{Q}_4 - \mathbf{A}_{22}, && \mathbf{C}_{11} = \mathbf{Q}_1 - \mathbf{Q}_7, \\ &\mathbf{Q}_6 = \mathbf{Q}_5^{-1}, && \mathbf{C}_{22} = -\mathbf{Q}_6. \end{aligned}$$

]

17. Show that the Karatsuba–Makarov method requires $n^3 - 2n^2$ additions when used recursively to compute the product of two $n \times n$ matrices. [*Solution.* Let $a(m, k)$ denote the algorithms which multiply matrices of size $m2^k$, where $a(m, 0)$ is the standard algorithm for matrix multiplication (with m^3 multiplications and $m^2(m-1)$ additions) and

$a(m, k)$, $k > 0$ is a recursive algorithm. To multiply two $n \times n$ matrices, where $n = m2^k$, Karatsuba–Makarov's algorithms require eight multiplications (M) of matrices of size $m2^{k-1}$, so that we can write

$$Ma(m, k) = 8Ma(m, k-1) = 8^k Ma(m, 0)$$
$$= 8^k m^3 = n^3.$$

Similarly for the additions (S)

$$Sa(m, k) = 8(m2^{k-1})^2 + 8Sa(m, k)$$

since the algorithms require eight additions of matrices of size $m2^{k-1}$ and eight multiplications of matrices of size $m2^{k-1}$ and the latter in turn require $Sa(m, k-1)$ additions each. Hence the additions recurrence gives

$$Sa(m, k) = 8^k m^3 - 2(4^k)m^2 = n^3 - 2n^2. \qquad]$$

18. Solve exercise 13 using Strassen's method.
19. Compute the product

$$\begin{bmatrix} 3 & 9 \\ 2 & 7 \end{bmatrix} \quad \begin{bmatrix} 1 & 6 \\ 4 & 5 \end{bmatrix}$$

using (a) Strassen's algorithm with seven multiplications and 18 additions, (b) Winograd's variant with seven multiplications and 15 additions and (c) Karatsuba–Makarov algorithm with eight multiplications and 12 additions.
20. Consider the following matrix decomposition:

$$\mathbf{A} = \begin{bmatrix} \mathbf{A}_{11} & \mathbf{A}_{12} \\ \mathbf{A}_{21} & \mathbf{A}_{22} \end{bmatrix} = \begin{bmatrix} \mathbf{I} & 0 \\ \mathbf{A}_{21}\mathbf{A}_{11}^{-1} & \mathbf{I} \end{bmatrix} \begin{bmatrix} \mathbf{A}_{11} & 0 \\ 0 & \nabla \end{bmatrix} \begin{bmatrix} \mathbf{I} & \mathbf{A}_{11}^{-1}\mathbf{A}_{12} \\ 0 & \mathbf{I} \end{bmatrix}$$

where $\nabla = \mathbf{A}_{22} - \mathbf{A}_{21}\mathbf{A}_{11}^{-1}\mathbf{A}_{12}$.

(a) Using the given formula derive an expression for computing the inverse $\mathbf{A}^{-1}$.

(b) Assuming that $\mathbf{A}$ is a general matrix of size n, show that computing the inverse by the recursive use of the process suggested in (i) is an algorithm of order $M(n)$ where $M(n)$ denotes the number of operations required for multiplication of two $n \times n$ matrices.

(c) Explain why the algorithm always works (theoretically) in the case when the original matrix is upper or lower triangular but may fail in the case of a general matrix.
21. Consider the polynomial $P(x) = \Sigma_{i=0}^{n} a(i)x^i$ where n is a perfect square. The Borodin–Munro algorithm for polynomial evaluation at two points, x_1 and x_2, is given as follows:

Write

$$\mathbf{A} = \begin{bmatrix} a(1) & a(2) \ldots & a(\sqrt{n}) \\ a(\sqrt{n}+1) & & \ldots \\ \ldots & & \ldots \\ \ldots & & a(n) \end{bmatrix} \quad \text{and} \quad \mathbf{X} = \begin{bmatrix} x_1 & x_2 \\ x_1^2 & x_2^2 \\ & \\ x_1^{\sqrt{n}} & x_2^{\sqrt{n}}, \end{bmatrix}$$

then compute $Y = \{y_{ik}\} = \mathbf{AX}$
and

$$P(x) = a(0) + y_{1k} + \sum_{i=2}^{\sqrt{n}} y_{ik} x_k^{(k-1)\sqrt{n}}, \qquad k = 1, 2.$$

(a) Estimate the number of multiplications required by the algorithm.
(b) By extending the algorithm for $\sqrt{n}$ points, $x_1, \ldots, x_{\sqrt{n}}$, show that a polynomial of degree n may be evaluated at $\sqrt{n}$ points in $O(n^{\log 7/2})$ arithmetic operations.

5

The Fast Fourier Transform

5.1 Introduction

In solving a mathematical problem, it is often convenient to transform the function, work with the transform, and then 'untransform' the result, in order to deduce non-obvious properties of the original function.

Among others the Fourier transform is of significant importance in practical work. Simultaneous visualization of a function and its Fourier transform is often the key to successful problem-solving.

Probably the best known application of this mathematical technique is the analysis of linear time-invariant systems. But it has also long been a principal analytical tool in such diverse fields as optics, theory of probability, quantum physics, antennas and signal analysis.

Such a wide and successful application was, however, an attribute of the continuous Fourier transform, i.e. the transform of the function of a variable which takes values in a continuum.

On the other hand, in numerical mathematics and data analysis, where one usually deals with ordered sets of numbers, functions of a discrete variable, which are also finite in extent, the discrete form of the Fourier transform is a more convenient tool. Unfortunately, until the mid-1960s the computation necessary to produce the discrete Fourier transforms were known to be notoriously time-consuming even with the tremendous computing speeds available on modern computers.

In 1965 Cooley and Tukey published a new method for computing the Fourier transforms which offered remarkable savings in computing time compared with all previously known methods. It became known as the Fast Fourier Transform or the FFT.

Because of the exceptional importance of the discrete Fourier transform (DFT) in practical applications, discovery of the FFT must be rated as one of the most significant achievements of computational mathematics in recent time.

With the Cooley and Tukey publication it became apparent that the technique of the FFT has been used *ad hoc* here and there, and, in fact, the algorithm has a fascinating history. Its general approach may be traced back to Runge and König (1924), Danielson and Lanczos (1942), and Good (1958, 1960). However, only with the brilliant contribution by Cooley and Tukey was the full potential of the approach realized.

Our concern in this chapter will be the computation of the discrete Fourier transform using the FFT algorithm. Accordingly we shall only briefly introduce the continuous Fourier transform and then move on to discuss the discrete Fourier transform. We shall then proceed to outline the FFT algorithm and consider its relative efficiency, storage requirements and error control.

5.2 The Continuous Fourier Transform

The Fourier Transform of a Function of One Variable and Its Inverse

The following formulae yield mutual and dual relations between a pair of functions.

If for a given function $f(v)$ we set

$$F(u)=\frac{1}{\sqrt{2\pi}}\int_{-\infty}^{\infty} f(v)e^{-iuv}\,dv, \quad i=\sqrt{-1}, \tag{5.2.1}$$

then the following holds:

$$f(v)=\frac{1}{\sqrt{2\pi}}\int_{-\infty}^{\infty} F(u)e^{+viu}\,du, \quad i=\sqrt{-1}, \tag{5.2.2}$$

where the integrals are interpreted as

$$\lim_{l\to\infty}\int_{-l}^{l} g(y)\,e^{+ixy}\,dy \tag{5.2.3}$$

with appropriate plus or minus sign used in the exponential function.

The function $F(u)$ is called *the Fourier transform* of $f(v)$ and is normally denoted by $\mathscr{F}[f]$, i.e. $\mathscr{F}[f]=F(u)$. If the function $f(v)$ is integrable in the interval $(-\infty,+\infty)$, then the function $F(u)$ exists for every value of v. Typically $f(v)$ is referred to as a function of the variable time and $F(u)$ is referred to as a function of the variable frequency.

The functions $F(u)$ and $f(v)$ are called the Fourier transform pair.

One also distinguishes between the Fourier transform, $F(u)$, and its inverse, $f(v)$, so that we have $\mathscr{F}[f(v)]=F(u)$ and $\mathscr{F}^{-1}[F(u)]=f(v)$.

Elementary Properties of the Fourier Transform

(a) The Fourier transform is a linear operator.

An operator P being linear on a specified set of functions means that

$$P(\alpha f+\beta g)=\alpha Pf+\beta Pg$$

for any two functions f and g of the set with α and β arbitrary constants. Now, if $f(v)$ and $g(v)$ have the Fourier tranforms $F(u)$ and $G(u)$, respectively, then the sum $f(v)+g(v)$ has the Fourier transform $F(u)+G(u)$, and in general the following holds:

$$\mathscr{F}[\alpha f+\beta g]=\alpha\mathscr{F}[f]+\beta\mathscr{F}[g].$$

(b) The Fourier transform is symmetric.
If $f(v)$ and $F(u)$ are a Fourier transform pair then $F(v)$ and $f(-u)$ are a Fourier transform pair.

(c) Scaling of the Fourier transform.
If the Fourier transform of $f(v)$ is $F(u)$ then the Fourier transform of $f(kv)$, where k is a real constant, is given as:

$$\frac{1}{|k|}F\left(\frac{u}{k}\right).$$

If the inverse Fourier transform of $F(u)$ is $f(v)$, then the inverse Fourier transform of $F(ku)$, where k is a real constant, is given by:

$$\frac{1}{|k|}F\left(\frac{v}{k}\right).$$

(d) The property of parity.
If $f(v)$ is an even function, i.e. $f(v)=f(-v)$, then its Fourier transform is an even function and is real.
If $f(v)$ is an odd function, i.e. $f(v)=-f(-v)$, then its Fourier transform is an odd and imaginary function.

The Convolution of Two Functions

Let functions $F(u)$ and $G(u)$ be the Fourier transforms of the functions $f(v)$ and $g(v)$ respectively. Formally, if $H(u)=F(u)G(u)$ then we can write

$$\begin{aligned}h(v)&=\frac{1}{\sqrt{2\pi}}\int_{-\infty}^{\infty}F(u)G(u)e^{ivu}\,du\\&=\frac{1}{\sqrt{2\pi}}\int_{-\infty}^{\infty}F(u)e^{ivu}\left\{\frac{1}{\sqrt{2\pi}}\int_{-\infty}^{\infty}g(\tau)e^{-iu\tau}d\tau\right\}du\\&=\frac{1}{\sqrt{2\pi}}\int_{-\infty}^{\infty}g(\tau)f(v-\tau)d\tau,\end{aligned}\tag{5.2.4}$$

that is, the functions

$$H(u)=F(u)G(u)\text{ and }h(v)=\frac{1}{\sqrt{2\pi}}\int_{-\infty}^{\infty}g(\tau)f(v-\tau)d\tau\tag{5.2.5}$$

form a Fourier pair.

The function $h(v)$ is called *the convolution* of the functions $f(v)$ and $g(v)$.

We thus have the important result that *the convolution of two functions is equal to the inverse Fourier transform of the product of the Fourier transforms of the functions.*

5.3 The Discrete Fourier Transform

The Fourier transform given by formulae (5.2.1) to (5.2.5) is called the infinite continuous Fourier transform because it involves continuous functions defined over the infinite domain. These transforms are widely used. However there is a wide range of problems where one works with the data which are given as sets of discrete quantities, and so the finite discrete Fourier transform is a more useful device when the mathematics of such a process is elaborated.

The finite discrete Fourier transform may be visualized as a special case of the infinite continuous Fourier transform when the latter is amenable to machine computation. (For a somewhat different interpretation of the DFT, as a special case of the Fourier series of a periodic function, see Cooley, Lewis and Welch, 1977.)

Definition of the DFT

Let $f(v)$, $v = 0, 1, \ldots, N-1$ be a sequence of N complex numbers. The finite discrete Fourier transform (DFT) of $f(v)$ is defined as

$$F(u) = \sum_{u=0}^{N-1} f(v) w_N^{-uv}, \quad u = 0, 1, \ldots, N-1, \tag{5.3.1}$$

where here and in sequel $w_N = \exp(2\pi i/N)$ and $i = \sqrt{-1}$. The values $F(u)$, $u = 0, 1, \ldots, N-1$, are also called the Fourier coefficients.

The inverse Fourier transform is defined as

$$f(v) = \frac{1}{N} \sum_{u=0}^{N-1} F(u) w_N^{uv}, \qquad v = 0, 1, \ldots, N-1, \tag{5.3.2}$$

which can also be written in the form

$$f(v) = \frac{1}{N} \left[\sum_{u=0}^{N-1} F^*(u) w_N^{-uv} \right]^*, \quad v = 0, 1, \ldots, N-1. \tag{5.3.3}$$

Formula (5.3.3) reads 'the complex conjugate of the Fourier transform of the complex conjugate of $F(u)$ divided by N gives $f(v)$'.

The formulae (5.3.1) and (5.3.2) define a transform pair:

$$\begin{aligned} f(v) &= \frac{1}{N} \sum_{u=0}^{N-1} F(u) w_N^{uv} = \frac{1}{N} \sum_{u=0}^{N-1} \left[\sum_{k=0}^{N-1} f(k) w_N^{-uk} \right] w_N^{uv} \\ &= \frac{1}{N} \sum_{k=0}^{N-1} f(k) \sum_{u=0}^{N-1} w_N^{u(v-k)} = f(v), \end{aligned} \tag{5.3.4}$$

since the exponential function w_N^{uv} satisfies the following orthogonality relationships

$$\sum_{u=0}^{N-1} w_N{}^{\theta u} w_N{}^{-ku} = \sum_{u=0}^{N-1} e^{i(2\pi/N)(v-k)u} = \begin{cases} N, & \text{if } k = v \bmod N, \\ 0, & \text{otherwise.} \end{cases} \tag{5.3.5}$$

Note that in the finite discrete Fourier transform both functions, $f(v)$ and $F(u)$, are defined over finite sets of discrete points only, i.e. $f(v)$ is given as $\{f(0), f(1), \ldots, f(N-1)\}$ and $F(u)$ is given as $\{F(0), F(1), F(2), \ldots, F(N-1)\}$. Consequently, we deal with finite sums. Hence, the name.

Periodicity of the DFT

The exponential function $w_N{}^{uv}$, as a function of u and v, is periodic of period N, i.e.

$$w_N{}^{uv} = w_N{}^{(u+N)v} = w_N{}^{u(v+N)}.$$

It follows immediately that the sequences $\{F(u)\}$ and $\{f(v)\}$ as defined by (5.3.1) and (5.3.2) are periodic of period N. For the theorems of the infinite continuous Fourier transform to be applied to the finite discrete transform the operation of integration must be replaced by the operation of summation over the discrete set of data points and taken modulo N, where N is the number of discrete data. One useful interpretation of this operation is to think of the sequence $\{f(v)\}$ as a function defined on equispaced integer points of a circle with circumference N (Gentleman and Sande, 1966). We can also consider $F(u)$ and $f(v)$ as defined for all integers, that is,

$$F(u), \quad u = 0, +1, +2, \ldots \qquad \text{and} \qquad f(v), \quad v = 0, +1, +2, \ldots$$

where

$$F(u) = F(kN + u) \qquad \text{and} \qquad f(v) = f(kN + v), \quad k = 0, +1, +2, \ldots$$

The periodic extension (5.3.5) of the finite sequences to infinite sequences makes for an easier visualization of the properties of the discrete Fourier transform and, in some cases, for their use in computational work. The finite sequences can, of course, be recovered by considering the values of the finite sequences at any N consecutive points.

Elementary Properties of the DFT

The DFT possesses the same properties as the continuous Fourier transform, the properties being appropriately modifed to suit the discrete and periodic character of the DFT. For example, the fundamental property of linearity of the transform is preserved, that is if

$$f(v), \quad v = 0, 1, \ldots, N-1 \qquad \text{and} \qquad F(u), \quad u = 0, 1, \ldots, N-1$$

and

$$g(v), \quad v = 0, 1, \ldots, N-1 \qquad \text{and} \qquad G(u), \quad u = 0, 1, \ldots, N-1 \tag{5.3.6}$$

are two DFT pairs then for any complex constants c and d,

$$cf(v) + \mathrm{d}g(v) \qquad \text{and} \qquad cF(u) + dG(u) \tag{5.3.7}$$

is a DFT pair.

The properties of symmetry, scaling and parity of the transform are similarly preserved in the DFT.

Convolution Theorem and Term-by-Term Products of Sequences

Let $f(v)$ and $g(v)$, and $F(u)$ and $G(u)$ be two sequences and their DFTs, respectively, defined as in (5.3.6). The term-by-term product of the sequences $F(u)$ and $G(u)$ is the sequence whose kth term is $F(k)G(k)$. The inverse transform of this product, in similarity to the convolution property of the continuous Fourier transform, turns out to be a convolution of the sequences $f(v)$ and $g(v)$. More explicitly, the sequences

$$h(v) = \sum_{\tau=0}^{N-1} f(\tau)g(v-\tau) = \sum_{\tau=0}^{N-1} f(v-\tau)g(\tau), \quad v = 0, 1, \ldots, N-1 \tag{5.3.8}$$

and $H(u) = F(u)G(u), \quad u = 0, 1, \ldots, N-1$

form a discrete Fourier transform pair.

Similary,

$$k(v) = f(v)g(v), \quad v = 0, 1, \ldots, N-1$$

$$\text{and } K(u) = \frac{1}{N}\sum_{\tau=0}^{N-1} F(\tau)G(u-\tau) = \frac{1}{N}\sum_{\tau=0}^{N-1} F(u-\tau)G(\tau), \quad u = 0, 1, \ldots, N-1 \tag{5.3.9}$$

form a discrete Fourier transform pair.

The convolutions defined by (5.3.8) and (5.3.9) are cyclic; that is, when one sequence moves over the end of the other, it does not encounter zeros, but rather the periodic extension of the sequence. This is consistent with the periodic extension property of the DFT which was noted earlier.

The convolution theorem corresponding to (5.2.4) can thus be stated as

$$\sum_{\tau=0}^{N-1} f(\tau)g(v-\tau) = \frac{1}{N}\sum_{u=0}^{N-1} F(u)G(u)w_N^{uv} \tag{5.3.10}$$

where, as before, $w_N = \exp(2\pi \mathrm{i}/N)$ with $\mathrm{i} = \sqrt{-1}$.

That is, we have a theorem which states that the convolution of two discrete functions is equal to the inverse Fourier transform of the product of the discrete Fourier transforms of the functions. The convolution theorem plays a fundamental role in applications of the discrete Fourier transform.

Proof of the relations given by (5.3.8), (5.3.9), and (5.3.10) is easily obtained using the orthogonality relationships (5.3.5).

The DFT and Operations on Polynomials

We shall now treat the subject of a rather remarkable relationship between the DFT and operations on polynomials. These results will subsequently be used in some applications of the DFT (see Chapter 6).

Let $(a_q, q=0, \ldots, N-1)$ be a sequence of real or complex numbers. The discrete Fourier transform of the sequence is defined as

$$b_p = \sum_{q=0}^{N-1} a_q w_N^{-pq}, \quad p=0,1,\ldots,N-1. \tag{5.3.11}$$

where, as before $w_N = \exp(2\pi i/N)$.

Now, denote the sequences $(a_q, q=0, \ldots, N-1)$ and $(b_p, p=0, \ldots, N-1)$ by two vectors **a** and **b**, respectively, of length N each. Furthermore, let **A** be an $N \times N$ matrix such that

$$\mathbf{A} = \{\alpha_{pq}\} = \{w_N^{-pq}\}. \tag{5.3.12}$$

Then the vector

$$b = \mathbf{Aa} \tag{5.3.13}$$

whose pth component $\mathbf{b}_p$ as given in formula (5.3.11) is called the DFT of vector **a**. The matrix **A** is non-singular and thus the inverse $\mathbf{A}^{-1}$ exists. Using the orthogonality relationships (5.3.5) it can be easily shown that inverse $\mathbf{A}^{-1}$ has the simple form given by

$$\mathbf{A}^{-1} = \{\beta_{pq}\} = \{w_N^{pq}\}. \tag{5.3.14}$$

On the basis of definitions (5.3.11) and (5.3.12) a close relationship can be derived between the discrete Fourier transform and polynomial evaluation and interpolation.

Let

$$P(x) = \sum_{k=0}^{N-1} a_k x^k \tag{5.3.15}$$

be an $(n-1)$st degree polynomial. This polynomial can be uniquely represented in one of two ways: either by a set of its coefficients $(a_k, k=0, \ldots, N-1)$, or by a set of its values at N distinct points $(P(x_j), j=0, \ldots, N-1)$. (Given the values of $P(x)$ at $x_0, x_1, \ldots x_{N-1}$, the process of finding the coefficient representation of a polynomial is called interpolation. When dealing with polynomials it is often desirable to be able to quickly convert from one form of representation to another.)

We shall now show that computing the DFT $\{b_p\}$ of vector $\mathbf{a} = (a_0, a_1, \ldots, a_{N-1})$:

$$b_p = \sum_{q=0}^{N-1} a_q w_N^{-pq}, \quad p=0,1,\ldots,N-1,$$

is equivalent to converting the coefficient representation of the polynomial

(5.3.15) to its representation at the N points $x_0 = w_N^0$, $x_N = w_N^1, \ldots, x_{N-1} = w_N^{N-1}$, where $w_N = \exp(-2\pi i/N)$.

Let the polynomial $P(x)$ be represented by its coefficients, $\{a_k, k = 0, 1, \ldots, N-1\}$. We wish to convert this representation to the representation $\{P(x_j), j = 0, 1, \ldots, N-1\}$. To do this we choose N arbitrary distinct values of the argument x and evaluate the polynomial at these points. The result will be the set of N points $\{P(x_j)\}$ which uniquely represent the polynomial. Now, suppose that the points x_j are chosen as follows:

$$x_j = w_N^j = \exp(-2\pi ij/N), \quad j = 0, 1, \ldots, N-1, \quad i = \sqrt{-1}. \tag{5.3.16}$$

We thus have

$$\begin{aligned} P(x_0) &= a_{N-1}(w_N^0)^{N-1} + a_{N-2}(w_N^0)^{N-2} + \ldots + a_0, \\ P(x_1) &= a_{N-1}(w_N^1)^{N-1} + a_{N-2}(w_N^1)^{N-2} + \ldots + a_0, \\ &\vdots \\ P(x_{N-1}) &= a_{N-1}(w_N^{N-1})^{N-1} + a_{N-2}(w_N^{N-1})^{N-2} + \ldots + a_0. \end{aligned} \tag{5.3.17}$$

The matrix form of (5.3.17) is

$$\begin{bmatrix} P(x_0) \\ P(x_1) \\ \vdots \\ P(x_{N-1}) \end{bmatrix} = \begin{bmatrix} 1 & w_N^0 & (w_N^0)^2 & \ldots & (w_N^0)^{N-1} \\ 1 & w_N^1 & (w_N^1)^2 & \ldots & (w_N^1)^{N-1} \\ & & & & \\ 1 & w_N^{N-1} & (w_N^{N-1})^2 & \ldots & (w_N^{N-1})^{N-1} \end{bmatrix} \begin{bmatrix} a_0 \\ a_1 \\ \vdots \\ a_{N-1} \end{bmatrix} \tag{5.3.18}$$

Comparing (5.3.18) with (5.3.11) we see that the vector

$$\mathbf{P}(\mathbf{x}) = (P(x_0), P(x_1), \ldots, P(x_{N-1}))$$

is the DFT of the vector $\mathbf{a} = (a_0, a_1, \ldots, a_{N-1})$.
A similar statement can be proved for the inverse DFT, that is that the inverse Fourier transform

$$a_q = \frac{1}{N} \sum_{p=0}^{N-1} b_p w_N^{pq}, \quad q = 0, 1, \ldots, N-1$$

is equivalent to computing an interpolating polynomial over the set of points, $x_k = w_N^k$, $k = 0, 1, \ldots, N-1$.

The set of points over which the DFT evaluates a polynomial could be chosen other than the $\{w_N^k\}$. For example, the integers $1, 2, \ldots, N$ could be used for this purpose (Aho, Hopcroft and Ullman, 1974, 1983).

The Convolution of Two Vectors and Multiplication of Two Polynomials

One of the important applications of the discrete Fourier transform is computing the convolution of two vectors.

Let $\mathbf{a} = (a_0, a_1, \ldots, a_{N-1})^T$ and $\mathbf{b} = (b_0, b_1, \ldots, b_{N-1})^T$ be two column vectors. We define the convolution of **a** and **b** as the vector $\mathbf{c} = (c_0, c_1, \ldots, c_{2N-2})^{\mathrm{T}}$ where

$$c_i = \sum_{j=0}^{N-1} a_j b_{i-j} \tag{5.3.19}$$

and $a_k = b_k = 0$ if $k < 0$ or $k \geqslant N$.
Thus, for example, for $N = 3$ we have

$$\begin{aligned} c_0 &= a_0 b_0, \\ c_1 &= a_0 b_1 + a_1 b_0, \\ c_2 &= a_0 b_2 + a_1 b_1 + a_2 b_0, \\ c_3 &= a_1 b_2 + a_2 b_1, \\ c_4 &= a_2 b_2. \end{aligned} \tag{5.3.20}$$

Now, consider the representation of a polynomial by its coefficients. The product of two $(N-1)$st degree polynomials

$$P(x) = \sum_{j=0}^{N-1} a_j x^j \quad \text{and} \quad Q(x) = \sum_{j=0}^{N-1} b_j x^j \tag{5.3.21}$$

is the $(2N-2)$nd degree polynomial

$$P(x)Q(x) = \sum_{k=0}^{2N-2} \left(\sum_{j=0}^{k} a_j b_{k-j} \right) x^k. \tag{5.3.22}$$

The coefficients of the product polynomial are exactly the components of the convolution of the coefficient vectors **a** and **b** of the original polynomials. Hence, if the two $(N-1)$st degree polynomials are represented by their coefficients then the coefficient representation of their product can be obtained by computing convolutions of the coefficient vectors of the two polynomial factors.

If $P(x)$ and $Q(x)$ are represented by their values at some N points, then to compute the value representation of their product we multiply pairs of values at the corresponding points.

The properties of the discrete Fourier transform discussed in this Section are used in the derivation of the main result in Chapter 6.

5.4 The Fast Fourier Transform

We now turn to the discussion of a highly efficient method for calculating the discrete Fourier transform of a given function $f(v)$. First, recall the definition of the finite DFT, i.e. given $f(v)$, $v = 0, 1, \ldots, N-1$, its DFT is defined as

$$F(u) = \sum_{v=0}^{N-1} f(v) w_N^{-vu}, \quad u = 0, 1, \ldots, N-1 \tag{5.4.1}$$

where

$$w_N = \exp(2\pi \mathrm{i}/N) \quad \text{with } i = \sqrt{-1}.$$

In practical problems, which use the DFT, the number of points, N, may easily be anything between several thousand and a million.

For decades discrete Fourier transforms were computed by a method based on direct implementation of the definition, thus requiring N^2 operations of multiplication and addition. This fact explains the reputation acquired by the DFT for being lengthy time-consuming calculations.

The Cooley and Tukey algorithm, the Fast Fourier Transform, reduces this number of operations to the number proportional to $N \log N$.

We shall now study this remarkable algorithm in greater detail.

The FFT Algorithm: Elementary Derivation

First, we shall give an elementary derivation of the Fast Fourier Transform (FFT) algorithm. This elementary derivation reveals the essential idea behind the algorithm. In later sections a more detailed derivation will be given. Consider formula (5.4.1). For the detailed mathematical analysis which follows, it is convenient to slightly modify notation in the formula: we denote the exponential function

$$w_N^{-vu} \quad \text{by} \quad e\left(\frac{vu}{N}\right), \quad \text{so that}$$

$$e\left(\frac{vu}{N}\right) = w_N^{-vu} = \exp(-2\pi i vu/N). \tag{5.4.2}$$

We note two important properties of $e(x)$, namely

$$e(r+s) = e(r)e(s)$$

and

$$e(rN) = 1, \quad \text{if } r \text{ is an integer.} \tag{5.4.3}$$

Suppose now that N is a product of two factors, r and s, so that $N = rs$. Then the parameters u and v can be expressed in the form:

$$\begin{aligned} u &= p_0 + p_1 r, \quad 0 \leqslant p_0 < r, \quad 0 \leqslant p_1 < s, \\ v &= q_0 s + q_1, \quad 0 \leqslant q_0 < r, \quad 0 \leqslant q_1 < s. \end{aligned} \tag{5.4.4}$$

Thus for (5.4.1) we have

$$\begin{aligned} F(u) &= \sum_{q_0=1}^{r-1} \sum_{q_1=0}^{s-1} f(v)\, e\left(\frac{(q_0 s + q_1)(p_0 + p_1 r)}{N}\right) \\ &= \sum_{q_1=0}^{s-1} \sum_{q_0=0}^{r-1} f(v)\, e\left(\frac{q_0 p_0}{r}\right) e\left(\frac{q_1 p_0}{rs}\right) e\left(\frac{q_1 p_1}{s}\right), \end{aligned}$$

since $e(q_0 p_1) = 1$.

Furthermore we can rewrite the expression as

$$F(u) = \sum_{q_1=0}^{s-1} e\left(\frac{q_1 p_1}{s}\right)\left\{e\left(\frac{q_1 p_0}{rs}\right) \sum_{q_0=0}^{r-1} f(v)\, e\left(\frac{q_0 p_0}{r}\right)\right\}. \tag{5.4.5}$$

Hence the following procedure will compute the Fourier coefficients given by (5.4.5):

Compute

$$c_p(q_1) = e\left(\frac{q_1 p_0}{rs}\right) \sum_{q_0=0}^{r-1} f(v)\, e\left(\frac{q_0 p_0}{r}\right), \tag{5.4.6}$$

$$q_1 = 0, 1, \ldots, s-1, \quad p_0 = 0, 1, \ldots, r-1.$$

Compute

$$F(u) = F(p_0 + p_1 r) = \sum_{q_1=0}^{s-1} e\left(\frac{q_1 p_1}{s}\right) c_{p_0}(q_1), \tag{5.4.7}$$

$$p_0 = 0, 1, \ldots, r-1, \quad p_1 = 0, 1, \ldots, s-1.$$

Example 5.4.1

Consider an example for $N = 12 = 4 \times 3 = r \times s$.
Thus we have the following set of points:

$$f(0), f(1), f(2), f(3), f(4), f(5), f(6), f(7), f(8), f(9), f(10), f(11).$$

This set is first partitioned into three subsets of four points each:

$$\{f(0), f(3), f(6), f(9)\}\ \{f(1), f(4), f(7), f(10)\}$$
$$\{f(2), f(5), f(8), f(11)\},$$

and the following sets of values are computed:

$$c_{p_0}(0) = e(0)\left[f(0) + f(3)e\left(\frac{p_0}{4}\right) + f(6)e\left(\frac{2p_0}{4}\right) + f(9)e\left(\frac{3p_0}{4}\right)\right], \quad p_0 = 0, 1, 2, 3,$$

$$c_{p_0}(1) = e\left(\frac{p_0}{12}\right)\left[f(1) + f(4)\, e\left(\frac{p_0}{4}\right) + f(7)e\left(\frac{2p_0}{4}\right) + f(10)\, e\left(\frac{3p_0}{4}\right)\right], \quad p_0 = 0, 1, 2, 3,$$

$$c_{p_0}(2) = e\left(\frac{2p_0}{12}\right)\left[f(2) + f(5)e\left(\frac{p_0}{4}\right) + f(8)\, e\left(\frac{2p_0}{4}\right) + f(11)e\left(\frac{3p_0}{4}\right)\right], \quad p_0 = 0, 1, 2, 3.$$

The twelve points $c_p(q_1)$ are then partitioned into four subsets of three points each

$$\{c_0(0), c_1(0), c_2(0)\} \quad \{c_0(1), c_1(1), c_2(1)\}$$
$$\{c_0(2), c_1(2), c_2(2)\} \quad \{c_0(3), c_1(3), c_2(3)\},$$

and the Fourier coefficients are computed to give:

$$F(0+4p_1) = \sum_{q_1=0}^{2} e\left(\frac{q_1 p_1}{3}\right) c_{q_1}(0)$$

$$= c_0(0) + e\left(\frac{p_1}{3}\right) c_1(0) + e\left(\frac{2p_1}{3}\right) c_2(0), \quad p_1 = 0, 1, 2,$$

$$F(1+4p_1) = \sum_{q_1=0}^{2} e\left(\frac{q_1 p_1}{3}\right) c_{q_1}(1)$$

$$= c_0(1) + e\left(\frac{p_1}{3}\right) c_1(1) + e\left(\frac{2p_1}{3}\right) c_2(1), \quad p_1 = 0, 1, 2,$$

$$F(2+4p_1) = \sum_{q_1=0}^{2} e\left(\frac{q_1 p_1}{3}\right) c_{q_1}(2)$$

$$= c_0(2) + e\left(\frac{p_1}{3}\right) c_1(2) + e\left(\frac{2p_1}{3}\right) c_2(2), \quad p_1 = 0, 1, 2,$$

We thus have a recursive procedure which defines the larger Fourier transform in terms of smaller ones. If, following Cooley and Tukey, we use the term 'operation' to mean a complex multiplication followed by a complex addition then we can easily see that the total number of operations required by the FFT algorithm is obtained as follows.

To produce each $c_{p_0}(q_1)(r+1)$ operations are required and there are $N = rs$ such elements. Hence at this step $N(r+1)$ operations are required altogether.

To produce each $F(p_0 + p_1 r)$ s operations are required and there are $N = rs$ such coefficients. Thus at this step Ns operations are required.

The total number of operations is then given as

$$N(r+s+1). \tag{5.4.8}$$

The obtained number of operations should be contrasted with N^2 operations required by a method based on the direct implementation of the defintion of the discrete Fourier transform. For example, for $N = 10^3$ we may have $r_1 = 10$ and $r_2 = 100$ which gives 111000 multiplications compared with 10^6 operations using direct method, a saving of a factor of nearly 10.

5.5 The FFT Algorithm for a Non-uniform Factorization

Consider now the case when N can be factorized into three generally unequal factors, $N = r_1 r_2 r_3$.

First, we have as before

$$F(u) = \sum_{v=0}^{N-1} f(v) w_N{}^{-uv} = \sum_{v=0}^{N-1} f(v) e\left(\frac{uv}{N}\right), \quad u = 0, 1, \ldots, N-1 \qquad (5.5.1)$$

where $e\left(\frac{uv}{N}\right) = w_N^{-uv} = \exp(-2\pi \mathrm{i} uv/N)$ with $\mathrm{i} = \sqrt{-1}$.

As an obvious extension of the two-factor case we now express the parameters u and v in the form:

$$\begin{aligned} u = p_0 + p_1 r_1 + p_2 r_1 r_2, \quad & p_0 = 0, 1, \ldots, r_1 - 1, \\ & p_1 = 0, 1, \ldots, r_2 - 1, \\ & p_2 = 0, 1, \ldots, r_3 - 1, \end{aligned}$$

$$(5.5.2)$$

$$\begin{aligned} v = q_0 + q_1 r_3 + q_2 r_2 r_3, \quad & q_0 = 0, 1, \ldots, r_3 - 1, \\ & q_1 = 0, 1, \ldots, r_2 - 1, \\ & q_2 = 0, 1, \ldots, r_1 - 1, \end{aligned}$$

We shall now distinguish between the two FFT algorithms, generally known as the *decimation-in-time FFT*, (Cooley and Tukey, 1965) and the *decimation-in-frequency FFT* (Gentleman and Sande, 1966).

The basic features of both algorithms are the same, but the formulae differ somewhat, and this latter fact leads to different ways of indexing the intermediate iteration arrays. Since the FFT often handles quite large arrays of data, it is of importance to keep the indexing of all intermediate as well as the final results as systematic and as simple as possible.

Depending on the data storage environment, one way of indexing intermediate results may be preferable to the other.

The FFT Algorithm which Uses Auxiliary Storage

We shall study the indexing system of the Cooley-Tukey algorithm assuming that the data are given as a sequence stored in a serial order and that an auxiliary array of the same size, N, is available to store the output.

From (5.5.1) and (5.5.2) we have

$$F(p_0 + p_1 r_1 + p_2 r_1 r_2) = \sum_{q_0=0}^{r_3-1} \sum_{q_1=0}^{r_2-1} \sum_{q_2=0}^{r_1-1} f(q_0 + q_1 r_3 + q_2 r_2 r_3)$$

$$\times\, e\left(\frac{(p_0 + p_1 r_1 + p_2 r_1 r_2)(q_0 + q_1 r_3 + q_2 r_2 r_3)}{N}\right).$$

$$(5.5.3)$$

Factorizing the exponential function in the expression (5.5.3) in terms of the parameters q_i, $i = 0, 1, 2$, we obtain

$$F(u) = \sum_{q_0=0}^{r_3-1} e\left(\frac{q_0 u}{N}\right) \sum_{q_1=0}^{r_2-1} e\left(\frac{q_1(p_0 + p_1 r_1)}{r_1 r_2}\right) \sum_{q_2=0}^{r_1-1} e\left(\frac{q_2 p_0}{r_1}\right) f(q_0 + q_1 r_3 + q_2 r_2 r_3). \tag{5.5.4}$$

Here are used facts such as

$$e\left(\frac{q_1(p_0 + p_1 r_1 + p_2 r_1 r_2)}{r_1 r_2}\right) = e\left(\frac{q_1(p_0 + p_1 r_1)}{r_1 r_2}\right) e(q_1 p_2)$$

$$= e\left(\frac{q_1(p_0 + p_1 r_1)}{r_1 r_2}\right)$$

since $e(q_1 p_2) = e^{-i2\pi q_1 p_2} = 1$.

In (5.5.4) some further factorization of the exponential components gives:

$$F(u) = \sum_{q_0=0}^{r_3-1} e\left(\frac{q_0 p_2}{r_3}\right) e\left(\frac{q_0(p_0 + p_1 r_1)}{r_1 r_2 r_3}\right) \sum_{q_1=0}^{r_2-1} e\left(\frac{q_1 p_1}{r_2}\right)$$

$$\times e\left(\frac{q_1 p_0}{r_1 r_2}\right) \sum_{q_2=0}^{r_1-1} e\left(\frac{q_2 p_0}{r_1}\right) f(q_0 + q_1 r_3 + q_2 r_2 r_3). \tag{5.5.5}$$

Expression (5.5.5) suggests the following procedure to compute the Fourier coefficients:

Compute the set $c_{p_0}(q_0, q_1)$

$$= e\left(\frac{q_1 p_0}{r_1 r_2}\right) \sum_{q_2=0}^{r_1-1} e\left(\frac{q_2 p_0}{r_1}\right) f(q_0 + q_1 r_3 + q_2 r_2 r_3), \tag{5.5.6}$$

$$q_0 = 0, 1, \ldots, r_3 - 1, \quad q_1 = 0, 1, \ldots, r_2 - 1, \quad p_0 = 0, 1, \ldots, r_1 - 1.$$

This is an r_1-point Fourier transform of points spaced $r_2 r_3$ apart. The twiddle factor in front of the sum depends on q_1 and p_0.

The values obtained are stored at the serial locations denoted by $p_0 + q_0 r_1 + q_1 r_1 r_2$.

Compute the set $c_{p_0} + p_1 r_1 (q_0)$

$$= e\left(\frac{q_0(p_0 + p_1 r_1)}{r_1 r_2 r_3}\right) \sum_{q_1=0}^{r_2-1} e\left(\frac{q_1 p_1}{r_2}\right) c_{p_0}(q_0, q_1), \tag{5.5.7}$$

$$q_0 = 0, 1, \ldots, r_3 - 1, p_1 = 0, 1, \ldots, r_2 - 1, p_0 = 0, 1, \ldots, r_1 - 1.$$

This is an r_2-point Fourier transform of points spaced $r_1 r_3$ apart. The twiddle factor depends on q_0 and $p_0 + p_1 r_1$.

The values obtained are stored at the serial locations denoted by $p_0 + p_1 r_1 + q_0 r_1 r_2$.

Compute the set $F(p_0 + p_1 r_1 + p_2 r_1 r_2) = F(u)$

$$= \sum_{q_0=0}^{r_3-1} e\left(\frac{q_0 p_2}{r_3}\right) c_{p_0 + p_1 r_1}(q_0),$$

$$p_2 = 0, 1, \ldots, r_3 - 1, \quad p_1 = 0, 1, \ldots r_2 - 1, \quad p_0 = 0, 1, \ldots, r_1 - 1. \tag{5.5.8}$$

This is an r_3-point Fourier transform of points spaced r_1r_2 apart. There is no twiddle factor at this step. The final results are stored in serial order at points denoted by $p_0 + p_1r_1 + p_2r_1r_2$.

The total number of operations required by this algorithm is given by

$$(r_3 + 1)N + (r_2 + 1)N + r_1N = N(r_1 + r_2 + r_3 + 2).$$

The algorithm described performs efficiently in the cases when an auxiliary array is readily available. It may happen, however, that the input array is so large that provision of the necessary auxiliary storage may significantly slow down the process of computation. This would be the case, for example, when backing store facilities which are relatively slowly accessed have to be used. We shall now show that, in fact, the FFT may be carried out *in situ* (that is, without using any auxiliary storage) though at the price of giving up the serial order of storage of the final Fourier transforms.

The FFT Algorithm in Place

In the cases when the FFT algorithm has to be carried out without using an auxiliary storage, its variant developed by Gentleman and Sande is found to be more flexible and convenient to use.

To obtain this algorithm factorization of the exponential function in equation (5.5.3) is carried out in relation to the parameters p_i, i = 0, 1, 2. The equation then becomes:

$$F(u) = \sum_{q_0=0}^{r_3-1} e\left(\frac{p_2q_0}{r_3}\right) e\left(\frac{p_1q_0}{r_2r_3}\right) \sum_{q_1=0}^{r_2-1} e\left(\frac{p_1q_1}{r_2}\right) e\left(\frac{p_0(q_0 + q_1r_3)}{r_1r_2r_3}\right)$$
$$\sum_{q_2=0}^{r_1-1} e\left(\frac{p_0q_2}{r_1}\right) f(q_0 + q_1r_3 + q_2r_2 + r_3). \qquad (5.5.9)$$

Assume again that the input data is stored in sequential order. The algorithm for computing the Fourier coefficients given by (5.5.9) is then as follows.

Compute the set $$c_{p_0}(q_0 + q_1r_3) = e\left(\frac{p_0(q_0 + q_1r_3)}{r_1r_2r_3}\right)$$
$$\sum_{q_2=0}^{r_1-1} e\left(\frac{p_0q_2}{r_1}\right) f(q_0 + q_1r_3 + q_2r_2 + r_3),$$
$$p_0 = 0, 1, \ldots, r_1 - 1, \quad q_1 = 0, 1, \ldots, r_2, \quad q_0 = 0, 1, \ldots, r_3. \qquad (5.5.10)$$

This is an r_1-point Fourier transform of points spaced r_2r_3 apart. Consider the subset of values $\{c_p(q_0 + q_1r_3),\ 0 \leqslant p_0 \leqslant r_1 - 1, q_0 + q_1r_3 \text{ fixed}\}$. The subset is produced using the r_1 input data values $\{f(q_0 + q_1r_3 + q_2r_2r_3),\ 0 \leqslant q_2 \leqslant r_1 - 1\}$. These r_1 input data values will not be needed in further computations and so we can release their r_1 storage locations. The released locations have so far been denoted, along with other locations storing the input data, by $q_0 + q_1r_3 + q_2r_2r_3$. They can be used to store the subset of the values just computed. The new notation for these locations is then set as $q_0 + q_1 r_3 + p_0 r_2 r_3$ where $p_0 = 0, 1, \ldots, r_1 - 1$. This new notation for the subset is consistent and is not overlapping with the other locations since the summation on q_2 represents an r_1-point Fourier transform of values spaced $r_2 r_3$ locations apart and the twiddle factor $e(p_0(q_0 + q_1 r_3)/r_1 r_2 r_3)$ depends upon the parameters p_0 and $q_0 + q_1 r_3$. The complete set of N values computed may be denoted by

$$\{c_{p_0}(q_0 + q_1 r_3), \quad 0 \leqslant p_0 \leqslant r_1 - 1, \quad 0 \leqslant q_1 \leqslant r_2 - 1, \quad 0 \leqslant q_0 \leqslant r_3 - 1\}.$$

Compute the set $c_{p_0} + p_1 r_1 (q_0)$

$$= e\left(\frac{p_1 q_0}{r_2 r_3}\right) \sum_{q_1=0}^{r_2-1} e\left(\frac{p_1 q_1}{r_2}\right) c_{p_0}(q_0 + q_1 r_3), \qquad (5.5.11)$$

$$0 \leqslant p_0 \leqslant r_1 - 1, \quad 0 \leqslant p_1 \leqslant r_2 - 1, \quad 0 \leqslant q_0 \leqslant r_3 - 1.$$

Hence we have the summation on q_1 which represents an r_2-point Fourier transform of values placed $r_1 r_3$ locations apart. The twiddle factor depends on the parameters p_1 and q_0 only. If we consider the subset of values $\{c_{p_0} + p_1 r_1(q_0), 0 \leqslant p_1 \leqslant r_2 - 1\}$, in the manner similar to the above, we note that to produce this subset the r_2 values $\{c_p(q_0 + q_1 r_3)\}$ were used. These values will not be needed in further Fourier computations and so we can release their r_2 locations, which are denoted by $q_0 + q_1 r_3 + p_0 r_2 r_3$, where $q_1 = 0, 1, \ldots, r_2 - 1$. As before, the vacated locations are used to store the subset of the values just computed. The new notation for these locations is set to be $q_0 + p_1 r_3 + p_0 r_2 r_3$, $0 \leqslant p_1 \leqslant r_2 - 1$. The complete set of N computed values is denoted by

$$\{c_p + pr(q_0), \quad 0 \leqslant p_0 \leqslant r_1 - 1, \quad 0 \leqslant p_1 \leqslant r_2 - 1, \quad 0 \leqslant q_0 \leqslant r_3 - 1\}.$$

Compute the set

$$F(p_0 + p_1 r_1 + p_2 r_1 r_2) = \sum_{q_0=0}^{r_3-1} e\left(\frac{p_2 q_0}{r_3}\right) c_{p_0} + p_1 r_1 (q_0), \qquad (5.5.12)$$

$$p_2 = 0, 1, \ldots, r_3 - 1, \quad p_1 = 0, 1, \ldots, r_2 - 1, \quad p_0 = 0, 1, \ldots, r_1 - 1.$$

Here the summation over q_0 represents N contiguous r_3-point Fourier transform. The results are stored sequentially at the locations denoted by $p_2 + p_1 r_3 + p_0 r_2 r_3$.

This completes the Gentleman–Sande FFT in-place algorithm. The Cooley–Tukey algorithm can also be arranged in-place but its indexing of the intermediate results needs a careful handling.

The in-place FFT algorithm produces the Fourier transform

$\{F(u), u = p_0 + p_1r_1 + p_2r_1r_2\}$ which is stored in the array with the sequential locations denoted as $p_2 + p_1r_3 + p_0r_2r_3$. This storage scheme is called a storage with 'digit reversed subscripts'. Since quite frequently the Fourier transforms are computed as a part of a larger problem where they may be produced as an intermediate data it is only important to obtain the array in the usual serial order suitable for further use. The standard procedures for computation the in-place FFT contain as a rule a computational device for 'unscrambling' the Fourier transforms. An algorithm for unscrambling the FFT will be given later.

5.6 Optimal Factorization for the FFT

We have seen that various degrees of saving in computing time may be achieved using the FFT algorithm provided an appropriate partitioning of the data is applied. Denoting, as usual, the size of the input data by N, we shall now investigate the best way to factorize N so as to achieve the greatest possible saving, using the FFT.

Let N be factorized into k factors

$$N = r_1r_2r_3 \ldots r_k. \tag{5.6.1}$$

Using the FFT factorizing technique in the way similar to the two-factor and three-factor cases discussed earlier, it is easy to demonstrate that the required number of operations is equal to

$N\left(k - 1 + \sum_{j=1}^{k} r_j\right)$ complex multiplications and

$N\left(\sum_{j=1}^{k} (r_j - 1)\right)$ complex additions.

For brevity, we shall use the term 'operation' to mean, as before, a complex multiplication followed by a complex addition. The number of such operations may be given as

$$P = N\left(\sum_{j=1}^{k} r_j\right), \tag{5.6.2}$$

where the terms $N(k - 1)$ in the number of complex multiplications and the term Nk in the number of complex additions are ignored as they do not affect the optimality criteria.

Equations (5.6.1) and (5.6.2) yield

$$\log_2 N = \log_2 r_1 + \log_2 r_2 + \ldots + \log_2 r_k$$

and

$$\frac{P}{N} = r_1 + r_2 + \ldots + r_k,$$

giving

$$\frac{P}{N \log_2 N} = \left(\sum_{j=1}^{k} r_j\right) \Big/ \left(\sum_{j=1}^{k} \log_2 r_j\right)$$

which may conveniently be presented as

$$\frac{P}{N \log_2 N} = \frac{\sum_{j=1}^{k} \left(\frac{r_j}{\log_2 r_j}\right) \log_2 r_j}{\sum_{j=1}^{k} \log_2 r_j}. \tag{5.6.3}$$

We now assume some freedom in the selection of N and its factors, r_j, and determine choices to minimize (5.6.3) with respect to $r_j, j = 1, \ldots, k$.

If all factors of N are equal, $r_1 = r_2 = \ldots = r_k = r$, then (5.6.3) can be written as

$$\frac{P}{N \log_2 N} = \frac{r}{\log_2 r},$$

giving

$$P = \left(\frac{r}{\log_2 r}\right) N \log_2 N = rN \log_r N. \tag{5.6.4}$$

In this case, we say that we have a radix r, or base r, algorithm. If the factors are not all equal, as in (5.6.3), we say that we have a mixed radix algorithm. From (5.6.4) we can see that for a radix r the algorithm P is proportional to $N \log_2 N$ with the proportionality factor $r/\log_2 r$, whose values are listed in Table 5.6.1.

Notice that if N is a power of 3 then the number of operations, P, assumes its lowest value

$$P = 1.88\, N \log_2 N. \tag{5.6.5}$$

In this sense, the right-hand side of (5.6.5) may be interpreted as the lower bound on the number of operations required to compute the discrete Fourier transform on a set of N points, using the FFT algorithm.

If $N = 2^m$, we may take $r = 2$, obtaining

$$P = 2\, N \log_2 N, \tag{5.6.6}$$

and the same expression holds for P if $N = 4^n$. For given N, this number is quite close to the lower bound on the number of operations, given by (5.6.5).

Table 5.6.1

r	2	3	4	5	6	7	8	9	10	11	16
$\frac{r}{\log_2 r}$	2.00	1.88	2.00	2.15	2.31	2.49	2.67	2.82	3.01	3.18	4.00

The use of radices 2 and 4 (also 8 and 16) has the advantage that some of the powers of $w_N = \exp(2\pi i/N)$ are simple numbers like ± 1, $\pm i$, $(1 \pm i)/\sqrt{2}$, and multiplications can be avoided.

Bergland (1968) studied in detail the number of additions and multiplications required for radices 2, 4, 8, and 16 and for the program written as economically as possible, that is omitting multiplications by simple powers of w_N and combining terms with common factors. The results show some economy in the use of radices 2, 4, 8, and 16 but the savings for 8 and 16 are small compared to the increased complexity of the algorithm. In fact, the radix 2 and 4 algorithms are the ones most widely used, whenever a choice of appropriate factorizing on N presents itself. The FFT algorithm of mixed radices have also been programmed. In particular, the algorithm of mixed radix 4 + 2 (i.e. all factors equal to 4 except perhaps the last, which may be 2) is also widely used. From Table 5.6.1 it is seen that the use of factors other than 2 and 4 does not increase the time of calculation substantially.

5.7 The FFT of Radix 2

Because of its simplicity in programming, the base 2 algorithm has been used extensively in subroutines and in hardware implementations.

Let us consider how it works. We assume that $N = 2^m$ and as an illustration, take the case of $N = 2^4$. The algorithm of radix 2 is then derived by expressing the parameters v and u in binary number form,

$$v = 2^{m-1}q_0 + \ldots + 2q_{m-2} + q_{m-1}, \tag{5.7.1}$$

in our example

$$v = 2^3 q_0 + 2^2 q_1 + 2q_2 + q_3,$$

and

$$u = 2^{m-1}p_0 + \ldots + 2p_{m-2} + p_{m-1}, \tag{5.7.2}$$

in our example

$$u = 2^3 p_0 + 2^2 p_1 + 2p_2 + p_3,$$

where the p's and the q's are equal to 0 or 1 and are termed 'bits'. With this convention, the Fourier transform may be written as a function of its binary parameters:

$$\begin{aligned} F(u) &= c(p_0, p_1, p_2, p_3) \\ &= \sum_{q_3}\sum_{q_2}\sum_{q_1}\sum_{q_0} f(q_0, q_1, q_2, q_3)\, e\left(\frac{uq_0 2^3 + uq_1 2^2 + uq_2 2 + uq_3}{N}\right). \end{aligned} \tag{5.7.3}$$

Transferring certain factors outside the appropriate summations we obtain

$$F(u) = c(p_0, p_1, p_2, p_3)$$

$$= \sum_{q_3} e\left(\frac{uq_3}{2^4}\right) \sum_{q_2} e\left(\frac{uq_2}{2^3}\right) \sum_{q_1} e\left(\frac{uq_1}{2^2}\right) \sum_{q_0} e\left(\frac{uq_0}{2}\right) c_0(q_0, q_1, q_2, q_3) \tag{5.7.4}$$

where for uniformity we have denoted $f(q_0, q_1, q_2, q_3)$ by $c_0(q_0, q_1, q_2, q_3)$. We now observe that

$$e\left(\frac{uq_0}{2}\right) = e\left(\frac{p_3 q_0}{2}\right),$$

since

$$e\left(\frac{uq_0}{2}\right) = \mathrm{e}^{-\mathrm{i}(2\pi/2)(p_0 2^3 + p_1 2^2 + p_2 2 + p_0) q_0}$$

$$= \mathrm{e}^{-\mathrm{i}(2\pi/2)(p_0 2^2 + p_1 2 + p_2) 2 q_0} \mathrm{e}^{-\mathrm{i}(2\pi/2) p_3 q_0} = e\left(\frac{p_3 q_0}{2}\right),$$

as

$$e^{-\mathrm{i}2\pi(p_0 2^2 + p_1 2 + p_2) q_0} = 1.$$

It follows that the innermost sum of (5.7.4), over q_0, depends only on p_3, q_1, q_2, q_3 and can be written as

$$c_1(p_3, q_1, q_2, q_3) = \sum_{q0} e\left(\frac{uq_0}{2}\right) c_0(q_0, q_1, q_2, q_3). \tag{5.7.5}$$

There are 16 (or N in the general case) such quantities altogether. The Fourier transform formula (5.7.4) now has the form

$$c_1(p_0, p_1, p_2, p_3) = \sum_{q_3} e\left(\frac{uq_3}{2^4}\right) \sum_{q_2} e\left(\frac{uq_2}{2^3}\right) \sum_{q_1} e\left(\frac{uq_1}{2^2}\right) c_1(p_3, q_1, q_2, q_3). \tag{5.7.6}$$

Noting, again, that

$$e\left(\frac{uq_1}{2^2}\right) = e\left(\frac{(p_2 2 + p_3) q_1}{2^2}\right),$$

we infer that the innermost sum of (5.7.6), over q_1, depends only on p_3, p_2, q_2, q_3 and can be written as

$$c_2(p_3, p_2, q_2, q_3) = \sum_{q_1} c_1(p_3, q_1, q_2, q_3) e\left(\frac{uq_1}{2^2}\right). \tag{5.7.7}$$

There are, again, N such quantities altogether.
Carrying out two further steps in manner similar to that just decribed, we obtain the discrete Fourier transform as

$$F(u) = c(p_0, p_1, p_2, p_3) = c_4(p_3, p_2, p_1, p_0). \tag{5.7.8}$$

Note that, as before, the parameter u of $F(u)$ must have its binary bits put in reverse order to yield its index in the array c_4.
Note also that the computational process given by (5.7.3)–(5.7.8) requires $m = \log_2 N$ steps, and a set of N elements is produced at each step.

General Formula and a Flowgraph of the FFT Algorithm of Radix 2

From (5.7.3) to (5.7.8) the kth step of the base 2 algorithm may be expressed as

$$\begin{aligned} c_{k+1}(p_{m-1}, \ldots, p_{m-1-k}, q_{k+1}, \ldots, q_{m-1}) \\ = c_k(p_{m-1}, \ldots, p_{m-2-k}, 0, q_{k+1}, \ldots, q_{m-1}) \\ + e\left(\frac{p_{m-1} + 2p_{m-2} + \ldots + 2^{k-1}p_{m-k}}{2^{k+1}}\right) \\ \times c_k(p_{m-1}, \ldots, p_{m-2-k}, 1, q_{k+1}, \ldots, q_{m-1}) \end{aligned} \tag{5.7.9}$$

where $\{c_k(r), r = 0, \ldots, N-1\}$ is the set of N complex numbers,

$k = 0, \ldots, \log_2 N, \quad e(x) = \mathrm{e}^{-\mathrm{i}2\pi x}$, as before,

and all the p's and the q's are either 0 or 1.
Rewriting (5.7.9) in a more compact form we get

$$c_{k+1}(r) = \begin{cases} c_k(r) + e(z/2^{k+1})c_k(r + 2^{m-1-k}) & \text{if } p_{m-1-k} = 0, \\ c_k(r - 2^{m-1-k}) + e(z/2^{k+1})c_k(r) & \text{if } p_{m-1-k} = 1, \end{cases} \tag{5.7.10}$$

where

$$z = p_{m-1} + 2p_{m-2} + \ldots + 2^k p_{m-1-k} \tag{5.7.11}$$

and

$$\begin{aligned} r = q_{m-1} + 2q_{m-2} + \ldots + 2^{m-2-k}q_{k+1} \\ + 2^{m-1-k}p_{m-1-k} + \ldots + 2^{m-1}p_{m-1}. \end{aligned} \tag{5.7.12}$$

Note that the arguments $r + 2^{m-1-k}$ and $r - 2^{m-1-k}$ can be obtained from the binary expansion of the r in equation (5.7.12), by replacing p_{m-1-k} by its 1-complement. (The 1-complement of a binary number is the result of changing each zero to one and each one to zero). The set $(c_m(r), r = 0, \ldots, N-1, m = \log_2 N)$ represents the final set of the discrete Fourier transforms, $(F(u), u = 0, \ldots, N-1)$, though in a rearranged order, so that the Fourier transform $F(u)$, where $u = p_0 + 2p_1 + \ldots + 2^{m-1}p_{m-1}$, is given by $c_m(r)$, where $r = p_{m-1} + 2p_{m-2} + \ldots + 2^{m-1}p_0$, and all the p's are either 0 or 1.

Using the general formula (5.7.10) we can now present a flowgraph of the FT computation. Figure 5.7.1 illustrates a flowgraph of the actual computation for the case $N = 2^4$. As shown, the data vector or array $(c_0(r), r = 0, \ldots, N-1)$ is represented by a vertical column of nodes on the left of

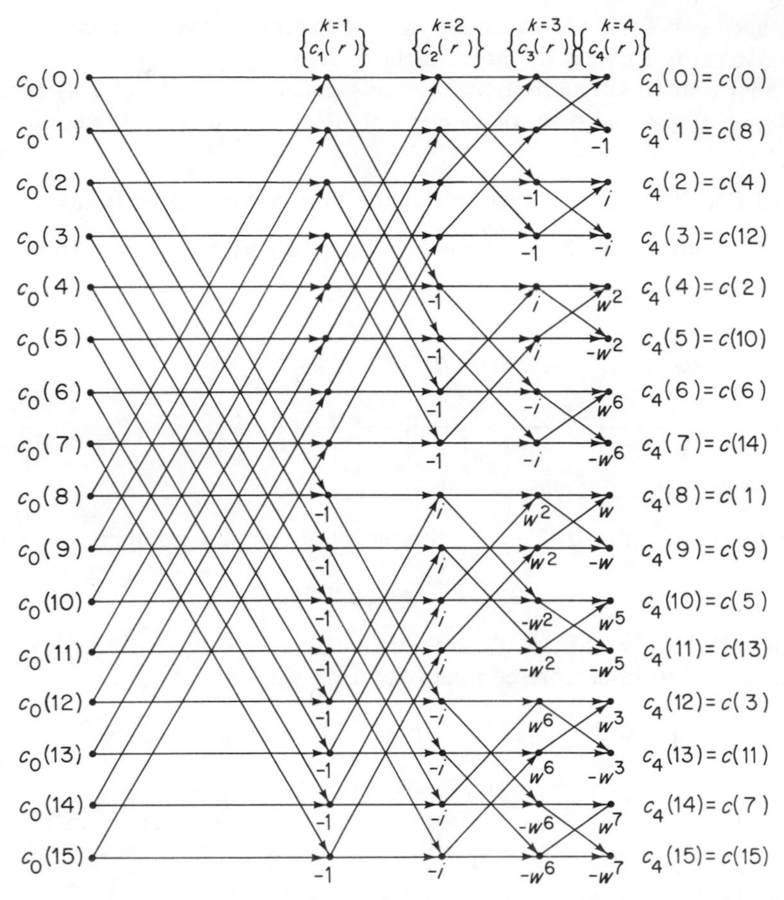

Figure 5.7.1 Flowgraph of the Cooley–Tukey FFT algorithm for $N = 2^4$ ($w_N^x = \exp(-2\pi ix/N)$, e.g. $w_N^8 = -1$, $w_N^4 = -i$, $w_N^{12} = -w_N^4 = i$, $w_N^0 = 1$). The weights for each branch, if other than unity, are shown immediately below the particular branch

the flowgraph. The second vertical array of nodes is the vector $c_1(r)$ computed in equation (5.7.10) for $k = 0$, etc. For general N there will be $m = \log_2 N$ computational arrays. The flowgraph is interpreted as follows. Each node is entered by two solid lines representing directed branches. A branch transmits a quantity from a node in one array, multiplies the quantity by the weight $w^x{}_N$, and inputs the result into the node in the next array. The weights, if other than unity, are shown for each branch. Results entering a node from the two branches are combined additively. According to the rules for interpreting the flowgraph we have, for example,

$$c_1(11) = c_0(3) + (-1)c_0(11),$$

which is determined by equations (5.7.10)–(5.7.12) for $k = 0$ and $r = (11)_2 =$

1011, the latter of which gives, in turn, $p_3 = 1, z = 1$ and

$$e\left(\frac{z}{2^{k+1}}\right) = e^{-2\pi i/2} = e^{-\pi i} = -1.$$

Each node of the flowgraph is expressed similarly.

We can see that the flowgraph is a concise method for representing the computation steps in the FFT algorithm.

Analysis of the FFT Flowgraph

We shall now study the flowgraph of Fig. 5.7.1 in some detail in order to emphasize some useful relations between the nodes. This will, in turn, allow us to simplify the computation formula (5.7.10) and, what is especially important, to show that the number of actual complex multiplications required to compute an $N-$ length array $(c_k(r), r = 0, \ldots, N-1)$ is equal to $N/2$ and not to N as it may seem from the formula (5.7.10)

Dual nodes

Consider the node pair $c_1(0)$ and $c_1(8)$. For their calculation only the nodes $c_0(0)$ and $c_0(8)$ are used; these latter nodes are not used in any other calculation in column $k = 1$.

Next consider the node pair $c_1(1)$ and $c_1(9)$. For their calculations only the nodes $c_0(1)$ and $c_0(9)$ are used. Again, these latter nodes are not used in any other calculation in column $k = 1$. Etc.

Similar observations are readily done in columns $k = 2$, $k = 3$ and $k = 4$.

We shall call the pairs $c_1(0)$ and $c_1(8)$, $c_1(1)$ and $c_1(9)$, etc., the dual nodes.

We further note that in the kth column the dual nodes are spaced at the distance $r = N/2^k = 2^{m-k}$ nodes, e.g. in column $k = 2$ the dual nodes $c_2(0)$ and $c_2(4)$, $c_2(1)$ and $c_2(5)$, etc., are spaced at the distance $r = 2^{m-k} = 4$ nodes.

These observations enable us to deduce the formula for calculation of the current column nodes in terms of the nodes of the immediately preceding column, as follows:

$$\begin{aligned} c_k(r) &= c_{k-1}(r) + w_N^{z'} c_{k-1}(r + 2^{m-k}) \\ c_k(r + 2^{m-k}) &= c_{k-1}(r) + w_N^{z''} c_{k-1}(r + 2^{m-k}) \end{aligned} \tag{5.7.13}$$

where the powers z' and z'' have to be determined.

To do this we note that the branches stemming from, say, node $c_1(12)$ are multiplied by factors $w_N^4 = e(4/N) = -\mathrm{i}$, and $w_N^{12} = e(12/N) = \mathrm{i}$, prior to input at nodes $c_2(8)$ and $c_2(12)$, respectively.

It is important to note that $w_N^4 = -w_N^{12}$ and, thus, in the formulae for computing $c_2(8)$ and $c_2(12)$ given by (5.7.13), only one complex multiplication is required since the same data, $c_1(12)$, is to be multiplied by these factors.

In general, if the weighting factor at one node is

$$w_N^x = e\left(\frac{x}{N}\right),$$

then the weighting factor at the dual node is

$$w_N^{x+(N/2)} = e\left(\frac{x+N/2}{N}\right) = e\left(\frac{x}{N}+\frac{1}{2}\right).$$

And since $e\left(\frac{x}{N}\right) = -e\left(\frac{x}{N}+\frac{1}{2}\right)$, only one multiplication is required in the computation of a dual node pair.

Formulae for computing a dual pair

We, thus, obtain the following formulae for calculating an array $(c_k(r), r = 0, \ldots, N-1)$:

$$c_k(r) = c_{k-1}(r) + e\left(\frac{z}{2^k}\right)c_{k-1}(r+2^{m-k}),$$

$$c_k(r+2^{m-k}) = c_{k-1}(r) - e\left(\frac{z}{2^k}\right)c_{k-1}(r+2^{m-k}), \tag{5.7.14}$$

where

$$z = p_{m-1} + 2p_{m-2} + \ldots + 2^{k-1}p_{m-k} \tag{5.7.15}$$

and

$$r = q_{m-1} + 2q_{m-2} + \ldots + 2^{m-1-k}q_k + 2^{m-k}p_{m-k} + \ldots + 2^{m-1}p_{m-1}, \tag{5.7.16}$$

and all the q's and the p's are either 0 or 1.
These formulae are equivalent to (5.7.10) to (5.7.12) but are more convenient for practical use. We note that the value of z is determined by

(a) writing the index r in binary form in m bits,

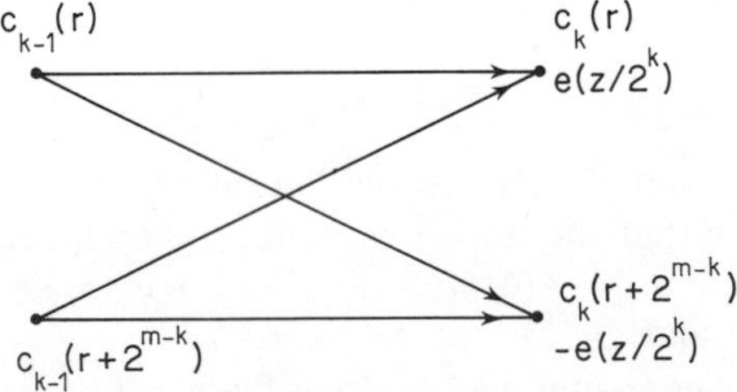

Fig. 5.7.2 Flowgraph for the basic 'butterfly' operation for the FFT algorithm of radix 2

(b) scaling or sliding this binary number $m-1$ bits to the right and filling in the newly opened bit position on the left with zeros, and
(c) reversing the order of the bits.

A flowgraph section of equation (5.7.14) is shown in Fig. 5.7.2. The diagram is referred to as the basic 'butterfly' operation for the radix 2 algorithm.

Skipping over the second nodes of the dual pairs

In computing an array ($c_k(r)$, $r=0, \ldots, N-1$) one normally begins with node $r=0$ and sequentially works down the array, computing the equation pair (5.7.14) to (5.7.16). Since the dual of any node in the kth column is always down $N/2^k$ in the column, it follows that one must skip some nodes after every $N/2^k$ node.

For example consider array $k=2$ of Fig. 5.7.1. We begin with node $r=0$. Its dual node is located at $r=N/2^2=4$. Next is node $r=1$, and its dual node is located at $r=5$, and so forth until we reach node $r=4$. At this point we have reached a set of nodes previously encountered, as these nodes are the duals for nodes $r=0, 1, 2$, and 3. So we must skip over the nodes $r=4, 5, 6$, and 7. We then carry out the computations for nodes $r=8, 9, 10$, and 11 and again skip over their duals, $r=12, 13, 14$, and 15. In general, if we work from the top down in array k, then we compute equations (5.7.14) to (5.7.16) for the first $N/2^k=2^{m-k}$ nodes, skip the next 2^{m-k}, etc., until we reach a node index greater than $N-1$.

The FFT radix 2 algorithm is thus basically concerned with computing the set of 'butterfly' operations given by (5.7.14) to (5.7.16), for $k=1, 2, \ldots, m$, $m=\log_2 N$.

In other words for every fixed k we compute the array $A(I)=c_k(I)$, $I=0, 1, \ldots, N-1$ using equations (5.7.14) to (5.7.16). Initially $A(I)$ is set to $c_0(I)$, $I=0, 1, \ldots, N-1$. For a fixed k, $1 \leqslant k \leqslant m$, equation (5.7.14) may be written in the form which is more convenient for the FFT specification:

$$A(I) \leftarrow A(I)+E \times A(K)$$
$$A(K) \leftarrow A(I)-E \times A(K)$$

where

$$I=0, 1, \ldots, N-1, \quad K=\mathrm{I}+2^k, \quad 1 \leqslant k \leqslant m,$$

and

$E=\exp(-2\pi\ iz_k/2^k)$ with z_k equal to the 'bit-reversed version' of the number I.

We shall next outline the bit reversing procedure and a sequence of steps required for unscrambling the Fourier transform sequence and then will give a complete procedure for the radix 2 FFT algorithm.

Unscrambling the FFT

Let K be a binary number $b_0b_1b_2b_3$, that is

$$K = 2^3 \times b_0 + 2^2 \times b_1 + 2 \times b_2 + 2^0 \times b_3 = (b_0b_1b_2b_3)_2. \quad (5.7.17)$$

We wish to compute the binary number R such that

$$R = (b_3b_2b_1b_0)_2 = 2^3 \times b_3 + 2^2 \times b_2 + 2 \times b_1 + 2^0 \times b_0. \quad (5.7.18)$$

The following procedure will produce the desired result:
bit b_3 is determined by

$$b_3 = K - (K \textit{ div } 2) \times 2, \quad (5.7.19)$$

the next bit b_2 is found by

$$\begin{aligned} b_2 &= K \textit{ div } 2 - ((K \textit{ div } 2) \textit{ div } 2) \times 2 \\ &= K \textit{ div } 2 - (K \textit{ div } 2^2) \times 2. \end{aligned} \quad (5.7.20)$$

The bits b_1 and b_0 are obtained in a similar manner. As soon as b_3, b_2, b_1, and b_0 are determined the number r is readily computed.

Example 5.7.1

Let $K = 14 = 2^3 \times 1 + 2^2 \times 1 + 2 \times 1 + 2^0 \times 0 = (1110)_2 = (b_0b_1b_2b_3)_2$
We wish to determine

$$R = (b_3b_2b_1b_0)_2 = (0111)_2 = 2^3 \times 0 + 2^2 \times 1 + 2 \times 1 + 2^0 \times 1 = 7.$$

$$\begin{array}{lll} 14 \textit{ div } 2 = 7; & 7 \times 2 = 14; & 14 - 14 = 0 \Rightarrow b_3 = 0. \\ 7 \textit{ div } 2 = 3; & 3 \times 2 = 6; & 7 - 6 \neq 0 \Rightarrow b_2 = 1. \\ 3 \textit{ div } 2 = 1; & 1 \times 2 = 2; & 3 - 2 \neq 0 \Rightarrow b_1 = 1. \\ 1 \textit{ div } 2 = 0; & 0 \times 2 = 0; & 1 - 0 \neq 0 \Rightarrow b_0 = 1. \end{array}$$

Collecting the bits sequentially from left to right as soon as they are determined we obtain

$$R = (b_3b_2b_1b_0)_2 = (0111)_2 = 7.$$

In the process of unscrambling the output array a situation occurs similar to the one which leads to skipping over some points in the computation of dual pairs, that is if we proceed down the array systematically interchanging $c_m(K)$ with the corresponding $c_m(I)$, eventually a node will be encountered which has already been interchanged. To eliminate the necessity to consider this node again we simply check to see if R (the integer obtained by bit reversing K) is less than K. If so, this implies that the node has already been interchanged in an earlier operation and should not be touched.

A different way to obtain the Fourier coefficients in a proper order at the end of the FFT process is to 'preprocess' the input data, that is to rearrange the input sequence $\{f(v)\}$ in the bit-reversed-index ordering at the outset of the FFT execution. The computed Fourier coefficients will then be output in normal serial order. For example, the input sequence

$f(0)\ f(1)\ f(2)\ f(3)\ f(4)\ f(5)\ f(6)\ f(7)\ f(8)\ f(9)\ f(10)\ f(11)\ f(12)\ f(13)\ f(14)$
$f(15)$

is first stored as the sequence

$f(0)\ f(8)\ f(4)\ f(12)\ f(2)\ f(10)\ f(6)\ f(14)\ f(1)\ f(9)\ f(5)\ f(13)\ f(3)\ f(11)\ f(7)$
$f(15)$.

The Fourier coefficients are computed and stored in a sequential order. Below is given an implementation of the FFT radix 2 algorithm computed in place.

Algorithm fft (A, m)

//$A(1:N)$ contains the input data, $N=2^m$. The Fourier coefficients are computed in place. Complex arithmetic is expressed explicitly and $\exp(-2\pi ij/N)$ is expressed in terms of sines and cosines.//

```
N := 2^m
J := 0
for I := 0 to N-1 do
//permute the input//
  if I < J then
  begin temp := A[J]; A[J] := A[I]; A[I] := temp end
  K := N div 2
  while K ⩽ J do begin J := J-K; K := k/2 end
  J := J+K
enddo
py := 3.14159265          //a constant//
for k := 1 to log2 N do
  power := 2^k; p := power/2
  r := 1.0 + i 0.0        //i = √-1//
  s := cos(py/p) - i × sin(py/p)        //i = √-1//
  for K := 1 to p do
    for I := K to n step power do
      index := I + p
      temp := A[index] × r
      A[index] := A[I] + temp
      A[I] := A[I] - temp
      enddo
    r := r × s
  enddo
enddo
```

5.8 DFT for Different Types of Data

Below we describe three procedures which permit one to use the complex FFT algorithm efficiently for special types of data. The procedures are given in such a form that one can use them by executing a complex DFT procedure with appropriate rearrangement of either the data or the output. The DFT subroutine is assumed to accept as input, an integer N and a sequence of complex numbers $f(v), v = 0, 1, \ldots, N-1$, and to yield, as output, the DFT sequence

$$F(u) = \sum_{v=0}^{N-1} f(v) w_N^{-uv}, \; u = 0, 1, \ldots, N-1. \tag{5.8.1}$$

Also useful is the inverse relationship to (5.8.1), i.e.

$$f(v) = \frac{1}{N} \sum_{u=0}^{N-1} F(u) w_N^{uv}, \; v = 0, 1, \ldots, N-1, \tag{5.8.2}$$

where $f(v)$ may be referred to as the inverse DFT (IDFT) of $F(u)$.

The Doubling Algorithm

Suppose that an N-point sequence, $f(v)$, $v = 0, 1, \ldots, N-1$, where N is even, is too long to compute in one application of the DFT algorithm while a sequence of $N/2$ points can be accommodated. In this case, define the two $N/2$-point sequences and their transforms:

$$f_1(v) = f(2v), \qquad v = 0, 1, \ldots, \frac{N}{2} - 1 \tag{5.8.3}$$

$$F_1(u) = \sum_{v=0}^{(N/2)-1} f_1(v) w_{N/2}^{-uv}, \qquad u = 0, 1, \ldots, \frac{N}{2} - 1,$$

and

$$f_2(v) = f(2v+1), \qquad v = 0, 1, \ldots, \frac{N}{2} - 1 \tag{5.8.4}$$

$$F_2(u) = \sum_{v=0}^{(N/2)-1} f_2(v) w_{N/2}^{-uv}, \qquad u = 0, 1, \ldots, \frac{N}{2} - 1.$$

Now consider the DFT, $F(u)$, of the complete N-point sequence, $(f(v))$:

$$F(u) = \sum_{v=0}^{N-1} f(v) w_N^{-uv}, \qquad u = 0, 1, \ldots, N-1 \tag{5.8.5}$$

from where, separating the odd- and even-indexed terms, we get

$$\begin{aligned} F(u) &= \sum_{v=0}^{(N/2)-1} f(2v) w_N^{-2uv} + \sum_{v=0}^{(N/2)-1} f(2v+1) w_N^{-(2v+1)u} \\ &= \sum_{v=0}^{(N/2)-1} f(2v) w_N^{-2uv} + \left[\sum_{v=0}^{(N/2)-1} f(2v+1) w_{N/2}^{-uv} \right] w_N^{-u}. \end{aligned} \tag{5.8.6}$$

The two sums are the $N/2$-point DFTs of $f_1(v)$ and $f_2(v)$, so

$$F(u) = F_1(u) + F_2(u)w_N{}^{-u}, \qquad u = 0, 1, \ldots, \frac{N}{2} - 1. \tag{5.8.7}$$

Substituting $u + N/2$ for u and using the fact that $F_1(u)$ and $F_2(u)$ are periodic with period $N/2$, we get

$$F\left(u + \frac{N}{2}\right) = F_1(u) - F_2(u)w_N{}^{-u}, \qquad u = 0, 1, \ldots, \frac{N}{2} - 1 \tag{5.8.8}$$

where we used the fact that $w_N{}^{-(u+N/2)} = -w_N{}^{-u}$.

Therefore, to obtain the DFT of $f(v)$ in this case, the procedure is as follows:

Form the sequences

$$f_1(v) = f(2v) \text{ and } f_2(v) = f(2v+1), \quad v = 0, 1, \ldots, \frac{N}{2} - 1.$$

Compute

$$F_1(u) = \sum_{v=0}^{(N/2)-1} f_1(v)w_{N/2}^{-uv}, \qquad u = 0, 1, \ldots, \frac{N}{2} - 1$$

and

$$F_2(u) = \sum_{v=0}^{(N/2)-1} f_2(v)w_{N/2}^{-uv}, \qquad u = 0, 1, \ldots, \frac{N}{2} - 1.$$

Compute

$$F(u) = F_1(u) + F_2(u)w_N{}^{-u}$$

and

$$F\left(u + \frac{N}{2}\right) = F_1(u) - F_2(u)w_N{}^{-u}, \qquad u = 0, 1, \ldots, \frac{N}{2} - 1.$$

The Inversion Algorithm

This algorithm assumes that a DFT sequence

$$F(u), \quad u = 0, 1, \ldots, N-1$$

is given as input and that one wishes to recover the sequence $f(v)$, $v = 0, \ldots, N-1$, that is to compute the inverse DFT

$$f(v) = \frac{1}{N} \sum_{u=0}^{N-1} F(u)w_N{}^{uv}. \tag{5.8.9}$$

Taking complex conjugates of both sides in (5.8.9) we get

$$\overline{f(v)} = \frac{1}{N} \sum_{u=0}^{N-1} \overline{F(u)}w_N{}^{-uv}, \quad v = 0, \ldots, N-1 \tag{5.8.10}$$

which may be interpreted as the DFT of the sequence

$$\left(\frac{\overline{F(u)}}{N}, \quad u = 0, \ldots, N-1\right).$$

(Here $\bar{z}$ denotes the complex conjugate of z.)

Therefore, the computational procedure to obtain (5.8.9) may be given as follows:

Form

$$\frac{\overline{F(u)}}{N}, \qquad u = 0, 1, \ldots, N-1$$

Compute

$$\overline{f(v)} = \sum_{u=0}^{N-1} \frac{\overline{F(u)}}{N} w_N^{-uv}, \qquad v = 0, 1, \ldots, N-1.$$

Form complex conjugates of $(\overline{f(v)})$, i.e.

$$f(v) = (\overline{\overline{f(v)}}), \qquad v = 0, 1, \ldots, N-1.$$

Two-at-a-time Algorithm for Real Sequences

Suppose that two data sequences $f_1(v)$ and $f_2(v)$, $v = 0, 1, \ldots, N-1$, are real, and one wishes to compute their respective DFTs. This can be done with some saving on the number of arithmetic operations by first forming a single complex sequence

$$f(v) = f_1(v) + if_2(v). \qquad v = 0, 1, \ldots, N-1, \tag{5.8.11}$$

from which we let

$$F(u) = \sum_{v=0}^{N-1} f(v) w_N^{-uv},$$

$$F_1(u) = \sum_{v=0}^{N-1} f_1(v) w_N^{-uv} \tag{5.8.12}$$

and

$$F_2(u) = \sum_{v=0}^{N-1} f_2(v) w_N^{-uv}, \quad u = 0, 1, \ldots, N-1,$$

and then observing that

$$F(u) = F_1(u) + iF_2(u), \qquad u = 0, 1, \ldots, N-1, \tag{5.8.13}$$

where $F_1(u)$ and $F_2(u)$ are conjugate even.

The complex function $f(z)$ is called *even* if $f(z) = f(-z)$; the complex function $f(z)$ is called *conjugate even* if $f(z) = \overline{f(-z)}$, which for the function periodic with period p also means that $f(z) = \overline{f(p-z)}$. In our case we, thus, have that $F_r(u) = \overline{F_r(N-u)}$ for $r = 1, 2$.

Next, taking the complex conjugate of the right-hand side of (8.5.13) and replacing u by $(N-u)$, we get

$$\begin{aligned}\overline{F_1(u)} - i\overline{F_2(u)} &= \overline{F_1(N-u)} - i\overline{F_2(N-u)} \\ &= \overline{F_1(u)} - i\overline{F_2(u)} = \overline{F(u)} = \overline{F(N-u)},\end{aligned}$$

that is

$$\overline{F(N-u)} = F_1(u) - iF_2(u), \quad u = 0, 1, \ldots, N-1. \tag{5.8.14}$$

Solving (8.5.13) and (8.5.14) for $F_1(u)$ and $F_2(u)$, we find

$$F_1(u) = \tfrac{1}{2}[\overline{F(N-u)} + F(u)]$$

$$F_2(u) = \tfrac{1}{2}[\overline{F(N-u)} - F(u)], \quad u = 0, 1, \ldots, \frac{N}{2}. \tag{5.8.15}$$

The computational procedure, therefore, may be given as:

Form

$$f(v) = f_1(v) + if_2(v), \qquad v = 0, 1, \ldots, N-1.$$

Compute the sequence

$$F(u) = \sum_{v=0}^{N-1} f(v) w_N^{-uv}, \qquad u = 0, 1, \ldots, N-1.$$

Form the sequence

$$\overline{F(N-u)}, \qquad u = 0, ,1, \ldots, \frac{N}{2}.$$

Compute

$$F_1(u) - \tfrac{1}{2}[\overline{F(N-u)} + F(u)],$$

$$F_2(u) = \tfrac{1}{2}[\overline{F(N-u)} - F(u)], \quad u = 0, 1, \ldots, \frac{N}{2}.$$

Values of $F_1(u)$ and $F_2(u)$ for $u > (N/2)$ need not be computed since the two functions are conjugate even.

5.9 Round-off Errors in the FFT

When programs of the FFT algorithm began to replace those using conventional methods, it was very soon noticed that not only was the speed improved by a factor proportional to $N/\log_2 N$, but, in general, the accuracy was much greater. The numerical stability of the FFT has been studied theoretically by several authors, all substantiating the empirical results. By comparing upper bounds, Gentleman and Sande (1966) have shown that accumulated floating-point round-off error is significantly less when one uses the FFT than when one computes the discrete Fourier transform directly, by using the DFT defining formula. Welch (1969) has derived approximate upper and lower bounds on the accumulated error in a fixed-point algorithm of radix 2. Weinstein (1969) has used a statistical model for floating-point round-off errors to predict the variance of the accumulated errors. Kaneko and Liu (1970) have also used a statistical approach and derived bounds for the mean squared error in a floating-point algorithm of radix 2. They treated both cases, the rounding and the chopping-off (truncation) of the excess digits in the value computed. Using the matrix form of the factorization, Ramos (1971) has derived approximate upper bounds for the ratios of the root-average-square (RAS) and the maximum round-off errors in the output to the RAS value of the output for both single and multidimensional transforms.

Following mainly Gentleman and Sande and to a certain extent Ramos we shall now derive upper bounds on the accumulated errors generated in computing the discrete Fourier transforms (a) using the standard technique and (b) using the FFT algorithm. For this we shall first need to express the FFT in matrix form.

Matrix Form of the FFT

Consider again the example for $N = 4 \times 3 = r \times s = 12$. We have the following input set:

$$f = \{f_0 f_1 f_2 f_3 f_4 f_5 f_6 f_7 f_8 f_9 f_{10} f_{11}\} \tag{5.9.1}$$

Partitioning the set into $s(=3)$ subsets gives

$$\{f_0 f_3 f_6 f_9\}\{f_1 f_4 f_7 f_{10}\}\{f_2 f_5 f_8 f_{11}\}. \tag{5.9.2}$$

The process by which a subset is obtained, that is starting with the q_0th element and adding to the subset every sth element thereafter, is called decimating by s. The matrix form of the set partitioned in this way is

$$\mathbf{c}^{(1)} = \begin{bmatrix} \mathbf{c}_0(q_0) \\ \\ \mathbf{c}_1(q_0) \\ \\ \mathbf{c}_2(q_0) \end{bmatrix} = \begin{bmatrix} 1 & 0 & 0 & 0 & 0 & 0 & 0 & 0 & 0 & 0 & 0 & 0 \\ 0 & 0 & 0 & 1 & 0 & 0 & 0 & 0 & 0 & 0 & 0 & 0 \\ 0 & 0 & 0 & 0 & 0 & 0 & 1 & 0 & 0 & 0 & 0 & 0 \\ 0 & 0 & 0 & 0 & 0 & 0 & 0 & 0 & 0 & 1 & 0 & 0 \\ 0 & 1 & 0 & 0 & 0 & 0 & 0 & 0 & 0 & 0 & 0 & 0 \\ 0 & 0 & 0 & 0 & 1 & 0 & 0 & 0 & 0 & 0 & 0 & 0 \\ 0 & 0 & 0 & 0 & 0 & 0 & 0 & 1 & 0 & 0 & 0 & 0 \\ 0 & 0 & 0 & 0 & 0 & 0 & 0 & 0 & 0 & 0 & 1 & 0 \\ 0 & 0 & 1 & 0 & 0 & 0 & 0 & 0 & 0 & 0 & 0 & 0 \\ 0 & 0 & 0 & 0 & 0 & 1 & 0 & 0 & 0 & 0 & 0 & 0 \\ 0 & 0 & 0 & 0 & 0 & 0 & 0 & 0 & 1 & 0 & 0 & 0 \\ 0 & 0 & 0 & 0 & 0 & 0 & 0 & 0 & 0 & 0 & 0 & 1 \end{bmatrix} \begin{bmatrix} f_0 \\ f_1 \\ f_2 \\ f_3 \\ f_4 \\ f_5 \\ f_6 \\ f_7 \\ f_8 \\ f_9 \\ f_{10} \\ f_{11} \end{bmatrix} \tag{5.9.3}$$

where $\mathbf{c}_0(q_0) = (f_0 f_3 f_6 f_9)^T$,

$\mathbf{c}_1(q_0) = (f_1 f_4 f_7 f_{10})^T$ and

$\mathbf{c}_2(q_0) = (f_2 f_5 f_8 f_{11})^T$.

The matrix in (5.9.3) is known as a permutation matrix (a permutation matrix is a square matrix of a particular type with unity and zero entries such that there is exactly one unity in each row and in each column). Let us denote this matrix by P and write

$$\mathbf{c}^{(1)} = \mathbf{Pf}. \tag{5.9.4}$$

To compute the first set of intermediate values, $c_{p_0}(q_1)$, which are given by (5.4.6):

$$c_{p_0}(q_1) = e\left(\frac{q_1 p_0}{rs}\right) \sum_{q_0=0}^{r-1} f(v) e\left(\frac{q_0 p_0}{r}\right),$$

we solve the following set of equations:

$$\begin{bmatrix} c_{p_0}(0) \\ c_{p_0}(1) \\ c_{p_0}(2) \end{bmatrix} = \begin{bmatrix} e\left(\frac{0 \times p_0}{12}\right) & & \\ & e\left(\frac{1 \times p_0}{12}\right) & \\ & & e\left(\frac{2 \times p_0}{12}\right) \end{bmatrix} \left(\begin{bmatrix} c_0(q_0) \\ c_1(q_0) \\ c_2(q_0) \end{bmatrix} \begin{bmatrix} E_0 & & \\ & E_1 & \\ & & E_2 \end{bmatrix} \right),$$

where

$$\mathbf{E}_{p_0} = \begin{bmatrix} e\left(\frac{0 \times p_0}{4}\right) & & & \\ & e\left(\frac{1 \times p_0}{4}\right) & & \\ & & e\left(\frac{2 \times p_0}{4}\right) & \\ & & & e\left(\frac{3 \times p_0}{4}\right) \end{bmatrix}$$

$p_0 = 0, 1, 2, \quad q_0 = 0, 1, 2, 3.$

The matrix form of the above set of 12 intermediate coefficients $c_{p_0}(q_1)$ can thus be given as

$$\mathbf{c}^{(2)} = \mathbf{D}_1 \mathbf{B}_1 c^{(1)} = \mathbf{D}_1 \mathbf{B}_1 \mathbf{P}_1 \mathbf{f},$$

where $\mathbf{c}^{(2)}$ is the column vector

$$(c_0(0) \ldots c_2(0) c_0(1) \ldots c_2(1) c_0(2) \ldots c_2(2) c_0(3) \ldots c_2(3))^T$$

$\mathbf{D}_1$ is the diagonal matrix of 12 twiddle factors,

$\mathbf{B}_1$ is the block-diagonal matrix whose blocks E_{p_0} are the Fourier transform of dimension $r = 4$,

$\mathbf{P}_1$ is the permutation matrix, and

$\mathbf{f}$ is column vector of input data.

The final set of the Fourier coefficients $\mathbf{F(u)}$ in our example is given as

$$\mathbf{F} = \mathbf{B}_2\mathbf{P}_2\mathbf{c}^{(2)} = (\mathbf{B}_2\mathbf{P}_2)(\mathbf{D}_1\mathbf{B}_1\mathbf{P}_1)\mathbf{f}. \tag{5.9.4}$$

In the general case of $N = r_1 \ldots r_m$ the matrix form of the discrete Fourier transform,

$$\mathbf{F}(u) = \mathbf{A}\mathbf{f}(v), \tag{5.9.5}$$

where the matrix $\mathbf{A} = \{a_{uv}\} = \mathrm{e}^{-\mathrm{i}2\pi uv/N}$, and $\mathbf{F}(u)$ and $\mathbf{f}(v)$ are column vectors of length N, is given by

$$\mathbf{F}(u) = \mathbf{P}_{m+1}(\mathbf{B}_m\mathbf{P}_m)(\mathbf{D}_{m-1}\mathbf{B}_{m-1}\mathbf{P}_{m-1}) \ldots (\mathbf{D}_1\mathbf{B}_1\mathbf{P}_1)\mathbf{f}(v) \tag{5.9.6}$$

where $\mathbf{P}_s, s = 1, \ldots, m+1$, are permutation matrices,

$\mathbf{D}_s, s = 1, \ldots, m-1$, are diagonal matrices of complex exponential twiddle factors, and

$\mathbf{B}_s, s = 1, \ldots, m$, are block-diagonal matrices whose blocks are complex Fourier transforms of dimension r_s, where $N = r_1 \ldots r_m$.

The formula (5.9.6) holds for the Gentleman–Sande algorithm as well as for the Cooley–Tukey algorithm but with different diagonal matrices, $\mathbf{D}_s$.

Error Analysis

Let $\mathbf{F}(u)$ be the *exact* Fourier transform and $\mathbf{F}(u)_{\mathrm{FT}}$ and $\mathbf{F}(u)_{\mathrm{FFT}}$ be the Fourier transforms *computed* using (5.9.5) and (5.9.6), respectively, then the errors incurred in computing the transforms by these algorithms may be given as

$$\mathbf{e}(u)_{\mathrm{FT}} = \mathbf{F}(u)_{\mathrm{FT}} - \mathbf{F}(u), \tag{5.9.7}$$

and

$$\mathbf{e}(u)_{\mathrm{FFT}} = \mathbf{F}(u)_{\mathrm{FFT}} - \mathbf{F}(u), \quad u = 0, 1, \ldots, N-1, \tag{5.9.8}$$

respectively.

We shall now show that the Euclidean norms of the vectors given by (5.9.7) and (5.9.8), i.e. the sums of squared errors, $|\mathbf{e}(u)|^2$, for $u = (0, \ldots, N-1)$, for algorithms of (5.9.5) and (5.9.6), are bounded from above as follows:

$$\frac{\|\mathbf{e}(u)_{\mathrm{FT}}\|_E}{\|\mathbf{F}(u)\|_E} \leqslant 1.06(2N)^{3/2}2^{-t} \tag{5.9.9}$$

and

$$\frac{\| \mathbf{e}(u)_{\mathrm{FFT}} \|_E}{\| \mathbf{F}(u) \|_E} \leqslant 1.06\sqrt{N}\left(\sum_s (2r_s)^{3/2}\right)2^{-t}, \tag{5.9.10}$$

where $\|\cdot\|_E$ denotes the Euclidean norm of a vector,

$N = r_1 \ldots r_m$, as before, and
t is the number of binary digits in the mantissa of the floating-point number.

Note that, in particular, from (5.9.10) it follows that if $N = r^m$, then

$$\frac{\| \mathbf{e}(u)_{\mathrm{FFT}} \|_E}{\| \mathbf{F}(u) \|_E} \leqslant 1.06\sqrt{N}m(2r)^{3/2}2^{-t}; \tag{5.9.11}$$

furthermore, if $N = 2^m$, then

$$\frac{\| \mathbf{e}(u)_{\mathrm{FFT}} \|_E}{\| \mathbf{F}(u) \|_E} \leqslant 8.5(\sqrt{N} \log_2 N)2^{-t}, \tag{5.9.12}$$

which can be compared with the corresponding case of formula (5.9.9), i.e. if $N = 2^m$ then

$$\frac{\| \mathbf{e}(u)_{\mathrm{FT}} \|_E}{\| \mathbf{F}(u) \|_E} \leqslant 3\sqrt{N}\, N 2^{-t}. \tag{5.9.13}$$

Proof of (5.5.9)

Equation (5.9.5) expressed in real arithmetic, may be written as

$$\begin{bmatrix} \mathrm{Re}(\mathbf{F}(u)) \\ \mathrm{Im}(\mathbf{F}(u)) \end{bmatrix} = \begin{bmatrix} \mathbf{C} & -\mathbf{S} \\ \mathbf{S} & \mathbf{C} \end{bmatrix} \begin{bmatrix} \mathrm{Re}(\mathbf{f}(v)) \\ \mathrm{Im}(\mathbf{f})(v)) \end{bmatrix}, \tag{5.9.14}$$

where **C** and **S** are real matrices with elements

$$c(u, v) = \cos\left(\frac{2\pi}{N}(u-1)(v-1)\right) \text{ and } s(u, v) = \sin\left(\frac{2\pi}{N}(u-1)(v-1)\right),$$

$u, v = 1, 2, \ldots, N$, and

$\mathrm{Re}(\mathbf{F}(u))$, $\mathrm{Im}(\mathbf{F}(u))$, $\mathrm{Re}(\mathbf{f}(v))$ and $\mathrm{Im}(\mathbf{f}(v))$ are the real and imaginary parts of the vectors $\mathbf{F}(u)$ and $\mathbf{f}(v)$.

Now, in terms of the norms we get

$$\| \mathbf{F}(u)_{\mathrm{FT}} - \mathbf{F}(u) \|_E = \|(\mathbf{Af}(v))_{\mathrm{FT}} - \mathbf{Af}(v)\|_E. \tag{5.9.15}$$

At this stage we need a lemma proved by Wilkinson which states that

If A is a real $p \times q$ matrix and $\mathbf{B}$ is a real $q \times r$ matrix then, for floating-point multiplication, the following holds

$$\| (\mathbf{AB})^{\mathrm{comp}} - (\mathbf{AB})^{\mathrm{exact}} \|_E \leqslant 1.06q2^{-t} \| \mathbf{A} \|_E \| \mathbf{B} \|_E, \tag{5.9.16}$$

where each term of the matrix product is the sum of q product terms. Applying the lemma to our problem and noting that $p = q = 3 = 2N$, and that the Euclidean norm of the $2N \times 2N$ real matrix

$$\begin{bmatrix} \mathbf{C} & -\mathbf{S} \\ \mathbf{S} & \mathbf{C} \end{bmatrix} \text{ is given by}$$

$$\left[\sum_{v=1}^{N}\sum_{u=1}^{N}\left(2\cos^2 (u-1)(v-1)\frac{2\pi}{N} + 2\sin^2 (u-1)(v-1)\frac{2\pi}{N}\right)\right]^{1/2} = \sqrt{2}N,$$

we get

$$\begin{aligned} \| \mathbf{e}(u)_{\mathrm{FT}} \|_E = \| \mathbf{F}(u)_{\mathrm{FT}} - \mathbf{F}(u) \|_E &\leqslant 1.06\sqrt{N}(2N)^{3/2}2^{-t}\,\|\mathbf{f}(v)\|_E \\ &\leqslant 1.06(2N)^{3/2}2^{-t}\,\|\mathbf{F}(u)\|_E, \end{aligned}$$

since

$$\|\mathbf{F}(u)\|_E = \left[\sum_{u=0}^{N-1}\left(\sum_{v=0}^{N-1} \mathrm{e}^{\mathrm{i}2\pi uv/N} f(v)\right)^2\right]^{1/2} = \sqrt{N}\,\|\mathbf{f}(v)\|_E \qquad (5.9.17)$$

This completes the proof of the error bound formula (5.9.9) for the ordinary DFT.

Proof of (5.9.10)

From equation (5.9.6) we have

$$\begin{aligned} \| \mathbf{F}(u)_{\mathrm{FFT}} - \mathbf{F}(u) \|_E = \| &\mathrm{fl}(\mathbf{P}_{m+1}\mathrm{fl}(\mathbf{B}_m\mathrm{fl}(\mathbf{P}_m\mathrm{fl}(\mathbf{M}_{m-1}\mathrm{fl}(\mathbf{M}_{m-2}\ldots\mathrm{fl}(\mathbf{M}_1\mathbf{f}(v))\underbrace{\ldots)}_{m+1 \text{ brackets}} \\ &- \mathbf{P}_{m+1}\mathbf{B}_m\mathbf{P}_m\mathbf{M}_{m-1}\mathbf{M}_{m-2}\ldots\mathbf{M}_1\mathbf{f}(v)\|_E, \end{aligned} \qquad (5.9.18)$$

where $\mathbf{M}_s$ denotes the matrix product $\mathbf{D}_s\mathbf{B}_s\mathbf{P}_s$, $s = 1, \ldots, m-1$ and fl(.) denotes as usual, the result computed in floating point arithmetic as opposed to the result computed exactly.

We shall now present relation (5.9.18) in the form:

$$\begin{aligned} \|\mathbf{e}(u)_{\mathrm{FFT}}\|_E = \|&\mathrm{fl}(\mathbf{P}_{m+1}\mathrm{fl}(\mathbf{B}_m\mathrm{fl}(\mathbf{P}_m\mathrm{fl}(\mathbf{M}_{m-1}\ldots\mathrm{fl}(\mathbf{M}_1\mathbf{f}(v))\underbrace{\ldots)}_{m+2 \text{ brackets}} \\ &- \mathbf{P}_{m+1}\mathrm{fl}(\mathbf{B}_m\mathrm{fl}(\mathbf{P}_m\mathrm{fl}(\mathbf{M}_{m-1}\ldots\mathrm{fl}(\mathbf{M}_1\mathbf{f}(v))\underbrace{\ldots)}_{m+1 \text{ brackets}} \\ &+ \mathbf{P}_{m+1}\mathrm{fl}(\mathbf{B}_m\mathrm{fl}(\mathbf{P}_m\mathrm{fl}(\mathbf{M}_{m-1}\ldots\mathrm{fl}(\mathbf{M}_1\mathbf{f}(v))\underbrace{\ldots)}_{m+1 \text{ brackets}} \\ &- \mathbf{P}_{m+1}\mathbf{B}_m\mathrm{fl}(\mathbf{M}_{m-1}\ldots\mathrm{fl}(\mathbf{M}_1\mathbf{f}(v))\underbrace{\ldots)}_{m \text{ brackets}} \\ &+ \ldots + \mathbf{P}_{m+1}\mathbf{B}_m\mathbf{P}_m\mathbf{M}_{m-1}\ldots\mathbf{M}_{s+1}\mathrm{fl}(\mathbf{M}_s\ldots\mathrm{fl}(\mathbf{M}_1\mathbf{f}(v))\underbrace{\ldots)}_{s \text{ brackets}} \end{aligned}$$

$$- P_{m+1}\mathbf{B}_m\mathbf{P}_m\mathbf{M}_{m-1}\ldots\mathbf{M}_{s+1}\mathbf{M}_s\mathrm{fl}(\mathbf{M}_{s+1}\ldots\mathrm{fl}(\mathbf{M}_1\mathbf{f}(v))\underbrace{\ldots)}_{s-1\text{ brackets}}$$

$$+\ \ldots + \mathbf{P}_{m+1}\mathbf{B}_m\mathbf{P}_m\mathbf{M}_{m-1}\ldots\mathbf{M}_s\ldots\mathbf{M}_2\mathrm{fl}(\mathbf{M}_1\mathbf{f}(v))$$

$$- \mathbf{P}_{m+1}\mathbf{B}_m\mathbf{P}_m\mathbf{M}_{m-1}\ldots M_s\ldots\mathbf{M}_2\mathbf{M}_1\mathbf{f}(v)\|_E. \qquad (5.9.19)$$

In order to proceed we shall need the following information:

(i) For any matrix $\mathbf{A}$ and any vector $\mathbf{x} \neq \mathbf{0}$ we have:

$$\|\mathbf{Ax}\|_E \leqslant \|\mathbf{A}\|_s \|\mathbf{x}\|_E \qquad (5.9.20)$$

where the subscripts E and s stand for the Euclidean vector and the spectral matrix norms, respectively. (The spectral norm of a matrix $\mathbf{A}$ is defined as $\|\mathbf{A}\|_s = [\text{modulus of the maximum eigenvalue of } (\mathbf{A}^*\mathbf{A})]^{1/2}$).

(ii) The spectral norms of $\mathbf{D}_s$, $\mathbf{B}_s$ and $\mathbf{P}_s$ are 1, $\sqrt{r_s}$ and 1, respectively, since $\mathbf{D}_s^*\mathbf{D}_s = \mathbf{I}$, $\mathbf{B}_s^*\mathbf{B}_s = r_s\mathbf{I}$ and $\mathbf{P}_s^*\mathbf{P}_s = \mathbf{I}$, where $\mathbf{I}$ is the $N \times N$ identity matrix and $r_s\mathbf{I}$ denotes the complex block in the block-diagonal matrix $\mathbf{B}_s$.

(iii) The matrices $\mathbf{P}_s$, $s = 1, \ldots, m+1$ are the permutation matrices and, as it has been noted earlier, simply reorder the vector values, introducing no round-off errors.

By virtue of (i), (ii), and (iii) we can rewrite expression (5.9.19) as

$$\|\mathbf{e}(u)_{\mathrm{FFT}}\|_E \leqslant \sqrt{r_1 r_2 \ldots r_m}\sum_{s=1}^{m}\|\mathrm{fl}(\mathbf{M}_s\mathrm{fl}(\mathbf{M}_{s-1}\ldots\mathrm{fl}(\mathbf{M}_1\mathbf{f}(v))\underbrace{\ldots)}_{s\text{ brackets}}$$

$$- \mathbf{M}_s\mathrm{fl}(\mathbf{M}_{s-1}\ldots\mathrm{fl}(\mathbf{M}_1\mathbf{f}(v))\underbrace{\ldots)}_{s-1\text{ brackets}}\|_E \qquad (5.9.21)$$

where matrix $\mathbf{M}_m$ denotes the product $\mathbf{B}_m\mathbf{P}_m$ and $r_1r_2\ldots r_m = N$, as before.

We now estimate the bound for the round-off error incurred during the calculations involving matrix $\mathbf{M}_s$, $s = 1, \ldots, m$, i.e.

$$\|\mathrm{fl}(\mathbf{M}_s\mathbf{z}_{s-1}) - \mathbf{M}_s\mathbf{z}_{s-1}\|_E$$

where $\mathbf{z}_{s-1}$ is an N-length vector.

We recall that matrix $\mathbf{M}_s = \mathbf{D}_s\mathbf{B}_s\mathbf{P}_s$ may be partitioned into N/r_s disjoint complex blocks and thus, the error from each block can be bounded separately. Since in real arithmetic each block is $2r_s$ square and since the Euclidean norm of each block is $\sqrt{2r_s}$, we get the error bound given by

$$1.06(2r_s)\sqrt{2\mathbf{r}_s}\,2^{-t}\|\mathbf{z}_{s-1,j}\|_E$$

where the vector $\mathbf{z}_{s-1,j}$ denotes the appropriate part of the vector $\mathbf{z}_{s-1}$.

For the complete matrix $\mathbf{M}_s$, it follows:

$$\|\mathrm{fl}(\mathbf{M}_s\mathbf{z}_{s-1}) - \mathbf{M}_s\mathbf{z}_{s-1}\|_E$$

$$\leqslant 1.06(2r_s)\sqrt{2r_s}\,2^{-t}(\|\mathbf{z}_{s-1,1}\|_E + \|\mathbf{z}_{s-1,2}\|_E + \ldots + \|\mathbf{z}_{s-1,N/r_s}\|_E)$$

$$= 1.06(2r_s)^{3/2}\|\mathbf{z}_{s-1}\|_E. \qquad (5.9.22)$$

Relation (5.9.22) may now be used to obtain the bounds for the right-hand side terms in (5.9.21). So, for the matrix $\mathbf{M}_1$ we get

$$\| \mathrm{fl}(\mathbf{M}_1\mathbf{f}(v)) - \mathbf{M}_1\mathbf{f}(v) \|_E \leqslant 1.06(2r_1)^{3/2} \| \mathbf{f}(v)\|_E. \tag{5.9.23}$$

Next, denoting $\mathbf{M}_1\mathbf{f}(v)$ by $\mathbf{z}_1$, for the matrix $\mathbf{M}_2$ we obtain:

$$\begin{aligned} \| \mathrm{fl}(\mathbf{M}_2\mathrm{fl}(\mathbf{z}_1)) - \mathbf{M}_2\, \mathrm{fl}(\mathbf{z}_1))\|_E &\leqslant 1.06(2r_2)^{3/2} \| \mathrm{fl}(\mathbf{z}_1)\|_E \\ &\leqslant 1.06(2r_2)^{3/2}\sqrt{r_1}\, \| f(v)\|_E, \end{aligned} \tag{5.9.24}$$

where we first assume that

$$\| \mathrm{fl}(\mathbf{z}_1)\|_E = \| \mathbf{z}_1 \|_E$$

and then use the bound

$$\| \mathbf{z}_1 \|_E \leqslant \|\mathbf{M}_1 \|_s \| \mathbf{f}(v)\|_E = \sqrt{r_1}\, \| \mathbf{f}(v)\|_E.$$

Continuing in this manner, we obtain estimates for each term of the sum on the right-hand side of the expression (5.9.21). Then, using these estimates we can write:

$$\begin{aligned} \| \mathbf{e}(u)_{\mathrm{FFT}} \|_E &\leqslant 1.06\sqrt{N} \left(\sum_{s=1}^{m} (2r_s)^{3/2} r_1 \ldots r_{s-1}\right) 2^{-t} \| \mathbf{f}(v)\|_E \\ &\leqslant 1.06\sqrt{N} \left(\sum_{s=1}^{m} (2r_s)^{3/2}\right) 2^{-t}\sqrt{N}\, \| \mathbf{f}(v)\|_E \\ &\leqslant 1.06\sqrt{N} \left(\sum_{s=1}^{m} (2r_s)^{3/2}\right) 2^{-t} \| \mathbf{F}(u)\|_E. \end{aligned} \tag{5.9.25}$$

This completes the proof of formula (5.9.10).

The upper bounds on the accumulated errors given by equations (5.9.9) and (5.9.10) indicate that, in general, the errors incurred in the final transforms computed using the FFT algorithm are smaller than the errors incurred in the computations involving the ordinary DFT algorithm.

A more careful analysis of the accumulated errors associated with the use of the FFT algorithm is due to Ramos (1970). In this analysis the errors in the elements of the matrices involved, i.e. the errors in the floating-point computations of sines and cosines, are considered directly. The upper bound on the errors derived is significantly lower than the bound given by (5.9.10). In terms of the Euclidean norms this bound may be expressed as

$$\frac{\| \mathbf{e}(u)_{\mathrm{FFT}} \|}{\| \mathbf{F}(u) \|_E} < \left[\sum_{s-1}^{m} 2\sqrt{r_s}(r_s + \gamma) + (m-1)(3 + 2\gamma)\right] 2^{-t}, \tag{5.9.26}$$

where γ is an absolute error constant, $\gamma \geqslant 0$, such that

$$\mathrm{fl}(\sin\, \mathrm{fl}(x)) = \sin \mathbf{x} + \gamma 2^{-t},$$
$$\mathrm{fl}(\cos\, \mathrm{fl}(\mathbf{x})) = \cos \mathbf{x} + \gamma 2^{-t}.$$

For $N = 2^m$ this bound is reduced to

$$\frac{\| \mathbf{e}(u)_{\text{FFT}} \|_E}{\| \mathbf{F}(u) \|_E} < [\sqrt{2} \log_2 N + (\log_2 N - 1)(3 + 2\gamma)] 2^{-t}. \tag{2.9.27}$$

In computational experiments reported by Ramos, the constant γ takes on the values in the range $1.5 < \gamma < 10$.

Kaneko and Liu (1970) studied the Gentleman–Sande FFT algorithm of radix 2 under the assumptions that the round-off errors are random variables uniformly distributed in the interval $(-2^{-t}, 2^{-t})$ and are independent of the numbers in the rounding of which they occur. Under these assumptions the authors established that the sum of the mean squared errors is bounded by

$$\frac{1}{2}\sqrt{\log_2 N}\, 2^{-t} < \frac{\left[\sum_{u=0}^{N-1} \mu[\,|\,\mathbf{e}(u)_{\text{FFT}}\,|^2]\right]^{1/2}}{\| \mathbf{F}(u) \|_E} < \sqrt{\log_2 N}\, 2^{-t}. \tag{2.9.28}$$

Formula (2.9.28) shows that the total relative mean square error in the Fourier transforms computed using the FFT radix 2 algorithm, may be expected to increase with N increasing, at most, as $m = \log_2 N$.

5.10 Conclusions

The FFT algorithm makes a remarkable improvement in the practicality with which one can apply discrete Fourier transforms in numerical calculation. In this chapter the algorithm has been discussed in detail. Its merits in terms of the numerical accuracy have also been exposed. Our aim was to present the basic principles of efficient computation of the DFT. With the material presented in this chapter, there should be little difficulty in programming an FFT radix 2 algorithm.

Exercises

1. Let $f(v)$ and $h(v)$ be periodic discrete functions:

$$f(v) = \begin{cases} 1, & v = 0, 4, \\ 2, & v = 1, 2, 3, \\ 0, & v = 5, 6, 7, \end{cases}$$

$$f(v + 8r) = f(v), \quad r = 0, \pm 1, \pm 2, \ldots,$$

$$h(v) = f(v),$$

$$h(v + 8r) = h(v), \quad r = 0, \pm 1, \pm 2, \ldots$$

Let further $g(v) = f(v) - h(v - 4)$.

(a) Compute the discrete Fourier transforms, $F(u)$, $H(u)$, and $G(u)$, of the functions $f(v)$, $h(v)$ and $g(v)$;

(b) Demonstrate the (discrete) convolution theorem using $f(v)$ and $h(v)$;
(c) Compute the inverse DFT of $F(u)$, $H(u)$ and $G(u)$.

2. Prove the following convolution properties:
   (a) convolution is commutative: $(h(v)f(v)) = (f(v)h(v))$;
   (b) convolution is associative: $h(v)[g(v)f(v)] = [h(v)g(v)f(v)]$;
   (c) convolution is distributive over addition:
   $h(v)[g(v)+f(v)] = h(v)g(v) + h(v)f(v)$.
3. In formula (5.4.1) assume that $N = r_1r_2r_3r_4$ and that the parameters u and v are expressed in the form

$$u = p_0 + p_1r_1 + p_2r_1r_2 + p_3r_1r_2r_3 \text{ with } p_0 = 0, 1, \ldots, r_1 - 1,$$
$$p_1 = 0, 1, \ldots, r_2 - 1,$$
$$p_2 = 0, 1, \ldots, r_3 - 1,$$
$$p_3 = 0, 1, \ldots, r_4 - 1,$$

$$v = q_0 + q_1r_4 + q_2r_4r_3 + q_3r_4r_3r_2 \text{ with } q_0 = 0, 1, \ldots, r_4 - 1,$$
$$q_1 = 0, 1, \ldots, r_3 - 1,$$
$$q_2 = 0, 1, \ldots, r_2 - 1,$$
$$q_3 = 0, 1, \ldots, r_1 - 1.$$

   Derive the FFT algorithm for the case where the exponential function is factorized in the first instance, in terms of the parameters p_j, $j = 0, 1, 2, 3$.
4. Write an algorithm *bitreverse* which would accept as input a number $K = 2^kb_0 + 2^{k-1}b_1 + \ldots + b_k = (b_0b_1 \ldots b_k)_2$ and compute the number $R = (b_kb_{k-1} \ldots b_0)_2 = 2^kb_k + 2^{k-1}b_{k-1} + \ldots + b_0$.
5. Consider two polynomials $P_n(x) = \sum_{k=0}^{n} a_{n-k}x^k$ and $Q_m(x) = \sum_{k=0}^{m} b_{m-k}x^k$ with real coefficients and of degree n and m, respectively. Show that the product $W_{n+m}(x) = P_n(x)Q_m(x)$ can be computed in $O((n+m)\log(n+m))$ arithmetic operations.
6. Derive the basic 'butterfly' operation for the Gentleman–Sande FFT algorithm.
7. Obtain an expression similar to (5.7.10) for the kth step of the Gentleman-Sande algorithm of radix 2.
8. Suppose that a real N-point sequence $f(v)$, $v = 0, 1, \ldots, N-1$ is given with N even. Show how the computation of the DFT sequence

$$F(u) = \sum_{v=0}^{N-1} f(v)w_N^{-uv}, \quad u = 0, 1, \ldots, N-1$$

   can be arranged with the use of the two-at-a-time and doubling algorithms and assert that in this way the computation time required is halved as compared with the use of direct DFT procedure.

9. Develop a flowgraph of the Gentleman-Snade FFT algorithm for $N = 2^3$.
10. Show that
    (a) The DFT of the real even sequence $f(v)$, $v = 0, 1, \ldots, N-1$, where N is an even integer, is equivalent to the cosine transform of the real sequence $f(v)$, $v = 0, 1, \ldots, N/2$;
    (b) The DFT in this case is also real and even and can be related to the coefficients of the cosine series

$$f(v) = \sum_{u=0}^{N/2} \alpha(u) \cos\left(\frac{\pi u v}{N}\right)$$

    by the formulae

$$\alpha(u) = F(u)/N, \qquad u = 0, N/2,$$
$$\alpha(u) = 2F(u)/N, \qquad u = 1, 2, \ldots, (N/2) - 1;$$

    (c) using the results obtained in (a) and (b), deduce a computational procedure for obtaining the DFT of the N-point real even sequence with N being an even integer.
11. Develop the FFT algorithm of radix '4 + 2' for the case $N = 8$.
12. Develop a computation procedure for DFT of a $2N$-point function by means of an N-point transform.
13. Find a polynomial whose values at $1, 2, 3, 4$ are, respectively, $1, 2, 2, 1$.
14. Let two $(N-1)$st-degree polynomials, $P(x) = \sum_{i=0}^{N-1} a_i x^i$ and $Q(x) = \sum_{i=0}^{N-1} b_i x^i$ be represented by their values at N points x_j, $j = 0, \ldots, N-1$. For their product $R(x) = P(x)Q(x)$ we then have $R(x_j) = \sum_{i=0}^{2N-2} c_i x_j^i = P(x_j)Q(x_j)$, $j = 0, \ldots, N-1$. Show that the vector $\mathbf{c} = (c_0, c_1, \ldots, c_{2N-2})$ can be obtained as the inverse Fourier transform of the componentwise product of the Fourier transforms of the vectors $\mathbf{a} = (a_0, a_1, \ldots, a_{N-1})$ and $\mathbf{b} = (b_0, b_1, \ldots, b_{N-1})$.(*Hint.* Consider $P(x)$ and $Q(x)$ as $(2N-2)$ st-degree polynomials where the coefficients of the $(N-1)$st highest powers of x are zero.)

6

Fast Multiplication of Numbers

Fast multiplication of numbers is another interesting problem in the context of the minimum computation time for functions. It is of practical and theoretical interest in the design of efficient algorithms. We shall consider the fastest available algorithm for multiplication of two numbers. The algorithm uses the convolution of two vectors which is obtained by means of the FFT. The development of the ideas on this algorithm follows mainly the exposition in Knuth (1969) and Borodin and Munro (1975).

6.1 On the Minimum Computation Time of Functions

Of the four basic arithmetic operations, addition, subtraction, multiplication and division, the latter two are the most time consuming. Given two integers of, say, length n and m, respectively, we can easily observe that using well-known conventional methods it takes longer to multiply or divide the integers than to add or subtract them.

Let the following primitive operations be defined:

(A) addition or subtraction of two one-digit integers, giving a one-digit answer and a carry,
(B) multiplication of a one-digit integer by another one-digit integer, giving a two-digit answer,
(C) division of a two-digit integer by a one-digit integer, provided that the quotient is a one-digit integer, and yielding also a one-digit remainder.

To add two integers using conventional methods requires the number of primitive operations proportional to the sum of the number of digits in the addends, $0(n+m)$, and to multiply two integers the number, proportional to the product of the number of digits in the two factors, $0(nm)$. Thus in terms of primitive operations it takes longer to multiply two numbers than to add them. It turns out however that the conventional method of multiplication is by far not the best possible.

Without loss of generality, assume that both integers to be multiplied are integers of the same length n, and consider the time function $T(n)$ of different multiplication algorithms. We shall use the term 'unit of time' in the sense of being equivalent to 'one primitive operation' of the stated type.

Toom (1963) and Schönhage (1966) have each developed a different multiplication algorithm which reduces the number of required primitive operations from n^2 (as in conventional method) to $n^{1+\varepsilon}$, for arbitrary small $\varepsilon > 0$. Schönhage has shown that his method can be executed (using a multitape Turing machine model) to multiply in a time proportional to

$$n^{1+(\sqrt{2}+\varepsilon)}/(\log_2 n)^{1/2},$$

and Cook (1966) proved the same result for Toom's method. Just how much further the bound can be reduced remains an open question.

Karatsuba and Ofman (1962) stated that multiplication of two n-bit binary numbers on an automata can be carried out in $O(n \log n)$, but they gave no proof of this statement.

We shall now show that for large enough n, the product of two n-digit numbers can be computed in $O(n \log_2 n \log_2 \log_2 n)$ time. For convenience, assume that we are working with the integers expressed in binary notation. Also, to complete the mathematical model which will be assumed throughout in the following, in addition to the operations (A), (B) and (C) above we define the fourth primitive operation:

(D) shifting a single binary digit left or right d positions, and $d > 0$.

An Initial Reduction

Given two binary numbers each of $n = 2q$ bits,

$$u = (u_{n-1}u_{n-2} \ldots u_0)_2 \text{ and } v = (v_{n-1}v_{n-2} \ldots v_0)_2,$$

to obtain their product by conventional method would take $O(n^2) = O(4q^2)$ primitive operations Now write

$$u = 2^q U_1 + U_0, \quad v = 2^q V_1 + V_0, \tag{6.1.1}$$

where $U_1 = (u_{n-1}u_{n-2} \ldots u_q)_2$ is the 'most significant half' of u and $U_0 = (u_{q-1}u_{q-2} \ldots u_0)_2$ is the 'least significant half' and, similarly,

$$V_1 = (v_{n-1}v_{n-2} \ldots v_q)_2 \text{ and } V_0 = (v_{q-1}v_{q-2} \ldots v_0)_2.$$

For the product we have

$$uv = (2^q U_1 + U_0)(2^q V_1 + V_0) = (2^{2q} + 2^q)U_1V_1 + 2^q(U_1 - U_0)(V_0 - V_1) + (2^q + 1)U_0V_0. \tag{6.1.2}$$

The computation has now been reduced to three multiplications of q-bit numbers $U_1V_1, (U_1 - U_0)(V_0 - V_1)$ and U_0V_0, plus some simple shifting, and adding operations. We thus have got an algorithm to multiply two binary numbers of $n = 2q$-bits, with the time function of $O(3q^2)$. The method was originated by Karatsuba and Ofman (1962) and Ofman (1962).

It is not difficult to see that formula (6.1.2) defines a recursive process for multiplication of two numbers which for large n is significantly faster than conventional methods.

If $T(n)$ is the number of primitive operations required to multiply two n-digit binary numbers then since shifting and adding can be done in $O(n)$ operations, by (6.1.2) we have

$$T(n) \leqslant 3T(n/2) + cn \text{ for some constant } c. \tag{6.1.3}$$

We shall now show by induction that (6.1.3) implies

$$T(2^m) \leqslant c(3^m - 2^m),\ m \geqslant 1. \tag{6.1.4}$$

Proof

Consider the multiplying scheme (6.1.2) applied to two 2-bit binary integers, $b = (b_1b_0)_2 = 2b_1 + b_0$ and $c = (c_1c_0)_2 = 2c_1 + c_0$, where b_1, b_0, c_1 and c_0 are equal to either 1 or 0. For their product we have

$$bc = (2^2 + 2)b_1c_1 + 2(b_1 - b_0)(c_0 - c_1) + (2 + 1)b_0c_0.$$

The computation formula calls for three primitive multiplications on b_1c_1, $(b_1 - b_0)(c_0 - c_1)$ and b_0c_0, and some shifting and primitive additions. At this stage we are not interested in the actual number of either shifting or addition operations, we simply denote their total number by, say, c. Hence we can write

$$T(2) \leqslant 3T(1) + c, \tag{6.1.5}$$

which shows that (6.1.3) holds for $n = 2$. Now choose constant c to be 'somewhat larger' than its original value in (6.1.5) and such that

$$T(2) \leqslant c. \tag*{(6.1.5)}$$

The inequality obtained can also be written as

$$T(2) \leqslant c(3^1 - 2^1).$$

Now assuming that

$$T(2^k) \leqslant c(3^k - 2^k)$$

holds for some k, $k \geqslant 1$, and using (6.1.3) with $n = 2^{k+1}$ we obtain

$$\begin{aligned} T(2^{k+1}) &\leqslant 3T(2^k) + c2^k \leqslant 3c(3^k - 2^k) + c2^k \\ &= c3^{k+1} - 3c2^k + c2^k = c(3^{k+1} - 2^{k+1}). \end{aligned} \tag{6.1.6}$$

This completes the proof.

Formula (6.1.6) can also be written as

$$T(n) \leqslant T(2^{\lceil \log n \lfloor}) \leqslant c(3^{\lceil \log n \rceil} - 2^{\lceil \log n \rceil}) \leqslant 3c3^{\log n} = 3cn^{\log 3} = 3cn^{1.585}, \tag{6.1.7}$$

and it shows that the recursive process (6.1.2) for multiplying two numbers is of $O(n^{1.585})$. For large n this is a considerable improvement over the standard method.

6.2 The Schönhage–Strassen Algorithm for Multiplying Two Integers

Generalization of the Initial Reduction

The time to multiply two numbers can be reduced still further as n approaches infinity. To see this, in the initial reduction method the operands u and v are split into $(r+1)$ parts of equal size q for a fixed $r \geqslant 1$. Then the time function $T(n)$ yields

$$T((r+1)q) \leqslant (2r+1)T(q) + cq \text{ for a fixed } r.$$

This general method is obtained as follows.
Let

$$u = (u_{(r+1)q-1}u_{(r+1)q-2} \ldots u_q u_{q-1} \ldots u_0)_2$$

and

$$v = (v_{(r+1)q-1}v_{(r+1)q-2} \ldots v_q v_{q-1} \ldots v_0)_2$$

be broken into $r+1$ parts,

$$\begin{aligned} u &= U_r 2^{qr} + U_{r-1}2^{q(r-1)} + \ldots + U_1 2^q + U_0 = \sum_{j=0}^{r} U_j 2^{qj}, \\ v &= V_r 2^{qr} + V_{r-1}2^{q(r-1)} + \ldots + V_1 2^q + V_0 = \sum_{j=0}^{r} V_j 2^{qj}, \end{aligned} \tag{6.2.1}$$

where each U_j and each V_j is a q-bit integer.
Form the polynomials

$$P_u(x) = \sum_{j=0}^{r} U_j x^j \quad \text{and} \quad P_v(x) = \sum_{j=0}^{r} V_j x^j, \text{ both of degree } r.$$

From (6.2.1) it is obvious that

$$u = P_u(2^q), \; v = P_v(2^q)$$

and therefore

$$w = uv = P_u(2^q)P_v(2^q). \tag{6.2.2}$$

The method we seek for computing uv may now be described with reference to formulae (6.2.1) and (6.2.2). We seek an efficient way to compute the coefficients of the polynomial

$$W(x) = P_u(x)P_v(x) = W_{2r}x^{2r} + \ldots + W_1 x + W_0. \tag{6.2.3}$$

First note that the coefficients W_s, $s = 0, 1, \ldots, 2r$ form a sequence of integers obtained from two sequences of integers $\{U_0, U_1, \ldots, U_r\}$ and $\{V_0, V_1, \ldots, V_r\}$:

$$
\begin{aligned}
W_0 &= U_0V_0, \\
W_1 &= U_0V_1 + U_1V_0, \\
W_2 &= U_0V_2 + U_1V_1 + U_2V_0, \\
&\vdots \\
W_r &= U_0V_r + U_1V_{r-1} + \ldots + U_{r-1}V_1 + U_rV_0, \qquad (6.2.4) \\
W_{r+1} &= U_1V_r + U_2V_{r-1} + \ldots + U_rV_1, \\
W_{r+2} &= U_2V_r + \ldots + U_rV_2, \\
&\vdots \\
W_{2r} &= U_rV_r
\end{aligned}
$$

We shall refer to the W_j's as the product coefficients.

Recalling the convolution theorem of Section 5.3 we see that it can be used to compute the product coefficients W_j and the computation process would entail:

(i) calculation of the respective Fourier transforms of the sequences $(U_0, \ldots, U_r)$ and $(V_0, \ldots, V_r)$,

$$
\text{i.e.}\quad F_k = \sum_{s=0}^{r} U_s \mathrm{e}^{-\mathrm{i}sk(2\pi/(r+1))},\ k = 0, \ldots, r,
$$

$$
G_k = \sum_{s=0}^{r} V_s \mathrm{e}^{-\mathrm{i}sk(2\pi/(r+1))},\ k = 0, \ldots, r,\ \mathrm{i} = \sqrt{-1}. \qquad (6.2.5)
$$

(ii) calculation of the product of the Fourier transforms obtained in (i), by componentwise multiplication,

i.e. $F_0G_0, F_1G_1, \ldots, F_rF_r$. (6.2.6)

(iii) calculation of the inverse Fourier transform of the products,

$$
\begin{aligned}
\text{i.e.}\quad W_0 &= F_0G_0\mathrm{e}^{\mathrm{i}t0(2\pi/1)}, && t = 0, \\
W_1 &= F_0G_0\mathrm{e}^{\mathrm{i}t0(2\pi/1)} + F_1G_1\mathrm{e}^{\mathrm{i}t1(2\pi/2)}, && t = 1, \\
&\vdots \\
W_r &= F_0G_0\mathrm{e}^{\mathrm{i}t0(2\pi/1)} + F_1G_1\mathrm{e}^{\mathrm{i}t1(2\pi/2)} + \ldots + F_rG_r\mathrm{e}^{\mathrm{i}tr(2\pi/(r+1))}, && t = r \\
W_{r+1} &= F_1G_1\mathrm{e}^{\mathrm{i}t1(2\pi/2)} + \ldots + F_rG_r\mathrm{e}^{\mathrm{i}tr(2\pi/(r+1))}, && t = r+1, \\
&\vdots \\
W_{2r} &= F_rG_r\mathrm{e}^{\mathrm{i}tr(2\pi/(r+1))} && t = 2r.
\end{aligned}
$$

(6.2.7)

Assuming that the FFT algorithm is used to compute Fourier transforms required in the above process we can expect that by virtue of the economies in the number of arithmetic operations associated with the use of the FFT the method outlined in (i) to (iii) is an efficient way to compute the product coefficients W_j's. However the use of the convolution theorem, the modular arithmetic and the properties of the integer numbers offers even more striking economies in computing the product coefficients. In the next section we discuss the concepts of modular arithmetic and its suitability for developing some fast algorithms.

Modular (or Residue) Arithmetic

Modular (or residual) arithmetic is an alternative to conventional arithmetic and in some situations it is preferable for use on large integer numbers. Instead of working directly with the number q, one uses several 'moduli' $p_1, p_2, \ldots, p_r$ which contain no common factors, and works with 'residues' $q \bmod p$, $q \bmod p_2, \ldots, q \bmod p_r$. We shall employ the following notation for the 'residues':

$$q_1 = q \bmod p_1, \ldots, q_r = q \bmod p_r.$$

The sequence $(q_1, q_2, \ldots, q_r)$ can easily be computed for an integer number q by means of divisions. Provided q is an integer within certain bounds determined by the capacity of a particular computer, no information is lost in the process of obtaining the residues as q can be computed from $(q_1, q_2, \ldots, q_r)$. More precisely, assuming q to be in the range $0 \leqslant q < p_1 p_2 \ldots p_r$ (sometimes it is more convenient to consider a completely symmetric range, $-\frac{1}{2}p_1 p_2 \ldots p_r < q < \frac{1}{2}p_1 p_2 \ldots p_r$), the representation $(q_1, \ldots, q_r)$ is unique and no other integer s, $0 \leqslant s < p_1 p_2 \ldots p_r$ will have the same residues as q. This fact is a corollary of the 'Chinese Remainder Theorem' which states that

> If $p = p_1 p_2 \ldots p_r$, where $p_1, p_2, \ldots, p_r$ are positive integers that are relatively prime in pairs, and $a, q_1, q_2, \ldots q_r$ are integers, then there is exactly one integer q that satisfies the conditions
>
> $$a \leqslant q < a + p, \text{ and } q \equiv q_j \bmod p_j \text{ for } 1 \leqslant j \leqslant r. \tag{6.2.8}$$

Proof

Assume that there are two integers q and s, $q \neq s \bmod p$ which satisfy the conditions

$$a \leqslant q < a + p,\ q \equiv q_j \bmod p_j,\ j = 1, \ldots, r$$

$$a \leqslant s < a + p,\ s \equiv q_j \bmod p_j,\ j = 1, \ldots, r.$$

Then $q = q_j + c_j p_j$, where c_j is some integer constant

$$\text{and } s \equiv (q - c_j p_j) \bmod p_j = q \bmod p_j.$$

Hence, $s-q$ is a multiple of p_j for all j and, since the $p_j, j=1, \ldots, r$, are relatively prime, $s-q$ is a multiple of $p=p_1p_2\ldots p_r$. It follows that $q=s \bmod p$. This contradicts the assumption made initially and shows that the problem (6.2.8) has at most one solution.

It remains to show the existence of at least one solution. We shall not dwell upon a strict proof of this second half of the theorem which may be found for example in Knuth (1969) but rather describe a practically usable method to convert from $(q_1, \ldots, q_r)$ to q. One such method was suggested by Garner.

Garner's method for conversion from $(q_1, q_2, \ldots, q_r)$ to q:

Compute constants c_{ij}, $1 \leqslant i < j \leqslant r$, such that

$$c_{ij}p_i \equiv 1 \bmod p_j. \tag{6.2.9}$$

The c_{ij} is called the multiplicative inverse of the p_i, modulo p_j and is sometimes expressed as

$$c_{ij} = p_i^{-1} \bmod p_j.$$

(For our purposes here we shall make use of only multiplicative inverses that are equal to unity and refer the reader for methods of computing c_{ij} in a general case to Knuth (1969).)

Set

$$\begin{aligned}
a_1 &= q_1 \bmod p_1,\\
a_2 &= (q_2 - a_1)c_{12} \bmod p_2,\\
a_3 &= ((q_3 - a_1)c_{13} - a_2)c_{23} \bmod p_3,\\
&\vdots\\
a_r &= (\ldots((q_r - a_1)c_{1r} - \ldots - a_{r-1})c_{r-1,r} \bmod p_r,
\end{aligned} \tag{6.2.10}$$

then

$$q = a_1p_{r-1}\ldots p_1 + a_{r-1}p_{r-2}\ldots p_1 + a_3p_2p_1 + a_2p_1 + a_1 \tag{6.2.11}$$

is the number satisfying the conditions

$$0 \leqslant q < p, \quad q \equiv q_j \bmod p_j: \quad 1 \leqslant j \leqslant r.$$

Example 6.2.1

Given is number $q=51$ and a set of mutually prime numbers

$$p_1 = 7,\ p_2 = 13,\ p_3 = 17 \text{ and } p_4 = 29.$$

Compute the residues

$$q_1 = 51 \bmod 7 = 2, \qquad q_3 = 51 \bmod 17 = 0,$$

$$q_2 = 51 \bmod 13 = 12, \qquad q_4 = 51 \bmod 29 = 22.$$

A 'modular representation' of the number 51 in this case is then given as (2, 12, 0, 22).
Suppose now we wish to recover the number 51 from its modular representation (2, 12, 0,22).

For this, we first compute the constants

$$c_{ij}, 1 \leqslant i < j \leqslant 4$$

from the equations

$$\begin{aligned}
&7c_{12} = 1 \bmod 13, \\
&7c_{13} = 1 \bmod 17, \quad 13c_{23} = 1 \bmod 17, \\
&7c_{14} = 1 \bmod 29, \quad 13c_{24} = 1 \bmod 29, \quad 17c_{34} = 1 \bmod 29,
\end{aligned}$$

and obtain

$$\begin{aligned}
&c_{12} = 2, \\
&c_{13} = 5, \quad c_{23} = 4, \\
&c_{14} = 25, \; c_{24} = 9, \; c_{34} = 12.
\end{aligned}$$

Next we compute

$$\begin{aligned}
&a_1 = 2 \bmod 7 - 2, \\
&a_2 = ((12 - 2)2) \bmod 13 = 7, \\
&a_3 = (((0 - 2)5 - 7)4) \bmod 17 = 0, \\
&a_4 = ((((22 - 2)25 - 7)9 - 0)12) \bmod 29 = 0.
\end{aligned}$$

Finally, we have

$$q = a_4 p_3 p_2 p_1 + a_3 p_2 p_1 + a_2 p_1 + a_1 = 51.$$

The sequence $(q_1, q_2, \ldots, q_r)$ may, thus, be considered as a new type of internal computer representation for the number q. It is called a 'modular representation'. Given q as $(q_1, q_2, \ldots, q_r)$ and s as $(s_1, s_2, \ldots, s_r)$ and assuming single-precision arithmetic, we have
for addition

$$(q_1, q_2, \ldots, q_r) + (s_1, s_2, \ldots, s_r) = ((q_1 + s_1) \bmod p_1, \ldots, (q_r + s_r) \bmod p_r) \tag{6.2.12}$$

for subtraction

$$(q_1, q_2, \ldots, q_r) - (s_1, s_2, \ldots, s_r) = ((q_1 - s_1) \bmod p_1, \ldots, (q_r - s_r) \bmod p_r) \tag{6.2.13}$$

for multiplication

$$(q_1, q_2, \ldots, q_r)(s_1, s_2, \ldots, s_r) = (q_1 s_1 \bmod p_1, \ldots, q_r s_r \bmod p_r), \tag{6.2.14}$$

where

$$(q_j + s_j) \bmod p_j = \begin{cases} q_j + s_j, & \text{if} \quad q_j + s_j < p_j, \\ q_j + s_j - p_j, & \text{if} \quad q_j + s_j \geqslant p_j, \end{cases} \tag{6.2.15}$$

and

$$(q_j + s_j) \bmod p_j = \begin{cases} q_j - s_j, & \text{if} \quad q_j - s_j \geqslant 0, \\ q_j - s_j + p_j, & \text{if} \quad q_j - s_j < 0, \end{cases} \qquad (6.2.16)$$

$q_j s_j \bmod p_j$ is formed by first multiplying $q_j s_j$ and then dividing by p_j.

It is usually convenient to let p_1 be the largest odd number that fits in a computer word, and p_2 be the largest odd number smaller than p_1, that is relatively prime to p_1, and so on until enough p_j's have been formed to give the desired range p.

In the sequel n-bit numbers, modulo $(2^n + 1)$, will be used. Such numbers lie in the range 0 to 2^n. The numbers in the range 0 to $2^n - 1$ each require n bits for their representation while the number 2^n needs $n + 1$ bits. For convenience this number will be represented by a special symbol, -1, that is

$$2^n = -1 \bmod (2^n + 1). \qquad (6.2.17)$$

This special case is easily handled by modular arithmetic. For example, if $q = 2^n$ and $s < 2^n$ then $(q + s) \bmod (2^n + 1)$ and $(q - s) \bmod (2^n + 1)$ are computed as in (6.2.15) and (6.2.16) with q replaced by -1, and $qs \bmod (2^n + 1)$ is obtained by computing $(2^n + 1 - s) \bmod (2^n + 1)$. Another useful formula for binary multiplication is

$$2^r(u_{n-1} \ldots u_0)_2 = [(u_{n-1-r} \ldots u_0(0 \ldots 0)_2 - (0 \ldots 0u_{n-1} \ldots u_{n-r})_2] \bmod (2^n + 1), \qquad (6.2.18)$$

for $0 \leqslant r \leqslant n$.

It shows that multiplication of a binary number by a power of 2 may be obtained using simple shift and subtraction/addition operations, only.

Our next step on the way to an efficient calculation of the production coefficients W_j's is to introduce the so called integer Fourier transform which is especially convenient for computations involving integers.

The Integer FT

The integer Fourier transform for a sequence of integers $\{b_s, s = 0, 1, \ldots, N - 1\}$ is defined as follows:

$$c_k = \left(\sum_{s=0}^{N-1} w^{ks} b_s \right) \bmod K, \qquad (6.2.19)$$

where w is an integer such that $w^N = 1 \bmod K$ and the computation of c_k is carried out modulo K. The integer FT may be treated in the same way as the finite FT except that all results are obtained modulo K.

The convolution theorem for the integer FT is particularly valid when N and w are powers of 2. Under these conditions on N and w the theorem renders a fast and easily programmable algorithm for computing the convolution of sequences of integers mod $(2^{N/2} + 1)$. The major phases of the algorithm are (i) an integer FT, (ii) componentwise multiplication and (iii) an inverse integer

FT. The convolution of two n-vectors with integer components is computed exactly, provided the components of the convolution are in the range 0 to $2^{N/2}$. If the components of the convolution are outside the range 0 to $2^{N/2}$ then they will be correct, modulo $(2^{N/2}+1)$. These results follow from the properties of modular arithmetic.

6.3 Multiplication of Two Integers Using Modular Arithmetic

Consider the problem given by (6.2.1) to (6.2.4) in the light of the computation process which uses modular arithmetic. There are five stages in this process at which the economies in computing time can be achieved. We shall consider each stage in turn.

Stage (a)

The two $n=(r+1)q$-bit integers, u and v, the product of which we seek to compute, are presented as

$$u = U_r 2^{rq} + U_{r-1} 2^{(r-1)q} + \ldots + U_0 = (U_r U_{r-1} \ldots U_0)_{2^q},$$
$$v = V_r 2^{rq} + V_{r-1} 2^{(r-1)q} + \ldots + V_0 = (V_r V_{r-1} \ldots V_0)_{2^q}, \qquad (6.3.1)$$

where U_j's and V_j's are q-bit integers each.
Since u and v are n-bit integers each, their values, in general, are in the range

$$0 \leqslant u, v < 2^n \text{ (with the convention } 2^n = -1 \bmod (2^n+1). \qquad (6.3.2)$$

Stage (b)

The product $w=uv$ in view of (6.3.1) is, generally, a $2n$-bit number defined in the range

$$0 = < w < 2^{2n} \qquad (6.3.3)$$

and represented as:

$$w = uv = W_{2r} 2^{2rq} + W_{2r-1} 2^{(2r-1)q} + \ldots + W_0 = (W_{2r} W_{2r-1} \ldots W_0)_{2^q}, \qquad (6.3.4)$$

where the product coefficients W_j's are $2q$-bit integers each and are expressed in terms of the U_k's and V_k's as given by (6.3.4).

Stage (c)

Now, suppose that all we want to do is to compute value of w, modulo (2^n+1). This value will then be in the range

$$0 \leqslant w \bmod (2^n+1) \leqslant 2^n. \qquad (6.3.5)$$

Here w can be uniquely determined using, say, the Garner method for conversion.

We can write

$$w \bmod (2^n + 1) = (uv) \bmod (2^n + 1).$$

This may be expressed as

$$\begin{aligned} w \bmod (2^n + 1) &= (R_r 2^{rq} + R_{r-1} 2^{(r-1)q} + \ldots R_1 2^q + R_0) \\ &= (R_r R_{r-1} \ldots R_0)_2. \end{aligned} \tag{6.3.6}$$

Using the fact that

$$2^{sq} \bmod (2^{(r+1)q} + 1) = \begin{cases} 2^{sq}, & \text{if} \quad s < r+1, \\ -2^{(s-r-1)q} & \text{if} \quad s \geqslant r+1, \end{cases}$$

the coefficients R_s in (6.3.6) are readily expressed as

$$\begin{aligned} R_s &= W_s - W_{r+s+1} \\ &= (U_s V_0 + U_{s-1} V_1 + \ldots + U_0 V_s) - (U_r V_{s+1} + \ldots + U_{s+1} V_r), \end{aligned} \tag{6.3.7}$$

where W_{2r+1} set equal to 0 (see also example 6.3.1).

We shall refer to the F_s's as the reduced product coefficients.

Example 6.3.1

Let

$$n = 2^2 q = (r+1)q$$

and

$$u = U_3 2^{3q} + U_2 2^{2q} + U_1 2^q + U_0, \quad v = V_3 2^{3q} + V_2 2^{2q} + V_1 2^q + V_0.$$

The product of u and v is given as

$$\begin{aligned} w = uv = U_3 V_3 2^{6q} &+ (V_3 U_2 + V_2 U_3) 2^{5q} + (V_3 U_1 + V_2 U_2 + V_1 U_3) 2^{4q} \\ &+ (V_3 U_0 + V_2 U_1 + V_1 U_2 + V_0 U_3) 2^{3q} \\ &+ (V_2 U_0 + V_1 U_1 + V_0 U_2) 2^{2q} + (V_1 U_0 + V_0 U_1) 2^q + V_0 U_0, \end{aligned}$$

and

$$\begin{aligned} w \bmod (2^{4q} + 1) &= (uv) \bmod (2^{4q} + 1) \\ &= (V_3 U_0 + V_2 U_1 + V_1 U_2 + V_0 U_3) 2^{3q} \\ &\quad + [(V_2 U_0 + V_1 U_1 + V_0 U_2) - U_3 V_3] 2^{2q} \\ &\quad + [(V_1 U_0 + V_0 U_1) - (V_3 U_2 + V_2 U_3)] 2^q \\ &\quad + [V_0 U_0 - (V_3 U_1 + V_2 U_2 + V_1 U_3)] \\ &= F_3 2^{3q} + F_2 2^{2q} + F_1 2^q + F_0, \end{aligned}$$

where we used the fact that

$$2^{6q} \bmod (2^{4q}+1) = (2^{4q}2^{2q}) \bmod (2^{4q}+1)$$
$$= 2^{4q} \bmod (2^{4q}+1)2^{2q} \bmod (2^{4q}+1) = -2^{2q},$$

and, similarly, that

$$2^{5q} \bmod (2^{4q}+1) = -2^{q},$$

$$2^{4q} \bmod (2^{4q}+1) = -1.$$

Recalling again that, generally, 2^p requires $p+1$ bits for its binary representation, we conclude that the product of two q-bit numbers must be less than 2^{2q} and, further, since W_s and W_{r+s+1} are sums of $s+1$ and $r-s$ such products, respectively where $0 \leqslant s \leqslant r, R_s$ must be in the range

$$(r-s)2^{2q} < R_s < (s+1)2^{2q}, \quad s = 0, 1, \ldots, r. \tag{6.3.8}$$

From (6.3.8) it follows that there are at most $(r+1)2^{2q}$ possible values which the R_s can assume.

Stage (d)

As soon as the values R_s are computed the product $w \bmod (2^n+1) = (uv) \bmod (2^n+1)$ given by (6.3.6) is computed in $(r+1)(2q+\log_2(r+1))$ additional primitive operations of shifting since the reduced coefficients lie in the range 0 to $(r+1)2^2q = 2^{\log_2(r+1)+2q}$ and in binary representation are the integers of length $(\log_2(r+1)+2q)$ bits at most. The $(r+1)$ such integers can be added together using just $(r+1)(\log_2(r+1)+2q)$ shifting operations.

Stage (e)

Noting that the time complexity function $(r+1)$ $(\log_2(r+1)+2q)$ is of order n and, further, recalling that the aim of our current analysis is to produce an algorithm to multiply two n-bit integers in significantly more efficient time than that of the standard multiplying algorithm which is of order n^2, we can see that the success of the analysis depends on fast computation of the reduced coefficients.

We now turn to discuss how these coefficients can be computed.

6.4 Computing the Reduced Product Coefficients Exactly

Since the reduced product coefficients are defined in the range

$$0 \leqslant R_s < (r+1)2^{2q}, s = 0, 1, \ldots, r, \tag{6.4.1}$$

we can carry out their computation, modulo $(r+1)2^{2q}$. This process of computing will, on the one hand, result in no loss of information on the coefficients, and on the other hand, enable us to carry out the computation rapidly, by computing the R_s's twice, once modulo $(r+1)$ and once modulo $(2^{2q}+1)$. Strictly speaking we should have the second computation, modulo 2^{2q} instead of modulo $(2^{2q}+1)$, but the latter is more convenient to perform and conversion to the exact values is then readily brought about, as we shall now see.

If we denote by

$$R_s^{(1)} = R_s \bmod (r+1) \quad \text{and} \quad R_s^{(2)} = R_s \bmod (2^{2q}+1), \tag{6.4.2}$$

then the exact values R_s can be obtained from the $R_s^{(1)}$ and $R_s^{(2)}$ by the formula

$$R_s = (2^{2q}+1)[R_s^{(1)} - R_s^{(2)} \bmod (r+1)] + R_s^{(2)}. \tag{6.4.3}$$

Formula (6.4.3) follows directly from the Garner method for conversion of a number to its full integer form from the modular representation, noting that

(i) $(r+1)$ and $(2^{2q}+1)$ are relatively prime and $(r+1) < (2^{2q}+1)$ by definition, and

(ii) since $(r+1) \leqslant 2^{2q}$ and $(r+1)$ is a power of 2, $(r+1)$ divides 2^{2q} and thus the multiplicative inverse of $(2^{2q}+1)$, modulo $(r+1)$ is equal to 1, i.e. $(2^{2q}+1) = 1 \bmod (r+1)$.

From (6.4.3) it follows that the three steps needed to compute the R_s are

Compute

$$R_s^{(1)} = R_s \bmod (r+1), \quad s = 0, 1, \ldots, r.$$

Compute

$$R_s^{(2)} = R_s \bmod (2^{2q}+1), \quad s = 0, 1, \ldots, r.$$

Let

$$R_s^{(3)} = (2^{2q}+1)[\mathrm{R}_s^{(1)} - R_s^{(2)} \bmod (r+1)] + R_s^{(2)} \text{ and}$$

Compute

$$R_s = \begin{cases} R_s^{(3)}, & \text{if} \quad R_s^{(3)} < (r+1)2^{2q}, \\ R_s^{(3)} - (r+1)(2^{2q}+1), & \text{if} \quad R_s^{(3)} \geqslant (r+1)2^{2q}, \end{cases}$$

$$s = 0, 1, \ldots, r.$$

This last step is introduced to satisfy the requirement that R_s should be taken modulo $(r+1)2^{2q}$ while $R_s^{(3)}$ is a number modulo $(r+1)(2^{2q}+1)$.

Next, the method to compute $R_s^{(1)}$'s and $R_s^{(2)}$'s will be outlined. It consists of two parts which will be referred to as Part One and Part Two. In the analysis to follow it is convenient to assume that the integer n and its factors $(r+1)$ and q, are all powers of 2, i.e.

$$n = 2^m = (r+1)q = 2^k 2^l,$$

so that

$$r+1=2^k, \quad q=2^l, \quad k+l=m. \tag{6.4.4}$$

The parameters k and l remain undetermined at this stage. The instructions on how to set their values in the best possible way will be given in Section 6.5.

Part One

$$\begin{aligned} R_s^{(1)} &= R_s \bmod (r+1) \\ &= [(U_sV_0+U_{s-1}V_1+\ldots+U_0V_s)-(U_rV_{s+1}+\ldots+U_{s+1}V_r)] \\ &\quad \bmod 2^k. \end{aligned}$$

Here we wish to carry out some computations modulo (power of 2) on binary integers. Such computations are simple and can be done quickly.

Write

$$\begin{aligned} R_s^{(1)} &= [(U_sV_0 \bmod 2^k + U_{s-1}V_1 \bmod 2^k + \ldots + U_0V_s \bmod 2^k) \\ &\quad -(U_rV_{s+1} \bmod 2^k + \ldots + U_{s+1}V_r \bmod 2^k)] \bmod 2^k \\ &= [((U_s \bmod 2^k)(V_0 \bmod 2^k)+(U_{s-1} \bmod 2^k)(V_1 \bmod 2^k)+\ldots \\ &\quad +(U_0 \bmod 2^k)(V_s \bmod 2^k))-((U_r \bmod 2^k)(V_{s+1} \bmod 2^k)+\ldots \\ &\quad +(U_{s+1} \bmod 2^k)(V_r \bmod 2^k))] \bmod 2^k \\ &= [(U_s'V_0'+U_{s-1}'V_1'+\ldots+U_0'V_s')-(U_r'V_{s+1}'+\ldots+U_{s+1}'V_r')] \\ &\quad \bmod 2^k, \end{aligned} \tag{6.4.5}$$

where we use the notation $U_j' = U_j \bmod 2^k$ and $V_j' = V_j \bmod 2^k$.
Since U_j' and V_i' are each k-bit long, their products, $U_j'\ V_j'$, are $2k$-bit long and of the convolutions are sums of at most $r+1=2^k$ such products, then each of the convolutions is at most $3k$-bit long. It follows that the convolutions $(U_s'V_0'+U_{s-1}'\ V_1'+\ldots+U_0'V_s')$ and $(U_r'V_{s+1}'+\ldots+U_{s+1}'V_r')$ can be 'read off' from the binary representation of the respective products

$$\begin{aligned} P_r^{(1)} &= (U_r'U_{r-1}'\ldots U_1'U_0')_{2^{3k}} \times (V_r'V_{r-1}'\ldots V_1'V_0')_{2^{3k}}, \\ P_r^{(2)} &= (U_0'U_1'\ldots U_r')_{2^{3k}} \times (V_0'V_1'\ldots V_r')_{2^{3k}}, \end{aligned} \tag{6.4.6}$$

where, as before, the bracket $(Y_q'Y_{q-1}'\ldots Y_0')_{2^{3k}}$ denotes the integer

$$Y_q'2^{3kq}+Y_{q-1}'2^{3k(q-1)}+\ldots Y_1'2^{3k}+Y_0'. \tag{6.4.7}$$

To illuminate the point let us consider a simple example, for $r+1=4=2^2$. We have the products

$$\begin{aligned} P_3^{(1)} &= (U_3'U_2'U_1'U_0')_{2^{3k}} \times (V_3'V_2'V_1'V_0')_{2^{3k}}, \\ P_3^{(2)} &= (U_0'U_1'U_2'U_3')_{2^{3k}} \times (V_0'V_1'V_2'V_3')_{2^{3k}}, \end{aligned}$$

and their respective binary representation may be given as

$$\begin{aligned}
P_3^{(1)} &= (U_3'2^{(3k)3} + U_2'2^{(3k)2} + U_1'2^{3k} + U_0') \\
&\quad \times (V_3'2^{(3k)3} + V_2'2^{(3k)2} + V_1'2^{(3k)} + V_0') \\
&= U_3'V'_3 2^{(3k)6} + (U_2'V_3' + U_3'V_2')2^{(3k)5} \\
&\quad + (U_1'V_3' + U_2'V_2' + U_3'V_1')2^{(3k)4} \\
&\quad + (U_0'V'_3 + U_1'V_2' + U_2'V_1' + U_3'V_0')2^{(3k)3} \\
&\quad + (U_0'V_2' + U_1'V_1' + U_2'V_0')2^{(3k)2} \\
&\quad + (U_0'V_1' + U_1'V_0')2^{3k} + U_0'V_0' \\
&= (P_6^{(1)}P_5^{(1)}P_4^{(1)}P_3^{(1)}P_2^{(1)}P_1^{(1)}P_0^{(1)})_{2^{3k}}, \text{ say} \\
P_3^{(2)} &= (U_0'2^{(3k)3} + U_1'2^{(3k)2} + U_2'2^{3k} + U_3') \\
&\quad \times (V_0'2^{(3k)3} + V_1'2^{(3k)2} + V_2'2^{(3k)} + V_3') \\
&= U_0'V_0'2^{(3k)6} + (U_1'V_0' + U_0'V_1')2^{(3k)5} \\
&\quad + (U_2'V_0' + U_1'V_1' + U_0'V_2')2^{(3k)4} \\
&\quad + (U_3'V_0' + U_2'V_1' + U_1'V_2' + U_0'V_3')2^{(3k)3} \\
&\quad + (U_3'V_1' + U_2'V_2' + U_1'V_3')2^{(3k)2} \\
&\quad + (U_3'V_2' + U_2'V_3')2^{3k} + U_3'V_3' \\
&= (P_6^{(2)}P_5^{(2)}P_4^{(2)}P_3^{(2)}P_2^{(2)}P_1^{(2)}P_0^{(2)})_{2^{3k}}, \text{ say}
\end{aligned}$$

Thus

$$\begin{aligned}
R_0' &= R_0 \bmod (2^{4q}+1) = P_0^{(1)} - P_2^{(2)} = P_6^{(2)} - P_4^{(1)}, \\
R_1' &= R_1 \bmod (2^{4q}+1) = P_1^{(1)} - P_1^{(2)} = P_5^{(2)} - P_5^{(1)}, \\
R_2' &= R_2 \bmod (2^{4q}+1) = P_2^{(1)} - P_0^{(2)} = P_4^{(2)} - P_6^{(1)}, \\
R_3' &= R_3 \bmod (2^{4q}+1) = P_3^{(1)} \qquad\qquad = P_3^{(2)}.
\end{aligned}$$

The products given by (6.4.6) can be computed using the multiplication method of Section 6.1 with the number of primitive operations of $0((r+1)^{1.585})$.

Part Two

$$R_s^{(2)} = R_s \bmod (2^{2q}+1), \quad s = 0, 1, \ldots, r.$$

This is the most time-consuming of the three stages. Savings on the number of operations required (as compared with the conventional procedure) are achieved due to the use of the integer convolution theorem. We shall first outline the sequence of computations performed in Part Two and then give a formal justification for these computations.

The computational procedure is as follows:

(a) Set $\psi = 2^{2q/(r+1)}$, so that $\psi^{r+1} = 2^{2q}$, and compute the integer Fourier transform with $w = \psi^2$, modulo $2^{2q}+1$, of the sequences

$$(U_0, \psi U_1, \psi^2 U_2, \ldots, \psi^r U_r) \text{ and } (V_0, \psi V_1, \ldots, \psi^r V_r),$$

i.e.

$$a_t = \left[\sum_{s=0}^{r} w^{st}(\psi^s U_s)\right] \text{mod } (2^{2q}+1)$$

and

$$b_t = \left[\sum_{s=0}^{r} w^{st}(\psi^s V_s)\right] \text{mod } (2^{2q}+1), \quad t = 0, \ldots, r. \tag{6.4.8}$$

(b) Compute the pairwise products of the Fourier transforms computed in (a), modulo $(2^{2q}+1)$, i.e.

$$c_t = a_t b_t \text{ mod}(2^{2q}+1), \quad t = 0, \ldots, r. \tag{6.4.9}$$

(c) Compute the inverse integer Fourier transform, modulo $2^{2q}+1$, of the sequence of pairwise products from step (b), i.e.

$$d_s = \left[\sum_{t=0}^{r} w^{-st} c_t\right] \text{mod } (2^{2q}+1), \; s = 0, \ldots, r. \tag{6.4.10}$$

The results of this computation will be

$$(d_0, d_1, \ldots, d_r) = (2^k R_0^{(2)}, 2^k \psi R_1^{(2)}, \ldots, 2^k \psi^r R_r^{(2)}) \text{ mod } (2^{2q}+1). \tag{6.4.11}$$

Note that

$$2^k = r+1. \tag{6.4.12}$$

Compute $R_s^{(2)}$ by multiplying d_s by $2^{-k}\psi^{-s}$ mod $(2^{2q}+1)$. In fact, since ψ is a power of 2, the $R_s^{(2)}$ may be obtained by an appropriate shifting operation analogous to (6.2.18).

We shall now justify the procedure given in Step 2.

By definition we have

$$R_s^{(2)} \equiv \sum_{0 \leqslant \alpha, \beta \leqslant r} U_\alpha V_\beta \text{ mod } (2^{2q}+1). \tag{6.4.13}$$

Multiplying (6.4.13) by ψ^s we get

$$\psi^s R_s^{(2)} \equiv \psi^s \left[\sum_{0 \leqslant \alpha, \beta \leqslant r} U_\alpha V_\beta \text{ mod } (2^{2q}+1)\right]$$

$$= \sum_{\substack{0 \leqslant \alpha, \beta \leqslant r \\ \alpha+\beta \equiv s \text{ mod } (r+1)}} (\psi^\alpha U_\alpha)(\psi^\beta V_\beta) \text{ mod } (2^{2q}+1). \tag{6.4.14}$$

For Step 2(c) we can write

$$d_s = \left[\sum_{t=0}^{r} \psi^{-2st} c_t\right] \bmod (2^{2q}+1)$$

$$= \left[\sum_{t=0}^{r} \psi^{-2st} a_t b_t \bmod (2^{2q}+1)\right] \bmod (2^{2q}+1)$$

$$= \left[\sum_{t=0}^{r} \psi^{-2st}\left(\sum_{j=0}^{r} \psi^{2jt}(\psi^j U_j)\right) \bmod (2^{2q}+1)\right.$$

$$\left.\times \left(\sum_{i=0}^{r} \psi^{2it}(\psi^i V^i)\right) \bmod (2^{2q}+1)\right] \bmod (2^{2q}+1)$$

$$= \left[\sum_{0 \leqslant j,i \leqslant r} (\psi^{2j}U_j)(\psi^{2i}V_i) \sum_{t=0}^{r} \psi^{2(-s+j+i)t}\right] \bmod (2^{2q}+1).$$

Now, assuming that (we shall later prove this statement)

$$\sum_{t=0}^{r} \psi^{2\gamma t} \equiv \begin{cases} r+1, & \text{if} \quad \gamma \bmod (r+1) = 0, \\ 0, & \text{if} \quad \gamma \bmod (r+1) \neq 0, \end{cases} \tag{6.4.15}$$

we obtain

$$d_s = \left[(r+1) \sum_{\substack{0 \leqslant j,i \leqslant r \\ 2(j+i-s) \equiv 0 \bmod (r+1)}} (\psi^{2j}U_j)(\psi^{2i}V_i)\right] \bmod (2^{2q}+1). \tag{6.4.16}$$

From (6.4.14), (6.4.12) and (6.4.16) it follows that

$$R_s^{(2)} = [2^{-k}\psi^{-s} \bmod (2^{2q}+1)]\, d_s, \; s = 0, \ldots, r,$$

which completes the justification of the computation procedure given in Step 2, except for a proof of statement (6.4.15).

In order to prove that

$$\left(\sum_{t=0}^{r} \psi^{2\gamma t}\right) \bmod (r+1) = \begin{cases} r+1 = 2^k, & \text{if} \quad \gamma \bmod (r+1) = 0, \\ 0, & \text{if} \quad \gamma \bmod (r+1) \neq 0, \end{cases}$$

we shall consider both cases in detail.

(i) If $\gamma \bmod (r+1) = 0$, then γ is a multiple of $(r+1)$ and we may set $\gamma = c(r+1)$, where c is a constant. We obtain

$$\psi^{2\gamma t} = \psi^{2c(r+1)t} = (\psi^{r+1})^{2ct} = (2^{2q})2^{ct} = (-1)^{2ct} \bmod (2^{wq}+1).$$

If follows

$$\left(\sum_{\substack{t=0 \\ \gamma \bmod (r+1) = 0}}^{r} \psi^{2\gamma t}\right) \bmod (r+1) = \left(\sum_{t=0}^{r} (-1)^{2ct}\right) \bmod (r+1) = r+1.$$

(ii) If $\bmod (r+1) \neq 0$, then γ must be an odd number, since $r+1$ is by definition a power of 2. Let $\gamma \bmod (r+1) = 2^{\varkappa}\lambda$, where λ is odd and $0 \leqslant \varkappa < k$,

$r+1=2^k$. Setting $T=2^{k-1-x}$, we get

$$\psi^{2\gamma T}=\psi^{2\gamma 2^{k-1-x}}=\psi^{\gamma 2^k 2^{-k}}=\psi^{2^k\lambda}$$
$$=(\psi^{(r+1)})^\lambda=(2^{2q})^\lambda\equiv -1 \bmod (2^{2q}+1).$$

Noting that

$$T=2^{k-1-x}=2^k 2^{-1-x}=(r+1)2^{-1-x}, \text{ or that } 2T=(r+1)2^{-x},$$

we can write

$$\left(\sum_{t=0}^{r}\psi^{2\gamma t}\right) \bmod (r+1)\equiv 2^x\sum_{t=0}^{2T-1}\psi^{2\gamma t}=2^x\sum_{t=0}^{T-1}(\psi^{2\gamma t}+\psi^{2\gamma(t+T)})$$
$$\equiv 2^x\sum_{t=0}^{T-1}(\psi^{2\gamma t}+\psi^{2\gamma t}(-1))\equiv 0.$$

This completes the proof.

6.5 Estimation of the Work Involved

The computational procedure is nearly complete. It remains only to specify the values k and l, where by definition $n=2^m=2^{k+1}$ (see (6.4.4)). To do this we shall first evaluate the total amount of work involved as a function of k and l, and then specify k and l in such a way as to minimize this function. We already know that the work required to compute the product $uv \bmod (2^n+1)$, after the reduced product coefficients have been computed, is of order n. We also know that Step 1 may be completed in $O(n)$ time, at most. We turn now to estimate the work involved in computation of Step 2. In this step three Fourier transforms, modulo $2^{2q}+1$, are used. Each of them consists of $k=\log_2(r+1)$ steps, and each step requires an amount of work of order n, since the operations involved are simple shifts of the $(6q+\log_2 q)$-bit numbers, modulo $(2^{2q}+1)$. Thus, each Fourier transform requires the work of $O(kn)$. Further, Step 2 also requires $r+1(=2^k)$ multiplications of $a_t b_t \bmod (2^{2q}+1)$, where the integers a_t and b_t are in the range 0 to 2^{2q}, and, thus, are of length $2q(=2^{l+1})$-bit each.

Now, let $T(n)$ denote the time it takes to multiply two n-bit numbers, modulo 2^n+1, by the method given. Let also $T'(n)=T(n)/n$. Then, for Step 2 we have

$$T(n)=2^k T(2^{l+1})+O(kn). \tag{6.5.1}$$

Dividing (6.5.1) by $n(=2^{k+1})$, we get

$$\frac{T(n)}{n}=\frac{2^k}{2^{k-1}}\,\frac{T(2^{l+1})}{2^{l+1}}+O\left(\frac{kn}{n}\right),$$

giving

$$T'(n)=2T'(2^{l+1})+O(k). \tag{6.5.2}$$

To yield the lowest possible time in (6.5.2), the l has to be chosen as small as possible. However, the choice of values for l is restricted by the relation between l and k which follows from the definition of ψ in Step 2(a). Namely, since

$$\psi = 2^{2q/(r+1)} = 2^{2^{l+1-k}}$$

should be a power of 2, then $l+1-k \geqslant 0$, or $l+1 \geqslant k$.
Condition (6.5.3) implies that the best we can do to minimize the $T'(n)$ in (6.5.2) is to set

$$l = \begin{cases} \dfrac{m}{2}, & \text{if } m \text{ is even,} \\ \dfrac{m-1}{2}, & \text{if } m \text{ is odd,} \end{cases} \tag{6.5.4}$$

and

$$k = \begin{cases} \dfrac{m}{2}, & \text{if } m \text{ is even,} \\ \dfrac{m+1}{2}, & \text{if } m \text{ is odd.} \end{cases} \tag{6.5.5}$$

The formulae (6.5.4) and (6.5.5) give the final specifications of l and k.

The conditions on l and k mean that there is a constant C such that

$$T'(n) \leqslant 2T'(2\sqrt{n}) + C \log_2 n \qquad \text{for all } n \geqslant 3, \tag{6.5.6}$$

since $2^l \leqslant 2^{m/2} = \sqrt{n}$ and $k = \log_2(r+1) < \log_2((r+1)q) = \log_2 n$.

We shall, finally, show that (6.5.6) implies that

$$T'(n) \leqslant C' \log_2 n \log_2 \log_2 n \text{ for suitable } C'. \tag{6.5.7}$$

In the proof we shall use the method of induction.

For $n = 4$ in (6.5.6) we have

$$T'(4) \leqslant 2T'(4) + 2C = 2(T'(4) + C). \tag{6.5.8}$$

Denoting $T'(4) + C$ by C', we can present (6.5.8) as

$$T'(4) \leqslant 2C' = C' \log_2 4 \log_2 \log_2 4.$$

Now assuming that (6.5.7) is valid for all $k \leqslant n-1$ we shall prove its validity for $k = n$.

Substituting $C' \log_2 2\sqrt{n} \log_2 \log_2 2\sqrt{n}$ for $T'(2\sqrt{n})$ in (6.5.6) we obtain

$$\begin{aligned} T'(n) &\leqslant 2C' \log_2 2\sqrt{n} \log_2 \log_2 2\sqrt{n} + C \log_2 n \\ &= 2C'(1 + \tfrac{1}{2}\log_2 n)\log_2(1 + \tfrac{1}{2}\log_2 n) + C \log_2 n \\ &= 2C' \log_2(1 + \tfrac{1}{2}\log_2 n) + C' \log_2 n \log_2(1 + \tfrac{1}{2}\log_2 n) \\ &\quad + C \log_2 n. \end{aligned}$$

For large n, $1+\frac{1}{2}\log_2 n \leqslant \frac{2}{3}\log_2 n$, and we can write

$$\begin{aligned}T'(n) &\leqslant 2C'(\log_2(\tfrac{2}{3})+\log_2\log_2 n)+C'\log_2 n(\log_2(\tfrac{2}{3})\\ &\quad +\log_2\log_2 n)+C\log_2 n\\ &= 2C'(1-\log_2 3)+2C'\log_2\log_2 n+C'\log_2 n(1-\log_2 3)\\ &\quad +C'\log_2 n\log_2\log_2 n+C\log_2 n\\ &= C'\log_2 n\log_2\log_2 n+2C'\log_2\log_2 n\\ &\quad +(C+C'-C'\log_2 3)\log_2 n+2C'(1-\log_2 3)\\ &= \mathrm{O}(\log_2 n\log_2\log_2 n).\end{aligned}$$

Finally, we have

$$T(n)=nT'(n)\leqslant \mathrm{O}(n\log_2 n\log_2\log_2 n). \tag{6.5.9}$$

This completes the estimation of the work involved in Step 2.

We note that Step 3 requires only a few operations of negligible cost, and we denote this cost by $\varepsilon(n)$.

The total amount of work involved in multiplication of two n-bit numbers, using the method outlined in Sections 6.3 and 6.4 is thus estimated by

$$W_{\text{total}}\leqslant \mathrm{O}(n)+\mathrm{O}(n)+\mathrm{O}(n\log_2 n\log_2\log_2 n)+\varepsilon(n), \tag{6.5.10}$$

which proves that two n-bit numbers can be multiplied in

$\mathrm{O}(n\log_2 n\log_2\log_2 n)$ time.

This rather remarkable result is due to Schönhage and Strassen (1971).

Exercises

1. Assuming that the Fourier transforms are computed using the FFT algorithm, estimate the total number of multiplication operations required by the computational process given by (6.2.1) to (6.2.7), for obtaining the product of two binary integers.
2. Give a 'modular representation' for one million with respect to the set of moduli $p_1=11$, $p_2=19$, $p_3=23$, $p_4=31$ and $p_5=43$.
3. Let the moduli be 2, 3, 5, 7. Given the modular representation (1, 2, 3, 4) recover the original number.
4. Find $(11101001100111)_2 \bmod (2^5+1)$.
5. Compute the product of the binary numbers $(1101010)_2$ and $(1000001)_2$ using the formulae (6.1.1) and (6.1.2).
6. Compute the product in exercise 5 using the procedure given by (6.2.1) to (6.2.7) and assuming that $q=1$ and $r=6$.
7. Write a program to find residues of a number modulo a collection of mutually prime moduli.

7

Internal Sorting

In this chapter we shall study some of the better known computer methods to sort in order a collection of n items. Sorting is one of the most frequently encountered tasks in data processing, and often takes up a significant proportion of the total computer time required by a particular processing system. Efficient sorting methods are therefore vital.

Two kinds of sorting are distinguished: internal sorting when the items are stored in the computer memory; and external sorting when there are more items than can be kept in the memory at once, and the sorting is done from tape or disc. Methods of internal sorting are not always readily extendable for external sorting since internal sorting allows flexibility in structuring and accessing the data, while external sorting often imposes strict constraints on accessing the data. As a result, different approaches are used for the development of algorithms suitable for a particular kind of sorting.

In the discussion which follows we refer to the entire collection of n items as an array or as a file. We assume that a key, $K(j)$, is associated with each item and that the file is sorted on this key. By sorting we understand the determination of a permutation $p_1, p_2, \ldots, p_n$ of the items which sets the keys in ascending order $K(p_1) \leqslant K(p_2) \leqslant \ldots \leqslant K(p_n)$. For notational clarity there will be no distinction between an item and its key—that is, we shall speak of an item K with key K.

One important class of sorting methods is based on comparison of the keys. For such methods the number of comparisons required is the characteristic on which the criteria of optimal sorting depend. Indeed, if we assume that a bounded number of arithmetic operations occur between execution of comparisons, then the total number of steps required by the method is proportional to the number of comparisons executed in a computation. Thus the number of comparisons affords a machine-independent criterion for the analysis of comparison sorts.

We shall first introduce examples of simple sorting techniques, then proceed to discuss more efficient algorithms and general efficiency bounds for comparison sorts. The most important part of the efficiency analysis is the study of the algorithm's performance on average on a randomly ordered input array.

7.1 Comparison Sorts

Consider a simple algorithm that finds the largest key on a given array of n keys. Given an array of unordered keys, $K(1), K(2), \ldots, K(n)$, not necessarily all distinct, find the largest key of the array, $K(k) = \max(K(j), 1 \leqslant j \leqslant n)$, such that k is as large as possible.

Algorithm findlargest

```
while n > 0 do
    K[n] := pivot
    for index := n - 1 downto 1 do
        if pivot < K[index] then pivot := K[index] endif
    enddo
enddo
```

An easily made observation suggests that the algorithm requires an amount of storage equal to thc lcngth of the input array and only a non-significant fixed amount of extra storage for storing the values of the pivot and the index. There is no need to analyse the memory requirements of the algorith. The run time analysis, on the other hand, amounts to counting the number of steps (or comparisons) in the algorithm and the number of times that each step is executed. The loop is executed $n - 1$ times. This quantity determines the time complexity of the algorithm because the only other quantity which depends on the size of the problem is the number of times that the value of the pivot is changed. The operation 'change the pivot value' is executed once or zero times per step of thc loop. Wc say that thc time complexity of the algorithm is of $O(n)$

However, the number of times of the pivot value is an interesting quantity and merits some further analysis. If the input array K is in non-decreasing order then the pivot's value will not changed at all after being initially set to $K(n)$. At the other extreme if the input array K is in non-increasing order then the pivot's value will be changed at most $n - 1$ times. Thus 0 and $n - 1$ are the minimum and maximum values for the pivot, respectively. A more interesting information pertains to the number of times that the pivot's value is changed on average.

To find this average number we set up a stochastic model of the computation. For example, let the keys be distinct values and let each of the $n!$ permutations of these values be equally likely. It is easy to see that the algorithm concerns the relative order of the keys and not their precise values. Consider the case of $n = 3$. Given a set of three different keys we find that whatever the actual values of the keys the following six possibilities are equally probable.

Permutation	Situation	The number of key moves
1 2 3	$K(1) < K(2) < K(3)$	0
1 3 2	$K(1) < K(3) < K(2)$	1
2 3 1	$K(2) < K(3) < K(1)$	1
3 1 2	$K(3) < K(1) < K(2)$	1
3 2 1	$K(3) < K(2) < K(1)$	2
2 1 3	$K(2) < K(1) < K(3)$	0

Under our assumption of all key permutations being equally likely, the probability that the key $K(n-1)$ is larger than $K(n)$ is 1/2. The probability that the key $K(n-2)$ is larger than either of the $K(n)$ and $K(n-1)$is 1/3. Continuing in this way we obtain the average number of pivot changes equal to the sum

$$1/2 + 1/3 + \ldots + 1/n = H_n - 1, \tag{7.1.1}$$

which for large n is approximately equal to

$$\log_e n - 0.423. \tag{7.1.2}$$

(Here $H_n = 1 + 1/2 + \ldots + 1/n$ is the notation used for the harmonic numbers and for large n $H_n = \log_e n + 0.577$.)

Algorithm *findlargest* can be used to sort in order an entire arbitrary array of keys.

7.2 Simple Selection Sort

Simple selection sort is one of the simplest sorting algorithms and works as follows.

Find the largest key in the array.
Exchange it with the key in the last position.
(This position is not considered again in the subsequent computation.)
Find the second largest key.
Exchange it with the key in the next to the last position.
(This position is not considered again in the subsequent computation.)
Continue this process until the entire array is sorted.

This algorithm is called selection sort because it works by repeatedly selecting the current largest key.

Algorithm selectsort

```
for index := n downto 2 do
    largest := index
    for loop := index - 1 downto 1 do
        if K[loop] > K[largest] then largest := loop endif
    enddo
```

```
        //exchange the positions of the two keys//
        temp := K[largest]
        K[largest] := K[index]
        K[index] := K[largest]
    enddo
```

To estimate the number of the key comparisons we note that the algorithm is comprised of two loops which are nested, and the comparison operation is executed once per step of the inner loop. Since in the inner loop $j-1$ steps are carried out per jth step of the outer loop and there are $n-1$ steps in the outer loop, the total number of comparisons required by the algorithm is given by

$$(n-1)+(n-2)+\ldots+2+1=n(n-1)/2. \tag{7.2.1}$$

This number is constant whatever the initial permutation of the keys.

In thc algorithm thc number of times that the key changes its position is given by the number of times, S_n, that the statement '*largest* := *loop*' is executed. The number S_n depends on the initial permutation of the keys and can therefore be described by its minimum, average and maximum values. If the initial array is in order then the simple selection sort requires no exchanges of the keys, the statement '*largest* := *loop*' is never executed and thus $S_{n,\min}=0$. If the initial array is in reverse order then at each step of the outer loop two keys change their positions and the length of the unordered subarray is reduced by one key at each end. This gives the maximum value for $S_{n,\max}$ as

$$(n-1)+(n-3)+\ldots+1=\begin{cases} n^2/4, & \text{if } n \text{ is even,} \\ (n^2+3)/4, & \text{if } n \text{ is odd.} \end{cases} \tag{7.2.2}$$

In order to determine the average number of key changes we note that

(a) If simple selection sort starts with a random permutation of $1, 2, \ldots, n$ then the first step of the outer loop yields a random permutation of $1, 2, \ldots, n-1$ followed by n, since the permutation $q_1, q_2, \ldots, q_{n-1}n$ is produced from each of the inputs

$$\begin{matrix} n & q_2 & g_3 & \ldots & q_{n-1} & q_n \\ q_1 & n & q_3 & \ldots & q_{n-1} & q_n \\ \vdots \\ q_1 & q_2 & q_3 & \ldots & q_{n-1} & n. \end{matrix}$$

(b) The occurrence of each permutation of $1, 2, \ldots, n-1$ in $K(1), K(2), \ldots, K(n-1)$ is equally likely, and the average number of times that the statement '*largest* := *loop*' is executed during the first step of the outer loop is given by $H_n - 1$.

In general, the average value of S_n, $S_{n,\text{aver}}$, satisfies the recurrence relation

$$S_{n,\text{aver}} = H_n - 1 + S_{n-1,\text{aver}}. \tag{7.2.3}$$

If follows that

$$S_{n,\text{aver}} = \sum_{j=2}^{n} H_j - n + 1 = (n+1)H_n - 2n. \tag{7.2.4}$$

(Here we have used the fact that $\sum_{j=1}^{n} H_j = (n+1) - n$, which can be proved by induction.)

In summary for the simple selection sort

the number of key comparisons is $n(n-1)/2 = O(n^2)$, and
the number of key changes is

$$\min = 0,$$

$$\text{aver} = (n+1)H_n - 2n = O(n \log n),$$

$$\max = \begin{matrix} n^2/4, & \text{if } n \text{ is even}, \\ (n^2+3)/4, & \text{if } n \text{ is odd}, \end{matrix} = O(n^2) \tag{7.2.5}$$

The overall complexity of the algorithm is of $O(n^2)$ and is determined by the number of key comparisons.

Simple selection sort works well for small arrays. It also turns out to be particularly suitable for one important application—for sorting arrays with very large records and small keys, when it is required that the records are actually to be rearranged—but for many other applications its quadratic time complexity is not very desirable property; the algorithm is too slow. Is it possible then to modify the method so that the number of comparisons required will be reduced?

We note that on the first scan of the array at least $n-1$ comparisons are needed as otherwise we cannot claim that the largest key has been found. However, on the subsequent scans the information about the relative order of the keys which has been obtained during the first scan can be utilized to reduce the number of key changes. Several algorithms have been designed to make the most of this information.

7.3 Heapsort—A Comparison Sort of Complexity *n* log *n*

An elegant and efficient algorithm which uses the idea of a repeated key selection is the Heapsort. The method was invented by Williams (1964) and Floyd (1964). Heapsort is a two-stage process: the first n steps build the array into a data structure 'heap', and the next n steps extract the keys in decreasing order and build the final ordered sequence, placed from right to left in an array.

At the heap creation stage the information about the relative order of the keys which is obtained after each key comparison is implemented in the form of a data structure 'heap'. In order to illucidate the concept of this important

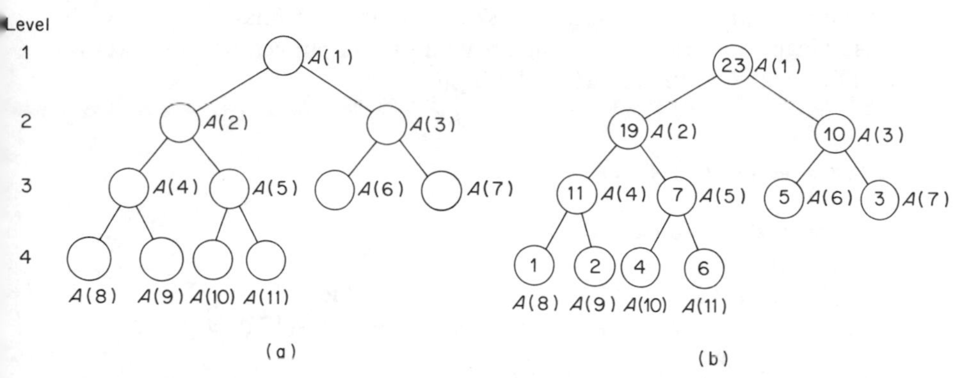

Figure 7.3.1 A binary tree structure for eleven keys: (a) a binary tree; (b) a heap

data structure we shall consider an array which we denote by $A(1{:}n)$. With this array we associate a binary tree, a data structure illustrated in Fig. 7.3.1(a).

A tree structure can be represented in many different ways. Graphical representation, as shown in Fig. 7.3.1, illustrates explicitly the branching relationships (which incidentally led to the very name 'tree'). The trees are usually drawn as if they 'grew' downwards, and the top node, that is $A(1)$ in Fig. 7.3.1, is called the root. A convenient way of numbering the nodes is to start at the root and to number the nodes from left to right on each level in turn so that if $A(k)$ is the root of a subtree than $A(2k)$ and $A(2k+1)$ are the two nodes immediately below it. The nodes $A(2k)$ and $A(2k+1)$ are called children of $A(k)$. Inversely, node $A(k)$ is said to be the parent of $A(2k)$ and $A(2k+1)$.

We define the root of a tree to be at level 1. If $A(k)$ is at level i then $A(2k)$ and $A(2k+1)$ are at level $i+1$. A node with no children is called a terminal node, a node which is not terminal is an internal node. The number of branches to be traversed in order to get from the root to a node $A(j)$ is called the path length of $A(j)$. The root has path length 1, its children have path lengths 2, and so on. In general a node at level i has the path length i.

Definition

A heap is a binary tree with nodes $A(1)$ to $A(n)$ inclusive which has all terminal nodes on the same level or on two adjacent levels at most, and which has the property that the value of a parent node is greater than or equal to the values of its children.

$A(k) \geqslant \max\ (A(2k),\ A(2k+1))$, for $1 \leqslant k \leqslant \lfloor n/2 \rfloor$ and $n \geqslant 2$. In FIg. 7.3.1(b) a heap is shown as a labelled tree. The sequence of keys on the path from the root to each terminal node is linearly ordered, for example, $23 > 19 > 7 > 4$. The largest key in a subtree is always at the root of that tree.

The Heapsort algorithm works as follows. Given an unordered array of n keys, $A(1:n)$, the algorithm first builds a proper heap and then systematically selects the current root key and places the key into its final position within array A. Heapsort uses no extra memory and is guaranteed to sort n keys in about $n \log n$ steps no matter what the input.

Algorithm createheap (A, n)

```
for k := n div 2 downto 1 do
    repeat
        if A[k] < max(A[2k], A[2k+1]) then
            index := the index of the largest of A[2k] and
            A[2k+1]
            //if only one child then index is the index of this child//
            temp := A[k]
            A[k] := A[index]
            A[index] := temp
            k := k div 2
        endif
    until k = 0
enddo
```

A heap formation process is illustrated in Example 7.3.1.

Example 7.3.1

	A(1)	A(2)	A(3)	A(4)	A(5)	A(6)	A(7)	A(8)	A(9)	A(10)	A(11)
	11	2	3	1	4	5	10	23	19	7	6
$k=5$	11	2	3	1	(*7*)	5	10	23	19	(*4*	6)
$k=4$	11	2	3	(*23*)	7	5	10	(1	19)	4	6
$k=3$	11	2	(*10*)	23	7	(5	*3*)	1	19	4	6
$k=2$	11	(*23*)	10	(*2*	7)	5	3	1	19	4	6
	11	23	10	(*19*)	7	5	3	1	(*2*	4)	6
$k=1$	(*23*)	(*11*	10)	19	7	5	3	1	2	4	6
	23	(*19*)	10	(*11*	7)	5	3	1	2	4	6
	23	19	10	(11)	7	5	3	(1	2)	4	6

The last row in Example 7.3.1 displays a heap created on eleven keys. The heap is also shown in Fig. 7.3.1(b).

At the second stage Heapsort extracts the largest key from the root node of the heap and stores it in $A(n)$ by interchanging the keys in $A(1)$ and $A(n)$. Thereafter location n of the array is no longer considered part of the heap. To rearrange the tree in locations $1, 2, \ldots, n-1$ into a heap, the new key in $A(1)$ is sunk as far down a path in the tree as necessary. The process of interchanging $A(1)$ and $A(n-1)$ is then repeated and thereafter the tree is considered to occupy locations 1, 2, . . . , $n-2$, etc. Selecting the largest key, 23, from the heap of Fig. 7.3.1(b) and the subsequent updating of the tree are illustrated in Fig. 7.3.2.

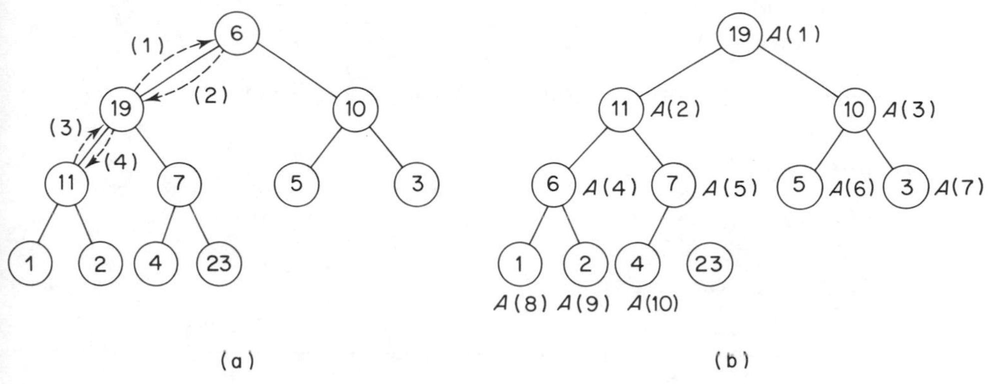

Figure 7.3.2 Selecting the largest key from the root node and updating the tree using Heapsort. (a) The keys 23 and 6 in the heap of Fig. 7.3.1(b) are interchanged. (b) The heap is rearranged while key 23 is no longer part of the heap

The array of Fig. 7.3.2(a) is

6	19	10	11	7	5	3	1	2	4	23

and the resulting array of Fig. 7.3.2(b) is

19	11	10	6	7	5	3	1	2	4	23

Algorithm updateheap(A, s)

//$A[1{:}s]$ is the current heapified array, $s > 1$.//
//Interchange the current largest key in $A[1]$ with the key in the right most position, $A[s]$.//

pivot := $A[1]$
$A[1]$:= $A[s]$
$A[s]$:= **pivot**
k := 1
while $2k < (s-1)$ **do**
index := *the index of the largest of* $A[2k]$ *and* $A[2k+1]$
//if only one child then index is the index of this child//
if $A[k] < A[index]$ **then**
//interchange $A[k]$ and $A[\text{index}]$//
pivot := $A[index]$
$A[index]$:= $A[k]$

```
                A[k] := pivot
                k := index
            else goto fin
            endif
        enddo
fin: return (A)
```

The complete Heapsort algorithm is implemented as follows.

Algorithm heapsort

```
    createheap (A, n)
    for s := n downto 2 do
        updateheap (A, s)
    enddo
```

The Number of Comparisons Required by Heapsort

Algorithm *createheap* comprises two loops which are nested. The outer loop consists of $n/2$ steps, and in the inner loop the number of steps is less than or equal to the depth of the heap tree (the depth of a tree is equal to the number of levels in the tree). Two comparisons at most are performed per step of the inner loop. It follows that if k denotes the depth of the heap then the number of comparisons required by the algorithm is of $O(nk)$.

Algorithm *undateheap* contains one loop. One comparison is executed per step of the loop and the loop executes at most the number of steps equal to the depth of the heap. This gives the number of comparisons as $O(k)$, where k again denotes the depth of the heap.

The total number of comparisons required by the complete algorithm *heapsort* is given by

$$O(nk) + (n-1)O(k) = O(nk).$$

Noting that a binary tree of n nodes and with all terminal nodes on at most two adjacent levels has the depth $k = ! \log_2 n!$, we obtain the number of comparisons required by the *heapsort* as $O(n \log n)$ whatever the input. This performance is a significant improvement on $O(n^2)$ of simple selection sort.

The number of key changes required by the heapsort for any input is also a quantity of $O(n \log n)$ as can easily be observed from the algorithm description.

The heapsort is of particular practical interest because the number of steps (comparisons) required by the algorithm to sort n keys is guaranteed to be proportional to $n \log n$ no matter what the input. Unlike the other methods that we have seen and will study further, there is no worst-case input that will make heapsort perform slower.

7.4 Sorting by Exchange: Bubble Sort

Another family of comparison sorting algorithms is based on the idea that each comparison of two keys is followed systematically by interchanging the pairs of keys that are out of order, until no more such pairs are left. Again, let us first look at a simple sorting algorithm which follows from a straightforward implementation of the idea, and then attend to the question of the optimal utilization of the same idea. One simple algorithm which uses an exchange of keys is the celebrated bubble sort. Actually we shall see that the method is very slow and expensive and the reason it is introduced is the fact that its analysis yields a very interesting mathematical discussion with the results of general and fascinating nature.

The algorithm works as follows. Given a sequence of unordered keys, $K(1)$, $K(2), \ldots, K(n)$, generally not all distinct, the bubble sort keeps scanning through the sequence, exchanging adjacent keys, if necessary; when no exchanges are required on some scan, the sequence is sorted.

Algorithm bubblesort

```
repeat
    pivot := A[1]
    for k := 2 to n do
        if A[k-1] > A[k] then
            pivot := A[k-1]
            A[k-1] := A[k]
            A[k] := pivot
        endif
    enddo
until pivot = A[1]
```

The algorithm is illustrated in Example 7.4.1

Example 7.4.1

Eleven keys are sorted using bubble sort.

```
3  4  2  1  7  6  5  10  11  19  23

3 ↔ 4 ↔ 2
    ┌────┘
3  2  4 ↔ 1
       ┌──┘
3  2  1  4 ↔ 7 ↔ 6
            ┌────┘
3  2  1  4  6  7 ↔ 5
               ┌──┘
3  2  1  4  6  5  7 ↔ 10 ↔ 11 ↔ 19 ↔ 23
```

Pass 1 at the end of the pass $pivot = 7$

$3 \leftrightarrow 2$

$2 \quad 3 \leftrightarrow 1$

$2 \quad 1 \quad 3 \leftrightarrow 4 \leftrightarrow 6 \leftrightarrow 5$

$2 \quad 1 \quad 3 \quad 4 \quad 5 \quad 6 \leftrightarrow 7 \leftrightarrow 10 \leftrightarrow 11 \leftrightarrow 19 \leftrightarrow 23$ Pass 2 at the end of the pass *pivot* = 6

$2 \leftrightarrow 1$

$1 \quad 2 \leftrightarrow 3 \leftrightarrow 4 \leftrightarrow 5 \leftrightarrow 6 \leftrightarrow 7 \leftrightarrow 10 \leftrightarrow 11 \leftrightarrow 19 \leftrightarrow 23$ Pass 3 at the end of the pass *pivot* = 3

$1 \leftrightarrow 2 \leftrightarrow 3 \leftrightarrow 4 \leftrightarrow 5 \leftrightarrow 6 \leftrightarrow 7 \leftrightarrow 10 \leftrightarrow 11 \leftrightarrow 19 \leftrightarrow 23$ No exchanges are required; the sequence is sorted.

Three quantities are involved in run time of the bubble sort: the number of comparisons, the number of key exchanges and the number of passes through the input data. In order to estimate these quantities we note that the algorithm is implemented using two loops which are nested. One step of the outer loop corresponds to one pass through the input array so that the total number of passes is given by the number of steps executed in the outer loop.

If the input sequence is in order then the first pass will also be the final pass as it will require no exchanges. Thus the minimum values for the number of comparisons, the number of exchanges and the number of passes are $n-1$, 0 and 1, respectively. If the input sequence is in reverse order then the first pass will 'bubble up' the largest key into its proper position; and exactly $n-1$ comparisons and $n-1$ exchanges will be used during the pass. The second pass will 'bubble up' the next to the largest key into its proper position and this by using $n-2$ each, comparisons and exchanges. Continuing the argument we find that in this case the bubble sort requires $n(n-1)/2$ each, comparisons and exchanges, as well as n passes. $O(n^2)$ is the maximum value on both the number of comparisons and exchanges required by the bubble sort.

Analysis of the values for the number of comparisons, exchanges and passes on average is not easily obtained by observing the algorithm's implementation. We shall study in detail one average number of exchanges, which for brevity will be denoted by E. The choice is partly due to the fact that the analysis of E involves using an interesting mathematical modelling technique, inversion tables. It was first beautifully expounded by Knuth (1968). The input sequence is assumed to be in random order and the keys are all distinct.

Definition

Let $a_1a_2 \ldots a_n$ be a permuation of the set $(12 \ldots n)$.

If $i < j$ and $a_i > a_j$, the pair (a_i, a_j) is called an 'inversion' of the permutation; for example, the permutation 3142 has three inversions: (31), (32), (42).

Each inversion is a pair of elements that is out of order, so the only permutation with no inversions is the sorted permutation $12 \ldots n$.

The inversion table $b_1 b_2 \ldots b_n$ of the permutation $a_1 a_2 \ldots a_n$ is obtained by letting b_j be the number of elements to the left of j that are greater than j. (In other words, b_j is the number of inversions whose second component is j).

For example the permutation

5 9 1 8 2 6 4 7 3

$b_1 b_2 b_3 b_4 b_5 b_6 b_7 b_8 b_9$

2 3 6 4 0 2 2 1 0 = 20 inversions in all

By definition we will always have

$$0 \leqslant b_1 \leqslant n-1, \quad 0 \leqslant b_2 \leqslant n-2, \ldots, \quad 0 \leqslant b_{n-1} \leqslant 1, \; b_n = 0$$

The most important fact about inversions is the observation that an inversion table uniquely determines the corresponding permutation. This correspondence is important because we can often translate a problem stated in terms of permutations into an equivalent problem stated in terms of inversion tables, and the latter problem may be easier to solve.

For our problem of obtaining the average of E we observe that if we interchange two adjacent elements of a permutation, it is obvious that the total number of inversions will increase or decrease by unity. Hence, *the number of exchanges in the bubble sort is equal to the total number of inversions in the given permutation*.

Now, denote by $I_n(k)$ the number of permutations of n elements which have exactly k inversions each.

Table 7.4.1 list the first few values of the function $I_n(k)$ for various n and k:

Table 7.4.1

n \ k	0	1	2	3	4	5	6	7	8	9	10
1	1	0	0	0	0	0	0	0	0	0	0
2	1	1	0	0	0	0	0	0	0	0	0
3	1	2	2	1	0	0	0	0	0	0	0
4	1	3	5	6	5	3	1	0	0	0	0
5	1	4	9	15	20	22	20	15	9	4	1
6	1	5	14	29	49	71	90	101	101	90	71

In Table 7.4.1, the first column, $I_n(0) = 1$, corresponds to a perfectly ordered permutation, i.e. $12 \ldots n$, and the second column, $I_n(1) = n-1$, to the case when just one pair is out of order (hence $n-1$ such permutations of n elements).

It is convenient to consider the generating function for the sequence ($I_n(k)$, $k = 0, \ldots m$).

Definition

Given a sequence of numbers $\{I_j, j = 0, 1, \ldots\}$ one can set up an infinite sum in terms of a parameter z,

$$G(z) = I_0 + I_1 z + I_2 z^2 + \ldots = \sum_j I_j z^j.$$

The function G is a single quantity which represents the whole sequence $\{I_j, j = 0, 1, \ldots\}$ and is called the generating function for the sequence. If the sequence is finite of size m its generating function becomes a polynomial of degree $m - 1$, $G(z) = \sum_{j=0}^{m-1} I_j z^j$.

The generating function possesses several important properties that make it easy to manipulate the function into form from which conclusions may be deduced concerning the number sequence itself. For our problem we let

$$G_n(z) = I_n(0) + I_n(1)z + \ldots + I_n(m)z^m = \sum_{k=0}^{m} I_n(k)z^k, \tag{7.4.1}$$

so that, say, for $n = 3$,

$$\begin{aligned} G_3(z) = 1 + 2z + 2z^2 + z^3 &= (1 + z + z^2)(1 + z) \\ &= (1 + z + z^2)G_2(z) \\ &\text{since } G_2(z) = 1 + z, \end{aligned}$$

for $n = 4$,

$$\begin{aligned} G_4(z) &= 1 + 3z + 5z^2 + 6z^3 + 5z^4 + 3z^5 + z^6 \\ &= (1 + z + z^2 + z^3)G_3(z), \text{ etc.,} \end{aligned}$$

and in general for our problem:

$$G_n(z) = (1 + z + z^2 + \ldots + z^{n-1})G_{n-1}(z). \tag{7.4.2}$$

(Proof of (7.4.2) can be obtained using an 'induction in reverse' approach.)

We shall now find an average number of inversions present in a given permutation assuming that each of the $n!$ permutations of n keys is equally likely.

The probability that a given permutation has exactly k inversions is

$$p_{nk} = \frac{\text{Number of permutations of } n \text{ elements with } k \text{ inversions}}{n!} = \frac{I_n(k)}{n!}. \tag{7.4.3}$$

Then the average number of inversions is

$$\sum_{k=0}^{m} k p_{nk}.$$

To determine p_{nk}, consider the generating function

$$\begin{aligned} g_n(z) &= \sum_{k=0}^{m} p_{nk} z^k = \frac{1}{n!} \sum_{k=0}^{m} I_n(k) z^k = \frac{1}{n!} G_n(z) \\ &= \frac{1+z+z^2+\ldots+z^{n-1}}{n} \frac{1+z+z^2+\ldots+z^{n-2}}{n-1} \cdots \frac{1+z}{2} \\ &= h_n(z) h_{n-1}(z) \ldots h_1(z). \end{aligned} \tag{7.4.4}$$

By virtue of the function definition and of expression (7.4.4):

$$\sum_{k=0}^{m} k p_{nk} = g_n{}'(1) = \frac{n-1}{2} + \frac{n-2}{2} + \ldots + \frac{1}{2} + 0 = \frac{1}{4} n(n-1). \tag{7.4.5}$$

This gives the average value for E equal to $n(n-1)/4$. We conclude that the time complexity of bubble sort on average is no better than $O(n^2)$.

7.5 Quicksort—A Comparison Sort of Complexity $n \log n$, on Average

The analysis of bubble sort shows that its time complexity is of order n^2 which is rather high for a sorting algorithm. We shall now study an algorithm which makes such efficient use of the basic comparison-exchange scheme that it is more popular than any other. The algorithm is called Quicksort and was invented by C. A. R. Hoare in 1962. Since the first publication of the basic algorithm it has been studied by many people, subjected to a thorough mathematical analysis and the analysis verified by extensive empirical experience. The algorithm works well in many diverse situations, requires only a small amount of extra memory, about $n \log n$ operations on average and is recursive.

Algorithm Quicksort

Assume an unordered array of keys, $K(1), K(2), \ldots, K(n)$. Quicksort works by partitioning the array into two parts, then sorting the parts independently. The exact position of the partition depends on the input, and in the basic algorithm the partitioning element is a key chosen from the input array, though there are modified versions of Quicksort which use more sophisticated mechanisms for the choice of the partitioning element. At the partitioning stage the array is rearranged so as to make the following three conditions hold:

the partitioning element, $K(k)$, is in its proper position in the array for some k,

all the elements in $K(1), \ldots, K(k-1)$ are less than or equal to $K(k)$,

all the elements in $K(k+1), \ldots, K(n)$ are greater than or equal to $K(k)$.

This can be easily implemented using the following strategy. Set two scan pointers, i and j, with $i = 1$ and $j = n$ initially. Compare $K(i)$ with $K(j)$, $K(j-1), \ldots, K(s)$ until finding a key $K(s)$ less than $K(i)$. Exchange the two keys, increase i by 1, and continue to compare and if in order to immediately increase i until another exchange takes place. After the exchange decrease j by 1, and continue to compare and if in order to immediately decrease j until a new exchange occurs. After the exchange increase i, and so on, until $i = j$. It so happens that for a better performance of the algorithm an exchange should take place even when $K(i)$ and $K(j)$, $i \neq j$, are equal values. When the scan pointers cross the partitioning element is found to be in its final place in the array. Example 7.5.1 illustrates the partitioning process for an array of eleven keys.

Example 7.5.1

This is sorting in order eleven keys using Quicksort.

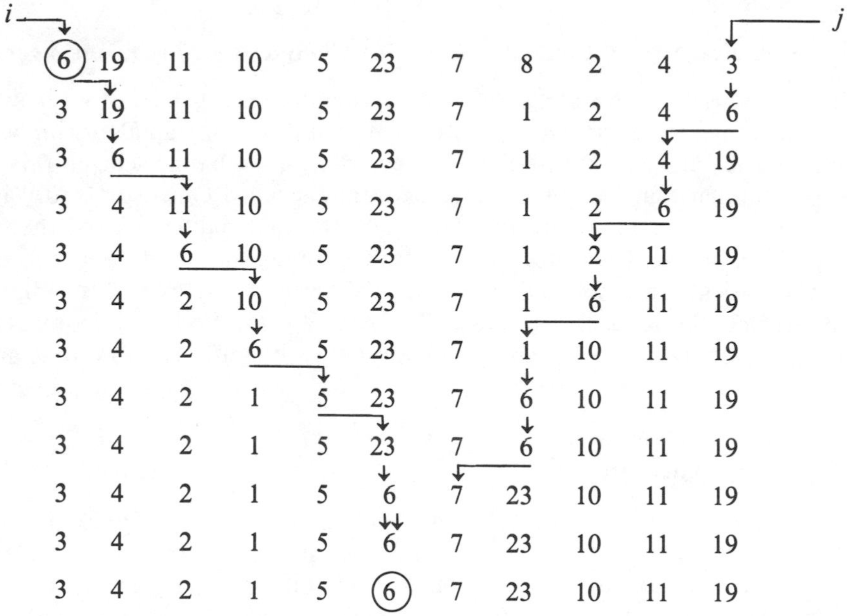

(6)	19	11	10	5	23	7	8	2	4	3
3	19	11	10	5	23	7	1	2	4	6
3	6	11	10	5	23	7	1	2	4	19
3	4	11	10	5	23	7	1	2	6	19
3	4	6	10	5	23	7	1	2	11	19
3	4	2	10	5	23	7	1	6	11	19
3	4	2	6	5	23	7	1	10	11	19
3	4	2	1	5	23	7	6	10	11	19
3	4	2	1	5	23	7	6	10	11	19
3	4	2	1	5	6	7	23	10	11	19
3	4	2	1	5	6	7	23	10	11	19
3	4	2	1	5	(6)	7	23	10	11	19

At the end of the process the partitioning element which is encircled in Example 7.5.1 occupies its proper position in the array. All keys on the left of the partitioning element have values less than the partitioning element, and all keys on the right have values greater than the paritioning element. Each of the two arrays (3 4 2 1 5) nd (7 23 10 11 19) is then sorted by Quicksort independently. This means that after each partitioning of the current array one of the arrays needs to be temporarily stored or 'remembered' while Quicksort works on the other array. Thus some auxiliary storage is required by the algorithm. We shall analyse the algorithm's requirements both in terms of time and space complexity but first a full implementation of the method is due.

Algorithm quicksort (r, s)

```
i := r
j := s + 1
pivot := K[i]        //the partitioning element is set//
if s > 1 then
    repeat
        repeat j := j - 1 until K[j] ≤ pivot
        repeat i := i + 1 until K[i] ≥ pivot
        if i < j then
            temp := K[i]; K[i] := K[j]; K[j] := temp
        endif
    until j ≤ i
    K[r] := K[j]; K[j] := pivot //the partitioning element belongs at
                                                   position j//
endif
if r < s then
    quicksort(r, j - 1)
    quicksort(j + 1, s)
endif
```

In this implementation the variable pivot holds the value of the partitioning element which is the first element in the input array, and i and j are the left and right scan pointers, respectively. The inner loop simply increments (decrements) a pointer and compares an array element against a fixed value. Subsequently the two subarrays are sorted recursively until the entire array is sorted.

Memory Requirements for Quicksort

In practice, after each partitioning the custom is to store the larger of the two arrays and work with the smaller array until a trivially short array is reached. This way, as we shall see below, about long$n$ arrays on average will have to be temporarily stored while awaiting their turn to be sorted.

Consider a randomly unordered array of n keys that we wish to quicksort.

For simplicity, assume that at each partitioning the larger arrays all come from one end.

After the first partitioning we have

$$K(1)\ldots K(p_1)s_1K(q_1)\ldots K(n),$$

where $p_1 \leqslant \dfrac{n}{2}$,

and the larger array is $\{K(q_1)\ldots K(n)\}$.

After the second partitioning (of the shorter array) we have

$$K(1)\dots K(p_2)s_2K(q_2)\dots K(p_1)$$

where $p_2 \leqslant \frac{p_1}{2} \leqslant \frac{1}{2}\left(\frac{n}{2}\right)$,

and the current larger array is $\{K(q_2)\dots K(p_1)\}$.

Repeating the process of partitioning in the above manner we eventually arrive at

$$p_k \leqslant \frac{1}{2}\, p_{k-1} \leqslant \dots \leqslant \frac{n}{2^k} < 2,$$

which yields $k > \log n - 1$, where k is the maximum number of stored arrays. (Note: the trivial array contains one element.)

In the general case of a randomly unordered initial array, the total number of subfiles created is bounded below by $\log n$. In practice, with some clever choice of the partitioning boundaries (see Hoare (1962), Singleton (1969), Frazer and McKellar (1970), Aho, Hopcroft, and Ullman (1974), Knuth (1969)), the actual number of the arrays can be made very close to this lower bound.

An array can be 'stored' by means of two words pointing to the ends of the array (the two words are stored on the stack). Hence Quicksort requires on average an extra space of about $2 \log n$ words. In some special cases this space may be reduced as, for example, when one can work on the array in such a way that the stored arrays all come from one end. The auxiliary space in this case is only half of that in the general case.

One drawback of the algorithm is that it runs very inefficiently on simple arrays. If the algorithm is applied on the input which is already sorted, then the partitions will degenerate. The algorithm will call itself n times, only deducting one element for each call. In this case the total number of arrays created by Quicksort is n and the space required to handle the recursion will be $2(n-1)$ words, which is extremely high. Furthermore, this means that each partitioning process will require the number of comparisons equal to the length of the current array, thus giving the number of comparisons as

$$n + (n-1) + \dots + 1n(n-1)/2 \qquad (7.5.1)$$

This time complexity immediately throws Quicksort into the class of inefficient sorting methods.

An initially sorted input is the worst case for Quicksort. Fortunately there are ways to prevent the worst case occurring in actual applications of the algorithm—see, for example, Sedgewick (1983). We shall mention one such way which suggests using a better partitioning element. The worst case can be avoided if a random element from the input array is used for a partitioning element, since the probability of the worst case occurring will then be negligible. Such a 'probabilistic approach' for the input data treatment is a useful tool in the design of algorithms, where one wishes to avoid any bias in the input. What happens in practice for Quicksort is that three elements from the array are chosen arbitrarily and then their median is used for the partitioning element. One way to choose the three elements is to take one from each left, middle and right of the array. This helps to make the worst case much more unlikely to occur in practical applications, since to 'quicksort' in $O(n^2)$ time, two out of three elements examined must be among the largest or the smallest elements in the array, and happening consistently through most of the partitions.

Average Time Required by Quicksort

In order to estimate the average performance of Quicksort, it is convenient to introduce a time unit as a unit which refers equally to one operation of either comparison or exchange. Quicksort, then lends itself to a straightforward analysis of the average time required to sort an array of n keys.

Denote the average time by $T(n)$.

Assume, for simplicity, that the array is some permutation of the numbers $(1, 2, \ldots, n)$, and that any permutation of these numbers is equally likely to occur.

Suppose that after the first partitioning we have created two arrays with the partitioning boundary s. Denote the time required to quicksort each of the arrays created, by $T(s)$ and $T(n-s)$, respectively.

Since s is equally likely to take on any value between 1 and n, and since the time expended on the first partitioning is cn, where c is a constant and $1 < c < 2$, (which follows from the observation that at the first partitioning n comparisons and $\leqslant n$ exchanges are required), we have

$$
\begin{aligned}
T(0) &= T(1) = 0, \\
T(n) &= \frac{1}{n} \sum_{s=1}^{n} [T(s) + T(n-s)] + cn, \\
&= \frac{1}{n} \left\{ \sum_{s=1}^{n} T(s) + \sum_{j=0}^{n-1} T(j) \right\} + cn \\
&= \frac{1}{n} \left\{ \sum_{s=1}^{n-1} T(s) + \sum_{j=1}^{n-1} T(j) + T(n) \right\} + cn \\
&= \frac{1}{n} \left\{ 2 \sum_{s=1}^{n-1} T(s) + T(n) \right\} + cn,
\end{aligned}
$$

or

$$
(n-1)T(n) = 2 \sum_{s=1}^{n=1} T(s) + cn^2, \quad 1 < c < 2. \tag{7.5.3}
$$

(7.5.3) is a recurrence relation. To solve it, we first get rid of the summation sign:

$$nT(n+1) = 2\sum_{s=1}^{n} T(s) + c(n+1)^2$$

$$(n-1)T(n) = 2\sum_{s=1}^{n-1} T(s) + cn^2$$

Subtracting the two equations we get

$$nT(n+1) - (n-1)T(n) = 2T(n) + c(2n+1),$$

or

$$nT(n+1) = (n+1)T(n) + c(2n+1). \tag{7.5.4}$$

Rewrite the recurrence relation (7.5.4) in the form

$$\frac{T(n+1)}{n+1} = \frac{T(n)}{n} + c\,\frac{2n+1}{n(n+1)}\,.$$

Now, recursively,

$$\begin{aligned}
\frac{T(n+1)}{n+1} &= \frac{T(n)}{n} + c\left[\frac{1}{n+1} + \frac{1}{n}\right]\\
\frac{T(n)}{n} &= \frac{T(n-1)}{n-1} + c\left[\frac{1}{n} + \frac{1}{n-1}\right]\\
&\vdots\\
\frac{T(2)}{2} &= \frac{T(1)}{1} + c\left[\frac{1}{2} + \frac{1}{1}\right],
\end{aligned}$$

which gives

$$\begin{aligned}
\frac{T(n+1)}{n+1} &= c\left[\left(\frac{1}{n+1} + \frac{1}{n} + \ldots + \frac{1}{2}\right) + \left(\frac{1}{n} + \frac{1}{n-1} + \ldots + 1\right)\right]\\
&= c\left[H_{n+1} - 1 + H_{n+1} - \frac{1}{n+1}\right]\\
&= 2cH_{n+1} - c\,\frac{n+2}{n+1}\,.
\end{aligned}$$

Hence

$$T(n) = 2cnH_n - c(n+1) \approx 2cn \log_e n - c(n+1) = O(n \ln n). \tag{7.5.5}$$

7.6 Insertion Sorts

Suppose that a sorted array of n distinct keys $K(1) < K(2) < \ldots < K(n)$ is given. We wish to insert X into the array by comparing it with the keys in the array and then placing X in its proper position among the keys in the array by moving larger keys one position to the right and inserting X into the vacated position. We shall give two simple algorithms which achieve just this.

Algorithm for sequential insertion compares X with $K(1), K(2), \ldots, K(n)$ in turn until one of the three possible cases result: $X < K(1)$, $K(i) < X < K(i+1)$ or $K(n) < X$. If $X < K(1)$ then the entire array is moved one position to the right and X becomes a new first element of the array. If $K(i) < X < K(i+1)$ then $(i+1)$ larger keys are moved one position to the right and X becomes the new $(i+1)$st element of the array, and if $K(n) < X$ then X is simply added to the array as its $(n+1)$st element. An implementation of the algorithm shown below in fact assumes that scanning of the array elements is done from right to left. This way the implementation looks neater. (For comparison, the reader is suggested to work out an implementation of the algorithm with the scan carried out from left to right.)

Algorithm seqinsert

```
pivot := X
i := n
if K[i] < pivot then K[i + 1] := pivot
else
    K[i + 1] := K[i]
    while K[i − 1] > pivot do K[i] := K(i − 1) enddo
endif
K[i − 1] := pivot
```

A careful check of the algorithm using, say, an example will disclose that it does not work properly because the while loop will run past the left end of the array if the pivot is smaller than $K(1)$. One common way to put this right is to set a 'sentinel' key in $K(0)$ which should be at least as small as the smallest element in the array; this is 'cheaper' than including a test (in our case for $i > 0$?) in a situation where the test almost always succeeds within the inner loop.

It is easy to see that algorithm *seqinsert* in the worst case requires n comparisons which is terribly expensive for such a simple process as the insertion of one new key into an array of sorted keys. It also requires up to n key shifts by one position to the right. An algorithm which inserts a new key with only about $\log n$ key comparisons is the algorithm for binary insertion. In this algorithm X is compared first with the key in the middle of the input array. If X is less than the middle element, only the lower half of the array needs to be considered in the subsequent comparisons and vice versa. The process is then repeated on the relevant half of the array, and so on, until the position for X is located. The larger keys are then shifted one position to the right and X is inserted in its proper place. It is obvious that similarly to the sequential insertion the maximum (worst case) number of key shifts for binary insertion remains linearly dependent on the size of the input .

Algorithm binaryinsert

```
i := 1
j := n
pivot := X
while j > k do
    index := (i + j) div 2
    if pivot  < K[index] then j := index - 1
    else i := index + 1
    endif
enddo
for k := n downto i do
K[k + 1] := K[k]
enddo
K[i] := pivot
```

Exercise 7.6.1

A simple method to sort n keys in order is to consider the keys one at a time, inserting each in its proper place among those already sorted. Write an algorithm insertsort based on (a) algorithm *seqinsert* and (b) algorithm *binaryinsert*.

Solution (a): Algorithm seqinsertsort

```
K[0] := minvalue             //sentinel element//
for k := 2 to n do
    pivot := K[k]
    j := k
    while K[j - 1] > pivot do
        K[j] := K[j - 1]
        j := j - 1
    enddo
    K[j] := pivot
enddo
```

In solution (a) algorithm *seqinsertsort* requires $n(n-1)/2$ key moves in the worst case. This high-time complexity seriously handicaps the algorithm. If we study the algorithm we can see that its slow performance is due to the fact that on each run of the inner loop only adjacent keys are exchanged. This results in many redundant moves of the keys, because if, for example, the smallest key happens to be at the wrong end of the array, it takes n steps to move this key into the position where it belongs.

Suppose we generalize the idea of key exchanges by allowing exchanges of keys that are far apart. To do this we introduce the increment, h, and in *seqinsertsort* replace '2' by '$h+1$' and every occurrence of '1' by 'h'. The resulting algorithm so rearranges an input sequence that taking every hth element (starting anywhere in the sequence) a sorted sequence is yielded. h is the distance between two keys that are compared and may possibly be exchanged; itis not fixed and can assume different values for each step of the for loop. In fact, using any sequence of values of h which ends in 1, the algorithm will produce a sorted array. By using some large values of h, elements in the array can be moved long distances, and in this way a reduction in the total number of key moves is achieved. The algorithm is called Shellsort.

Example 7.6.1

A sequence of 17 keys is sorted in order using Shellsort and the increments $h = 13$, 4 and 1.

1 19 15 18 20 9 14 7 5 21 1 13 16 12 29 5 4

$h = 13$

1 12 15 5 4 9 14 7 5 21 1 13 16 19 29 18 20

$h = 4$

1 4 9 5 7 5 14 1 13 16 12 15 18 19 29 21 20

$h = 1$

1 4 5 7 5 9 1 13 14 12 15 16 18 19 21 20 29

$h = 1$

1 4 5 5 7 1 9 13 12 14 15 16 18 19 20 21 29

$h = 1$

1 4 5 5 1 7 9 12 13 14 15 16 18 19 20 21 29

$h = 1$

1 4 5 1 5 7 9 12 13 14 15 16 18 19 20 21 29

$h = 1$

1 4 1 5 5 7 9 12 13 14 15 16 18 19 20 21 29

$h = 1$

1 1 4 5 5 7 9 12 13 14 15 16 18 19 20 21 29

7.7 Optimum Comparison Sorting

In this section we consider some optimal characteristics of comparison sorting. We assume that an array of n keys to be sorted has no known structure, and that the only operation that can be used to gain information about the array is the comparison of two keys. Under these assumptions, we pose the question: what are the minimum requirements on the number of comparisons between keys in the comparative sorting algorithms?

The problem is made more precise under further assumptions that all the keys to be sorted are distinct and that the sorting methods we have in mind are based solely on an abstract linear ordering relation '$<$' between the keys. Then, all n-key comparative sorting methods which satisfy the above constraints can be represented in terms of an extended binary tree structure such a shown in Fig. 7.7.1.

In the tree, each internal node contains two indices '$i:j$' for comparison of $K(i) vs. K(j)$. The left subtree of this node represents the subsequent comparisons to be made if $K(i) < K(j)$, and the right subtree the subsequent comparisons if $K(i) < K(j)$. Each terminal node contains permuation $q_1, q_2, \ldots, q_n$ of $(1, 2, \ldots, n)$ thus denoting the fact that the ordering $K(q_1) < K(q_2) < \ldots < K(q_n)$ has been established.

In Fig. 7.7.1 $K(1)$ is first compared with $K(2)$; if $K(1) > K(2)$ then (via the right subtree) $K(3)$ is compared with $K(4)$. If $K(3) < K(4)$ then $K(2)$ is compared with $K(3)$; if $K(2) < K(3)$ then we know that $K(2) < K(3) < K(4)$, and (via the left subtree) compare $K(1)$ with $K(4)$. With the outcome of $K(1) > K(4)$ the permutation $K(2) < K(3) < K(4) < K(1)$ is yielded.

A comparison of $K(i)$ with $K(j)$ in our tree always means the original keys $K(i)$ and $K(j)$, not the keys that might currently occupy the ith and jth positions of the array after the keys have been shuffled around. Generally some comparisons may be redundant and should be avoided. For example, in Fig. 7.7.2 comparison of $K(1)$ with $K(3)$ is unnecessary because at the stage it is done we already know that $K(1) > K(2)$ and $K(2) > K(3)$, and this implies $K(1) > K(3)$, that is there is no need to compare $K(1)$ and $K(3)$ directly.

Since we are interested in minimizing the number of comparisons, we assume that no redundant comparisons are made. All permutations of the input keys are possible and every permutation defines a unique path from the root to an external node; it follows that there are exactly $n!$ external nodes in a comparison tree which sorts n keys with no redundant comparisons.

Lower Bound on the Maximum Number of Comparisons

We are now able to show that any algorithm which sorts by comparisons must on some array of length n use at least $O(n \log n)$ comparisons. Indeed, if all the internal nodes of a comparison tree are at levels $<k$, then it is obvious that there can be at most 2^k external nodes in the tree. Hence

$$n! \leqslant 2^k, \tag{7.7.1}$$

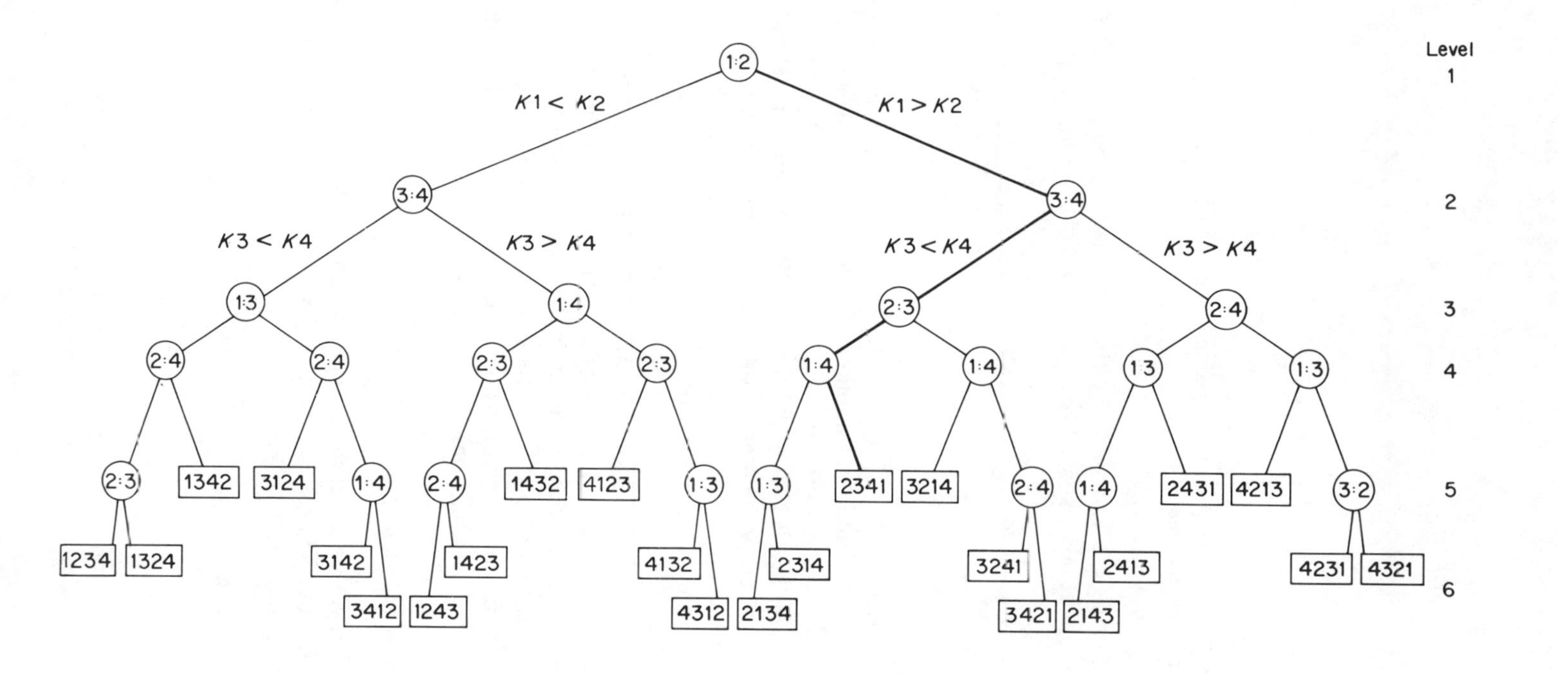

Figure 7.7.1 An extended binary tree for an array of size $n = 4$

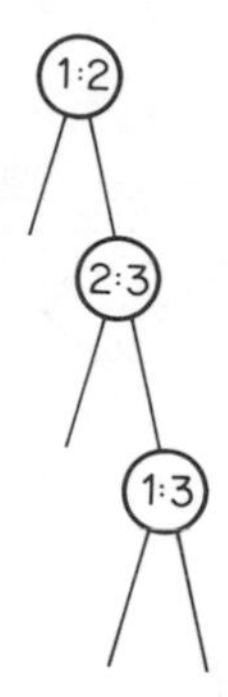

Figure 7.7.2 An example of a redundant comparison, $K(1):K(3)$, in an extended binary tree

where the inequality sign is explained by the fact that the number 2^k may include the redundant comparisons. From (7.7.1) it follows that

$$k \geqslant \log_2 n!, \tag{7.7.2}$$

and by Stirling's formula:

$$k \geqslant O(n \log n). \tag{7.7.3}$$

The extended binary tree of Fig. 7.7.1 shows that if no redundant comparisons are included then k is the minimum number of comparisons that suffices to sort any n keys. This completes the proof.

What we have established is that under the decision tree model it is impossible to design a general purpose comparison sort algorithm which would sort in less than $\log_2 n!$ comparisons. Only very few algorithms have been developed that attain this lower bound on comparisons. One such algorithm is the merge-insertion sort of Ford and Johnson (1959). It calls for no more than the fewest necessary number of comparisons to sort an array of size n, when $n \leqslant 11$ or when $n = 20$ or 21. For other instances of n the algorithm requires the number of comparisons slightly higher than the lower bound but fewer than all other known algorithms. The smallest instance of the problem on which the algorithm can be demonstrated is 5, that is any five keys can be sorted in order in seven comparisons by the merge-insertion.

Merge-Insertion Sort on Five Keys

Let $K1$, $K2$, $K3$, $K4$, $K5$ be an arbitrary permutation of five keys. Compare the two pairs $K1:K2$ and $K3:K4$ (leave out key $K5$). The result can be shown diagrammatically as follows:

x_1 x_2

y_1 y_2 y_3 (2 comparisons)

where x_1, x_2 and y_1, y_2 stand for larger and smaller keys of two sorted pairs, respectively; and y_3 stands for the key which was left out.

Compare the two larger keys of the pairs. The result will be:

(1 comparison)

indicating that $a < b < d, \quad c < d$.

Insert e among (a, b, d). This leads to one of the following situations:

(2 comparisons at most)

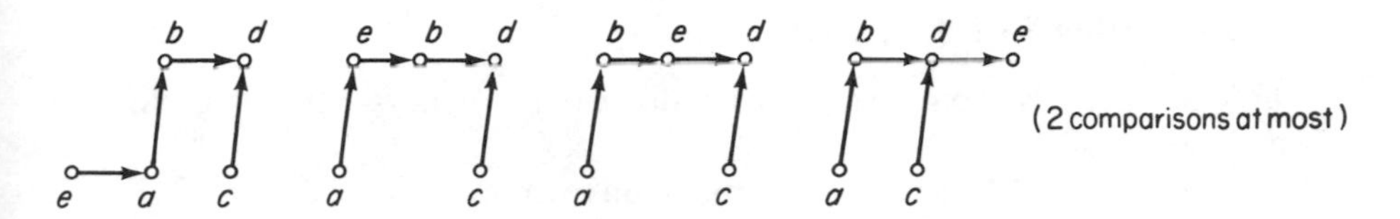

Insert c among the keys which are less than d. This will require two extra comparisons at most, giving seven comparisons in total.

Merge-insertion Sort on 11 Keys

Compare the five pairs $K(1):K(2)$, $K(3):K(4)$, $\ldots$, $K(9):K(10)$. (Leave out $K(11)$.). We get

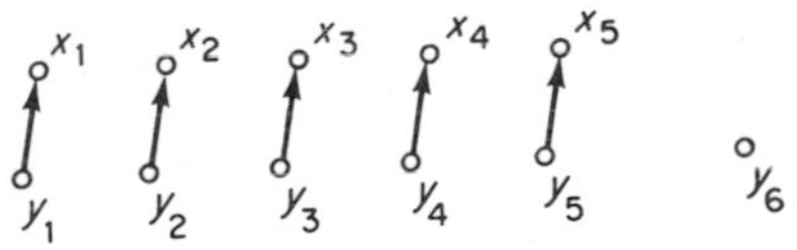

Sort the five larger keys of the pairs using merge-insertion. (Note the recursive use of the method.) A configuration obtained may be shown diagrammatically as

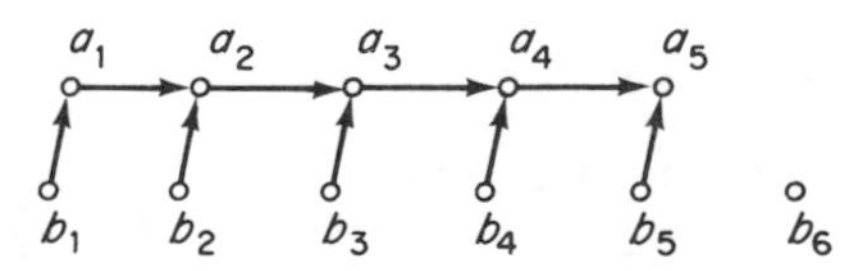

Insert b_3 among (b_1, a_1, a_2) then b_2 among the keys less than a_2. By now we arrive at the configuration

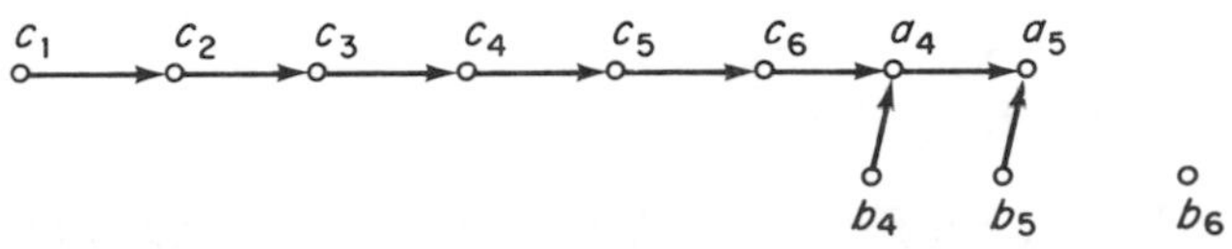

Insert b_5 among $(c_1, c_2, \ldots, c_6, a_4)$ using binary insertion—that is, first compare b_5 to c_4, then to c_2 or c_6, etc. This will take three comparisons at most.

Insert b_6 in its proper place among the sorted keys using binary insertion; hence there will be four comparisons at most.

The total number of comparisons is

$$5 + 7 + (2 + 2) + (3 + 3) + 4 = 26.$$

Merge-insertion Sort on n Keys

Carry out pairwise comparisons of n div 2 pairs of keys. (If n is odd, leave one key out.)

Sort the n div 2 larger keys by merge-insertion.
(Let the N keys now be denoted as

$$a_1, a_2, \ldots, a_{n \text{ div } 2}, b_1, \ldots, b_{n \text{ div } 2},$$

where $a_1 \leqslant a_2 \leqslant \ldots \leqslant a_{n \text{ div } 2}$

and $b_i \leqslant a_i$, $1 \leqslant i \leqslant n$ div 2.)

Call the sequence $a_1, a_2, \ldots a_{n \text{ div } 2}, b_1$ the 'main chain'.)
Insert the remaining b's into the main chain using binary insertion and in the following order

insert b_1

insert b_3 then b_2

insert b_5 then b_4

insert b_{11} then $b_{10}, b_9, b_8, b_7, b_6$

⋮

insert b_{t_k} then $b_{t_k - 1}, \ldots, b_{t_{k-1}+1}$

⋮

until there are no more keys.

The crucial point in the algorithm is the generating of the sequence $\{t_1, t_2, \ldots, t_k, \ldots\} = \{1, 3, 5, 11, \ldots\}$ where t_k is such that each of b_{t_k}, $b_{t_k - 1}, \ldots, b_{t_{k-1}} + 1$ can be inserted with at most k comparisons. An analysis of the merge-insertion shows that $t_k = (2^{k+1} + (-1)^k)/3$.

The number of comparisons required to sort n keys by merge-insertion is given as

$$\sum_{k=1}^{n} \left\lceil \log_2(3k/4) \right\rceil . \qquad (7.7.4)$$

For $n = 1$ to 22 the values of this sum and for the lower bound on comparisons are

$\log_2(3k/4)$	0	1	3	5	7	10	13	16	19	22	26	30	34	38	42	46	50	54	58	62	66	71
$\lceil \log_2 n! \rceil$	0	1	3	5	7	10	13	16	19	22	26	29	33	37	41	45	49	53	57	62	66	70

We see that merge-insertion is truly optimal for $1 \leqslant n \leqslant 11$ and $n = 20, 21$. For other values of n, though above the lower bound, the Ford–Johnson method uses the fewest number of comparisons among all commonly known comparison sorts. However, in 1975 Manacher offered an algorithm which uses fewer comparisons than the merge-insertion for infinitely many n; the smallest such value is $n = 189$. Manacher's algorithm employs a minimum comparison merging of sorted arrays method developed by Hwang and Lin (1972).

7.8 Lower Bound on the Average Number of Comparisons

We have studied the worst-case performance of sorting algorithms. The time performance of an algorithm on average is another important measure, particularly in real-life applications. We shall derive a lower bound on the average number of comparisons required by any comparison sort to sort in order an n-key array. The extended binary tree of Fig. 7.7.1 (page 233) shows that if the external path length of the tree is defined as the sum of the distances from the root to each of the external nodes then the average number of comparisons in a sorting method is equal to the external path of the tree divided by $n!$. The following theorem holds.

Theorem 7.7.1

Assuming that all permutations of a sequence of n keys are equally likely to appear as input the external path length of the tree is at least $n! \log(n!)$.

Proof

Let $D(m)$ be the external path length of an extended binary tree with m terminal nodes. We shall first show that

$$D(m) \geqslant m \log_2 m. \tag{7.8.1}$$

The case $m = 1$ is trivial. Now, assume that statement (7.8.1) is valid for all values of m less than M. Consider an extended binary tree T with M terminal nodes. The tree T consists of a root having a left subtree T_i with i terminal nodes and a right subtree T_{M-i} with $M - i$ terminal nodes for some $1 \leqslant i \leqslant M$. And so, we have

$$D(M) = i + D(i) + (M - i) + D(M - i). \tag{7.8.2}$$

Therefore, the minimum sum is given as

$$D(M) \geqslant \min_{1 \leqslant i \leqslant M} [M + D(i) + D(M - i)]. \tag{7.8.3}$$

Invoking the assumption (7.8.1) we can write

$$D(M) \geqslant M + \min_{1 \leqslant i \leqslant M} [i \log i + (M-i)\log(M-i)] \tag{7.8.4}$$

The expression in square brackets attains its minimum when $i = M/2$, and thus we obtain

$$D(M) \geqslant M + 2\,\frac{M}{2}\log\left(\frac{M}{2}\right) = M + M(\log M - 1) = M \log M. \tag{7.8.5}$$

Setting $M = n!$ we get a lower bound for the external path length of the tree, i.e.

$$D(n!) \leqslant n! \log n! \tag{7.8.6}$$

This completes the proof.

From the theorem it follows that the average number of comparisons in any comparison sort is greater than or equal to

$$\frac{n! \log_2 n!}{n!} = \log_2 n! \approx O(n \log n), \tag{7.8.7}$$

which is a tight lower bound on the average number of comparisons for large n.

7.9 Selection of the kth Largest of n

In some applications one is interested in selecting the kth largest element in a sequence of n elements. One way is to find the median of a sequence. The problem can of course be solved by first sorting thesequence in (non-increasing) order and then picking the kth element. Such a method, as we have already seen, would require at least $n \log_2 n$ comparisons. Can the selection problem be solved 'faster' than that? Suppose that the largest of n elements is required. In this case $n-1$ comparisons are sufficient for the solution. To find the second largest element means first finding both the largest and the second largest elements. This can be done in $n - 1 + \log_2(n-1)$ comparison (Schreier, 1932; Slupecki, 1951; Kislitsyn, 1964). For a general case of $k \leqslant \lceil n/2 \rceil$, it was found that $n - k + (k-1) \times \lceil \log_2(n-k+2) \rceil$ comparisons are sufficient (Hadian and Sobel, 1969). For $k = n/2$ this bound gives $O(n \log n)$. It was believed for a long time that to find the median involves sorting at least one-half of the array, that is that the time complexity of $O(n \log n)$ is asymptotically optimal for the problem. Then in 1973 Blum *et al.* found that by careful application of the divide-and-conquer strategy (that is, partitioning the problem) and recursion the kth largest element can be found in $O(n)$ time. The basic idea of the approach is as follows:

Let $K(1), K(2), \ldots, K(n)$ be a sequence of keys, all distinct. Select an arbitrary partitioning element, Q, among the keys. Compare Q to each of the $n-1$ other keys rearranging the keys (as in Quicksort) so that all keys greater than Q appear in positions $K(1), \ldots, K(t-1)$, to the left of q, and all keys less than Q appear in positions $K(t+1), \ldots, K(n)$, to the right of Q. Q is the tth largest. If $k = t$, the problem is solved. If $k < t$, use the same method to find the kth largest among the first $(t-1)$ keys. If $k > t$, use the same method to find the $(k-t)$th largest of the last $(t-1)$ keys.

An implementation of the selection algorithm is given below with the partitioning element chosen to be the right end key of the input array.

```
Algorithm select (K, r, k, s)
    i := r - 1
    j := s
    pivot := K[j]                    //the partitioning element//
    if s > 1 then
    //partitioning process//
        repeat
            repeat i := i + 1 until K[i] := pivot
            repeat j := j - 1 until K[j] := pivot
            if i < j then
                temp := K[j]; K[j] := K[i]; K[i] := temp
            endif
        until j ≤ i
        K[s] := K[i]; K[i] := pivot     //the partitioning element belongs
                                          at position i//
        case k of
            : k = i : return(i)          //K[i] is the kth largest//
            : k < i : select(K, r, k, i - 1)
            : k > i : select(K, i + 1, k - i, s)
        endcase
    endif
```

Bounds on the Maximum Number of Comparisons

The selection of the kth largest element is achieved in linear time provided that a partitioning element is found in linear time such that the given array is partitioned into two arrays $K(1), \ldots, K(t-1)$ and $K(t+1), \ldots, K(n)$ and each of the two arrays is no more than a fixed fraction of the size of the original array. The trick is in how the partition element Q is chosen.

To see how this can be done we shall now set up a partitioning process so as to satisfy as closely as possible the minimal requirements on the number of comparisons. The process is developed in four steps.

Step One

Partition the input array into arrays of r elements each, where r is a reasonably small number. For clarity we assume that in our exposition $r = 15$.

Sort each of the small arrays in (non-increasing) order, noting that any 15 elements can be sorted in order with 42 comparisons (for example, by merge-insertion).

The number of comparisons required by this step equals $42(n/15)$.

Now let $C(n)$ be the maximum number of comparisons needed to find the kth largest from scratch, and $C'(n)$ the maximum number of comparisons needed to find the kth largest among n elements after they have been partitioned into sorted arrays of length 15 each. We obtain the relation

$$C(n) = 42\left(\frac{n}{15}\right) + C'(n). \tag{7.9.1}$$

Step Two

From each sorted array select the median and form the array of medians. The median array contains $n/15$ elements.

Find the median of the array of medians.

Denote the number of comparisons required by this step by $C(n/15)$. After completion of Step Two the situation is as shown in Fig. 7.9.1. Here the rows represent $n/15$ arrays of length 15, each sorted in decreasing order. The central vertical strip is the array of medians and M is its median. The elements of A are known to be greater than M. The elements of D are known to be less than M. The relations to M of the elements in blocks B and C are unknown.

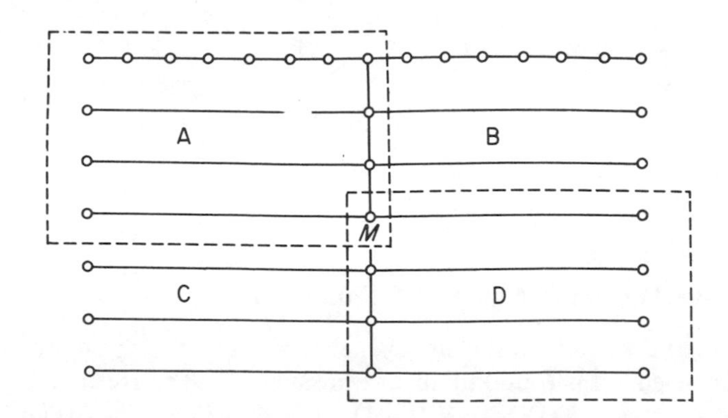

Figure 7.9.1 The rows represent $n/15$ arrays of length 15, each sorted in decreasing order. The central vertical strip is the array of medians and M is its median. The elements of A are known to be greater than M. The elements of D are known to be less than M. The relations to M of the elements in blocks B and C are unknown

Step Three

Carry out binary insertion of the median M into each of the rows in blocks C and B in order to find out exactly where among the elements of blocks B and C the median M belongs.

This step calls for $3(n/15)$ comparisons since each row of C and B contains seven elements and so an insertion into a row requires three comparisons at most.

Step Four

By this time the original array has been partitioned into two arrays, one on the left of M containng all elements greater than M, and one on the right of M containing all elements less than M.

Let $m-1$ be the number of elements which are greater than M. This makes M the mth element, and one of the following outcomes occurs. If $k=m$ then the solution is found; if $k>m$ then we need to find the $(k-m)$th largest of the $(n-m)$ small elements; if $k<m$ then we need to find the kth largest element of the m large elements.

The observation that allows savings in the total number of comparisons at this step is this. Since the relative order between M and the elements of A and D is known, each of the two arrays, $\{m-1$ large elements$\}$ and $\{n-m$ small elements$\}$, consists of no more than $B+C+\{$either A or $D\}$ blocks, that is of no more than $(3/4)n$ elements. We further note that of $(3/4)n$ elements, $n/2$ elements are sorted arrays of length 15 each and the remaining $n/4$ elements are sorted arrays of length 7 each. The process then proceeds to pairwisely merge the shorter arrays into arrays of length 15, with a dummy element is added after the merge of each pair. The process requires $13(n/60)$ comparisons since 13 comparisons at most are needed to merge two arrays of length seven each.

When the process of merging is complete the problem is reduced to the problem of finding an element of an arbitrary fixed rank, that is, either the kth largest of the $m-1$ large elements or $(k-m)$th of the $(n-m)$ small elements. In either case the array to be processed is at most $(3/4)n$ elements long and is structured as a set of the 15-element sorted arrays.

Denote by $C'(3n/4)$ the maximum number of comparisons required to find the kth or the $(k-m)$th largest of this new array.

We are now ready to complete the time analysis of the entire algorithm. Steps Two to Four yield the following relation

$$C'(n)=C\left(\frac{n}{15}\right)+3\left(\frac{n}{15}\right)+13\left(\frac{n}{60}\right)+C'\left(\frac{3}{4}n\right) \tag{7.9.2}$$

Substituting (7.9.1) into (7.9.2) we obtain

$$C'(n)=42\left(\frac{n}{225}\right)+C'\left(\frac{n}{15}\right)+3\left(\frac{n}{15}\right)+13\left(\frac{n}{60}\right)+C'\left(\frac{3n}{4}\right)$$

which gives

$$C'(n) = 0.6033n + C'\left(\frac{n}{15}\right) + C'\left(\frac{3n}{4}\right).\tag{7.9.3}$$

Assuming that $C'(n)$ is of the form $C'(n) = Kn$, we obtain

$$Kn = 0.6033n + \frac{Kn}{15} + \frac{3Kn}{4},$$

so that $K = 3.29$ and $C'(n) \leqslant 3.29n$. (7.9.4)

Substitution of $C'(n)$ into (7.9.1) yields

$$C(n) \leqslant 2.8n + 3.29n = 6.09n.\tag{7.9.5}$$

As shown by Blum *et al.* (1973) this boundary can be reduced further to $C(n) \leqslant 5.43n$ using the more scrupulous strategy of rejecting everything greater (or less) than the median M and reconstituting arrays of length 15 from the irregular arrays that remain.

The values for r other than 15 can be used as well. However one needs to check that the value of r is chosen so that the assumption $C'(n) = Kn$ in (7.9.4) holds, that is that the total length of the fraction-arrays processed in Steps One, Two and Three sum up to the array size which is less than the n.

A lower bound on $C(n)$ is asserted by the theorem which states that any algorithm which determines the kth largest of an unordered array of length n requires at least

$$n - 1 + \min\{k - 1, n - k\}$$

comparisons. (This gives $\frac{3}{2}n$ comparisons to find the median of the array.) An elegant proof of this theorem may be given using the concept of combinatories; see, for example, Yeh (1976).

Upper Bound on the Average Number of Comparisons

We assume a random order on the array and estimate the average number of comparisons required to find the kth largest of n elements.

Let $C_k(n)$ be the average number of comparisons required by the algorithm of the previous section, to find the kth largest element. Then the following relation is asserted:

$$C_k(n) = K_1 n + \frac{1}{n}\sum_{m=k+1}^{n} C_k(m) + \frac{1}{n}\sum_{m=1}^{k+1} C_{k-m}(n-m)\tag{7.9.6}$$

where $K_1 n$ is the number of comparisons required by the algorithm to establish the rank of the 'median of medians', M, among the elements of the array, that is, steps One to Three of the algorithm.

If $k > m$ then the average number of comparisons needed to find the kth largest of the array $K(1), \ldots, K(m)$ is $C_k(m)$ and if $k < m$, the average number of comparisons to find the $(k-m)$the largest of the array $K(m+1), \ldots, K(n)$ is $C_{k-m}(n-m)$. In (7.9.6) the second term on the right corresponds to the case of $k > m$ and the third term, to the case of $k < m$.

Relation (7.9.6) can be written as

$$C_k(n) = K_1 n + \frac{1}{n} \sum_{j=k}^{n-1} C_k(j) + \frac{1}{n} \sum_{j=n-k+1}^{n-1} C_{k+j-n}(j), \tag{7.9.7}$$

giving

$$C_k(n) \leqslant K_1 n + \max_k \left\{ \frac{1}{n} \left[\sum_{j=k}^{n-1} C_k(j) + \sum_{j=n-k+1}^{n-1} C_{k+j-n}(j) \right] \right\}. \tag{7.9.8}$$

We assume that $C_k(1) \leqslant K_1$ and then show that

$$C_k(n) \leqslant 4K_1 n \text{ for all } n \geqslant 2. \tag{7.9.9}$$

Proof

For $n = 2$ we have

$$C_k(2) \leqslant 2K_1 + \max_k \left\{ \frac{1}{2} \left[\sum_{j=k}^{1} C_k(j) + \sum_{j=3-k}^{1} C_{k+j-2}(j) \right] \right\} \leqslant 4(2K_1).$$

Assume that $C_k(n) \leqslant 4K_1 n$ holds for all $n \leqslant N-1$, then write (7.9.8) for $n = N$ as

$$C_k(N) \leqslant K_1 N + \max_k \left\{ \frac{4K_1}{N} \left[\sum_{j=k}^{N-1} j + \sum_{j=N-k+1}^{N-1} j \right] \right\}. \tag{7.9.11}$$

The summation in the square brackets yields

$$C_k(N) \leqslant K_1 N + \max_k \left\{ \frac{4K_1}{N} \left[(N-1)N - \frac{(k-1)k}{2} - \frac{(N-k)(N-k+1)}{2} \right] \right\},$$

which after the maximization of the function in the square brackets gives

$$C_k(N) \leqslant K_1 N + \frac{4K_1}{N} \left[\frac{3}{4} \left(N^2 - \frac{4}{3} N + \frac{1}{3} \right) \right] \leqslant 4K_1 N.$$

The proof is now complete.

Exercises

1. Rewrite the simple selection sort algorithm so that on each scan it would search simultaneously for the current largest and smallest keys and then place each of the keys into their proper positions within the array.

2. Sort in order the array 72 39 51 81 3 13 5 17 33 49 11 using algorithm *heapsort*.
3. Write an algorithm to build up a heap assuming that one element at a time is inserted into an already existing heap. Show that the performance of the algorithm is of $O(n \log n)$.

   [*Solution:*

   Algorithm buildheap (s)
   $s := s + 1$
   $K[s] := newkey$
   $pivot := K[s]$
   while $K[s \text{ div } 2] \leqslant pivot$ **do**
   $K[s] := K[s \text{ div } 2]$
   $s := s \text{ div } 2$
   enddo
   $K[s] := pivot$]

4. Show that in algorithm *heapsort* the initial heap is constructed in $O(n)$ time.
5. Prove that any comparison-based algorithm which finds the largest and second largest largest key in an array of n unordered keys requires $n + \lceil \log_2 n \rceil - 2$ comparisons.

   [*Solution*

   If the problem is solved using the notion of heap, then the input array is first rearranged into the heap using $n - 1$ comparisons. After removal of the largest element from the root node of the tree the updating of the tree takes $\log_2 - 1$ comparisons and this operation places the second largest key into the root node of the new heap. Hence the total number of comparisons to select the largest and the second largest keys is equal to $n - 1 + \lceil \log_2 n \rceil - 1$.]

6. Construct a binary tree of minimum height for finding the largest of six keys, proving the fact that the tree is of minimum height.
7. Find what permutations of 1 2 3 4 5 are transformed into 5 3 4 2 1 by the heap creation phase of the *heapsort*.
8. How many inversions has the permutation 7 2 3 8 9 1 5 6?
9. What is the maximum number of times that the largest key could be moved during the execution of *quicksort*?
10. One way to improve performance of *quicksort* is to use as the partitioning element the median of three elements which are chosen arbitrarily from the input sequence. Justify why or why not the performance of *quicksort* is improved in this way.
11. True or false: If the median-of-three approach improves performance of

quicksort then one would have thought that a further improvement may be achieved by using as the partitioning element the median of five or more elements chosen from the input sequence. Write a program and test the conjecture by some experiments on large and small arrays.

12. A momentary reflection suggests that there is a recursive element in the formulation of the binary insertion algorithm. Write a recursive implementation of the binary insertion, *bininsert*(k, s), and estimate the number of key shifts the algorithm would require in the worst case.
13. Write a basic algorithm shellsort following instructions given in the text.
14. Sort seven keys in order in 13 comparisons using merge-insertion algorithm.
15. Show that in merge-insertion algorithm the subsequences of the smaller keys to be inserted into the sorted sequence of larger keys if guided by the parameter t_k which is given as $(2^{k+1} + (-1)^k)/3$.
    Hint. First derive and then solve the recurrence $t_{k-1} + t_k = 2^k$.
16. Analyse the O(n) selection algorithm of Tarjan *et al* to find the kth largest of n keys for the cases of $r = 7, 9, 13, 17, 19$
17. Given an arbitrary array of n distinct keys where n is a perfect square the following method is used to select the largest key:
    divide the array into $\sqrt{n}$ groups of $\sqrt{n}$ keys each,
    find the largest key of each group, that is find the 'group leaders',
    find the largest key of the entire array by considering the 'group leaders' array and selecting its largest key.

    Extend the method to sort in order a complete array and give an estimate of the total number of comparisons required by this sort in the worst case.
18. Show that $\sum_{j=1}^{n} H_j = (n+1)H_n - n$.
19. The following are four versions of an algorithm that finds for a given value K and an array KEY of size n, the value of the element of KEY with the smallest subscript i such that KEY$(i) \geqslant K$. Study each of the versions and then explain precisely what each algorithm does and for each of the versions indicate the danger points and possible pitfalls due to which, for certain inputs, incomplete or altogether incorrect results may be produced.

```
(a) answer := 0
    for i := 1 to n do
        if KEY[i] >= K then
            answer := KEY[i]
        endif
    enddo
```

```
(c) found := false
    for i := to n do
        if KEY[i] >= K then
            if not found then
                found := true
                answer := KEY[i]
            endif
        endif
    enddo
```

```
(b) found := false
    for i := 1 to n do
        if KEY[i] ⩾ K then
            found := true
            answer := KEY[i]
        endif
    enddo
```

```
(d) found := false; i := 1
    while (i ⩽ n and
    not found) do
        if KEY[i] ⩾ K then
            found := true
            answer := KEY[i]
        end
        else i := i + 1
        endif
    enddo
```

[*Answer*:

(a) The algorithm may given an incorrect answer for the value of $K = 0$.
(b) The algorithm breaks down for the null input.
(c) In the algorithm the **for loop** uses redundant steps since the loop needs to be carried out fully no matter what input.
(d) The algorithm performs correctly for every input.]

20. Write an algorithm to rearrange an array of n keys so that all the keys equal to the median are in place, with larger keys to the left and smaller keys to the right.
21. Outline an algorithm which takes an array of $3n$ keys and rearranges the n largest keys in the first n positions, the next n in the next n positions, and the n smallest keys in the last n positions.
22. Suppose that an application requires selection of the kth largest element (for various arbitrary k) at large number of times on the same sequence. (a) Suggest an application with such a need; (b) suggest a method which would be best in this case.

8

External Sorting

External sorting means sorting an array under the assumption that the length of the array is larger than the computer can hold in its internal memory.

External sorting is quite different from internal sorting as the data are stored on comparatively slow peripheral memory devices (such as tapes, discs, drums) and thus the data structure has to be arranged in such a way that these devices can cope quickly with the requirements of the sorting algorithm. It is customary to call by a (sequential) file, an array which resides on disc or tape, the latter implying that at each moment one and only one of the file elements is directly accessible. Accordingly, external sorting is sometimes referred to as sorting of (sequential) files. The majority of internal sorting techniques—for example selection, exchange, insertion—are virtually useless for external sorting.

An exception is sorting by merging where two or more sorted files are combined into one sorted file. Sorting by merging is used in both internal and external sorts. The most external sorts are an in-core sort followed by 'external merge'.

In many data processing systems a large sorted data file (master file) is maintained and to it new entries are added on a regular basis. Instead of adding each new entry as soon as it is input into the system, a number of new entries are 'batched' into a separate file, which is then sorted, appended to the master file and the whole extended file resorted by merging the two files. Many other similar applications are well suited to be solved by merging. External merge can be done without difficulty on the least expensive external memory devices as this process uses only very simple data structures, such as linear lists which are traversed in a sequential manner as stacks or as queues. (A linear list is a set of elements with linear relationships between these elements, i.e. all elements except for the last and first are preceded by exactly one element and followed by exactly one element in the list. A stack and a queue are linear lists which work on the principle 'last-in-first-out' (LIFO) and 'first-in-first-out' (FIFO), respectively.)

We shall first consider the basic merge technique which merges two sorted files, a process called a two-way merge. Next ways will be discussed in which merging is adapted for the needs of the external sorting.

8.1 The Two-way Merge

The two-way merge is applied to two sorted input files of size n (input file N) and m (input file M) in order to produce an output file L of size $n + m$. Essential to the technique are three pointers, of which two, i and j, point to the current positions in files N and M and the third, k, to the next position in the output file L. It is assumed that all keys preceding i and j have already been merged into $k - 1$ positions of file L. Keys pointed to by i and j are compared and the smaller key is transferred to position k in L. The pointers are updated and the process is repeated.

The following is a direct implementation of this method.

Algorithm simple_2_way_merge

```
i := 1; j := 1
M[m + 1] := maxkey; N[n + 1] := maxkey //sentinel elements//
for k := 1 to m + n do
    if M[i] < N[j]
    then begin L[k] := M[i]; i := i + 1 end
    else begin L[k] := N[j]; j := j + 1 end
    endif
enddo
```

To simplify the implementation two sentinel keys with values larger than all the other keys are added, one in each array, M and N. When the M array is exhausted, the loop moves the rest of the N array into the L array. The time taken by the algorithm is obviously proportional to $m + n$.

We note that the process calls for an additional memory space of $m + n$ locations (records) to store the output file. The extra storage requirement is an essential feature of merge sorts in general. And though some reduction in the volume of additional storage may be possible due to various devices that make use of parts of files M and N which becomes vacant with the progress of merging, the merge sorts cannot be done *in situ*, since they require at least as many data moves as comparisons.

So, in merging two files, for each comparison made, at least one key goes into the output file; e.g. in the two-way merge there are $m + n$ transmissions of keys. In general, this means that in estimating the time complexity of merge algorithms, the operation of transmission of a key is as basic as that of comparison of keys.

Since additional storage is required for a practical implementation, one may consider ways to reduce its volume. If one link field is added to each of the keys in files M and N thus forming two sorted lists, M and N, then everything required by the merging algorithm can be done using simple link manipulations without moving the keys at all. This device is particularly beneficial when keys are long and occupy several words each.

Suppose that the files M and N are set up as two linked lists which are pointed at by two pointers, p and q, respectively. An auxiliary array LINK is also set up such that at the end of the merge the elements of LINK will contain the pointers to the elements of M and N so as to give an enumeration of the entire population of these elements as a sorted sequence.

A basic implementation of the algorithm is this.

Algorithm listmerge

```
i := p; j := q; k := 0
while i<>null j <>null do
    if i.key ⩽ j.key
    then begin LINK[k] := i; i := i.next end
    else begin LINK[k] := j; j := j.next end
    endif
    k := k + 1
enddo
repeat
    if i = 0
    then begin LINK[k] := j; j := j.next end
    else begin LINK[k] := i; i := i.next end
    endif
    k := k + 1
until (i = null and j = null)
```

In this implementation of *listmerge* the repeat loop transfers into LINK the pointers to the elements of the remaining list after one of the two lists has been exhausted. The last node of any list has its link field value set to null.

Comparisons in the Two-way Merge

The maximum and minimum number of comparisons required in the two-way merge are readily obtained and are given as $m+n-1$ and $\min\{m,n\}$, respectively.

However, estimation of the average number of comparisons calls for a more careful analysis.

Denotig by $C(n,m)$ the number of comparisons required to merge two sorted files of length m and n respectively, we can write

$$C(n,m) = m + n - S, \tag{8.1.1}$$

where S is the number of keys transmitted after exhaustion of one of the files, i.e. if the file M is exhausted after comparing $K(m)$ of the file M with $K(j)$ of the file N, then the keys $K(j), \ldots, K(n)$ of the file N are simply transmitted to the file C and thus

$$S = n - j + 1. \tag{8.1.2}$$

From (8.1.1) it follows that to find the average value of C, we have to find the average value of S.

Now, let s be some fixed integer such that $1 \leqslant s \leqslant m + n$. Then the probability that $S \geqslant s$ is given as

$$P(S \geqslant s) = p_s = \frac{\begin{pmatrix}\text{The number of possible}\\ \text{arrangements of } m \text{ keys}\\ \text{of file } M \text{ among first}\\ n - s \text{ keys of file } N\end{pmatrix} + \begin{pmatrix}\text{The number of possible}\\ \text{arrangements of } n \text{ keys}\\ \text{of file } N \text{ among the first}\\ m - s \text{ keys of file } M\end{pmatrix}}{\begin{matrix}\text{The total number of possible}\\ \text{arrangements of } m \text{ keys of file}\\ M \text{ among } n \text{ keys of file } N\end{matrix}}$$

or

$$p_s = \frac{\binom{m+n-s}{m} + \binom{m+n-s}{n}}{\binom{m+n}{m}}, \quad 1 \leqslant s \leqslant m + n, \tag{8.1.3}$$

where $\binom{X}{Y}$ stands for the binomial coefficient.

The mean and the variance of S

$$\text{Let } q_s = P(S = s) \text{ for } 1 \leqslant s < m + n. \tag{8.1.4}$$

Then, the mean of S, $\mu_{mn}[S]$, and the variance of S, $\sigma_{mn}{}^2[S]$, can be defined as

$$\mu_{mn}[S] = \sum_{s=1}^{m+n} s q_s \tag{8.1.5}$$

and

$$\sigma_{mn}{}^2[S] = \sum_{s=1}^{m+n} s^2 q_s - \mu_{mn}{}^2[S], \tag{8.1.6}$$

respectively.

Now,

$$p_s = P(S \geqslant s) = P(S = s) + P(S = s + 1) + \ldots + P(S = m + n)$$
$$= q_s + q_{s+1} + \ldots + q_{M+N}$$

and, similarly,

$$p_{s+1} = P(S \geqslant s + 1) = q_{s+1} + q_{s+2} + \ldots + q_{m+n}$$

Subtracting the last two expressions we get

$$p_s - p_{s+1} = q_s. \tag{8.1.7}$$

Substitution of (8.1.7) into (8.1.5) and (8.1.6) yields

$$\mu_{mn}[S] = \sum_{s=1}^{m+n} sq_s = \sum_{s=1}^{m+n} p_k \tag{8.1.8}$$

and

$$\sigma_{mn}{}^2[S] = \sum_{s=1}^{m+n} s^2 q_s - \mu_{mn}{}^2[S] = \sum_{k=1}^{m+n} (2k-1)p_k - \mu_{mn}{}^2[S]. \tag{8.1.9}$$

Using (8.1.3) the equation (8.1.8) can be written as

$$\mu_{mn}[S] = \sum_{s=1}^{m+n} \frac{\binom{m+n-s}{m} - \binom{m+n-s}{n}}{\binom{m+n}{m}}$$

$$= \sum_{s=1}^{n} \frac{\binom{m+n-s}{m}}{\binom{m+n}{m}} + \sum_{s=1}^{m} \frac{\binom{m+n-s}{n}}{\binom{m+n}{n}}, \tag{8.1.10}$$

since

$$\binom{m+n-s}{m} = 0 \quad \text{for } s > n,$$

$$\binom{m+n-s}{n} = 0 \quad \text{for } s > m$$

and

$$\binom{m+n}{m} = \binom{m+n}{n}.$$

Further, by virtue of the formula

$$\sum_{k=0}^{p} \binom{r+k}{k} = \binom{r+p+1}{p},$$

relation (8.1.10) can be brought to the form

$$\mu_{mn}[S] = \frac{\binom{m+n}{n-1}}{\binom{m+n}{m}} + \frac{\binom{n+m}{m-1}}{\binom{m+n}{n}},$$

giving

$$\mu_{mn}[S] = \frac{n}{m+1} + \frac{m}{n+1}. \tag{8.1.11}$$

In a similar manner, for the variance (8.1.9) we obtain

$$\sigma_{mn}^2[S] = \frac{n(2n+m)}{(m+1)(m+2)} + \frac{m(2m+n)}{(n+1)(n+2)} - \mu_{mn}^2[S]. \tag{8.1.12}$$

Hence the number of comparisons in the two-way merge is given as

$$C(n,m) = \begin{cases} \text{min } \min(n,m), \\ \text{ave } m+n-\mu_{mn}, \text{ dev } \sigma_{mn}, \\ \text{max } n+m-1. \end{cases} \tag{8.1.13}$$

where μ_{mn} and σ_{mn}^2 are determined by (8.1.11) and (8.1.12), respectively.

8.2 Merge Sorting

Two commonly known sorting methods based on two-way merge are the balanced two-way merge sort and the natural two-way merge sort. In both methods the input file is examined from both ends working towards the middle. However the two sorts differ in the way in which the intermediate sorted subfiles (or runs) are brought about. In the balanced two-way merge sort a file of n keys one starts with n runs of lengths 1 and merges them into runs of length 2, except possibly the last run. Next the sorted two-element runs are merged into runs of length 4, and so on until the complete file is output in order. The method is illustrated in Example 8.2.1. The first two runs of length 1 are the first and the last keys of the file, and they are merged and placed in the first two locations of the output file. The next two runs of length 1 are the second and the penultimate keys of the input file, and they are merged and placed in the last two locations of the output file, and so on. After k complete passes over the data the generated runs have each length 2^k, except possibly for the last run. The balanced two-way merge sort is best suited for $n = 2^m$, but it can be used for n other than a power of 2.

In the natural two-way merge sort the lengths of the initial runs are decided by the presence of 'naturally sorted' runs in the data file. Suppose we start at the left end of the input file. The first j keys form a sorted sequence while the $(j+1)$st key is the first out-of-order key. It means that the initial run on the left end is of length j. Similarly at the right end of the data file the initial run is formed by $K(n), \ldots, K(n-t)$. This situation is shown in Fig. 8.2.1.

The two 'natural' runs $\{K(1), \ldots, K(j)\}$ and $\{K(n), \ldots, k(n-t)\}$ are then merged and placed at the left end in the output file. The next pair of natural runs is generated in a similar manner, merged into one sorted run and placed at the right end in the output file, and so on until the complete file has been sorted.

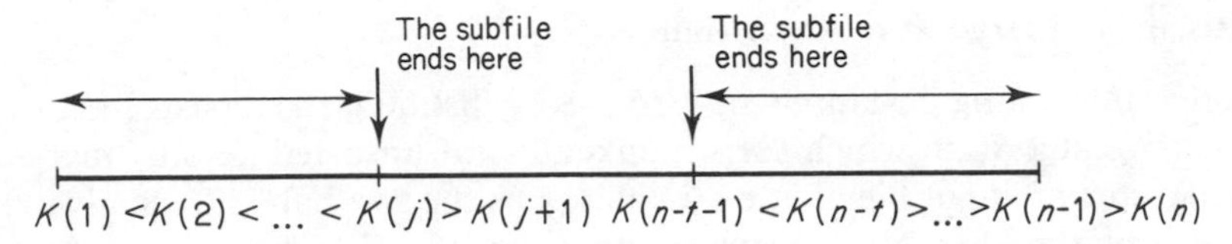

Figure 8.2.1 Generation of 'natural' subfiles in natural two-way merge sort

Example 8.2.1

This is a balanced two-way merge sort on eleven keys.

	$\overrightarrow{46}$	$\overrightarrow{2}$	$\overrightarrow{31}$	$\overrightarrow{54}$	$\overrightarrow{10}$	$\overrightarrow{25}$	$\overleftrightarrow{14}$	$\overleftarrow{72}$	$\overleftarrow{18}$	$\overleftarrow{39}$	$\overleftarrow{74}$	$\overleftarrow{91}$	$\overleftarrow{61}$
Pass 1	46	61	31	74	10	18	14	72	25	54	39	91	2
Pass 2	2	46	61	91	10	18	25	72	14	74	54	39	31
Pass 3	2	31	39	46	54	61	74	91	72	25	18	14	10
Pass 4	2	10	14	18	25	31	39	46	54	61	72	74	91

Maximum Number of Comparisons in the Balanced Two-way Merge Sort

Let the data file be of size n. At the start of the process there are n runs of length 1, and in the first pass they are sorted into $n/2$ runs of length 2 in

$$1 \times \left\lceil \frac{n}{2} \right\rceil \text{comparisons.}$$

In the second pass $n/2$ runs of length 2 are sorted into runs of length 4, except possibly the last run, in

$$3 \times \left\lceil \frac{n}{2^2} \right\rceil \text{comparisons.}$$

In general in the rth pass there are $n/2^{r-1}$ runs of length 2^{r-1}, except possibly the last run. These are sorted into runs of length 2^r in

$$(2^r - 1) \times \left\lceil \frac{n}{2^r} \right\rceil \text{comparisons.}$$

Thus we have

$$1 \times \left\lceil \frac{n}{2} \right\rceil + 3 \times \left\lceil \frac{n}{2^2} \right\rceil + \ldots + (2^r - 1) \times \left\lceil \frac{n}{2^r} \right\rceil + \ldots + (2^k - 1) \times \left\lceil \frac{n}{2^k} \right\rceil$$

$$= nk - \left\lceil \frac{n}{2} \right\rceil \sum_{s=0}^{k-1} \frac{1}{2^s}, \quad \text{where } 2^k < n \leqslant 2^{k+1}. \tag{8.2.1}$$

This gives the maximum number of comparisons as $(0(n \log n)$. The estimate is exact if n is a power of 2, and a slight overestimate if n is not an exact power of 2.

A Recursive Merge Sort Algorithm

Another interesting development is to use the listmerge as a basis to construct a recursive algorithm which sorts a linked list of unsorted keys by merging *in situ*. To sort an input file divide it in half, sort the two halves (recursively) and merge them together. No temporary nodes, no extra lists are needed: the merge sorting is done by rearranging the nodes within the data list. The following algorithm is a direct implementation to recursively merge sort a linked list. The algorithm takes a pointer, r, to an unsorted list as an input and outputs the value of the pointer to the sorted version of the list.

```
Algorithm recmergesort (r, n)
    if r.next = null then return (r)
    else
        p := r
        for i := 2 to n div 2 do r := r.next enddo
            q := q.next
            q.next := null
            listmerge(recmergesort(p, ndiv2),recmergesort(q, n - (ndiv2)))
    endif
```

The algorithm sorts an input list of size n by splitting it into two halves, pointed to by p and q, sorting the two halves recursively, then 'listmerging' them to produce the final sorted list. Again the convention that the last node of any list is set to null is adhered to here. The data structures assumed in the algorithm are depicted in Fig. 8.2.2.

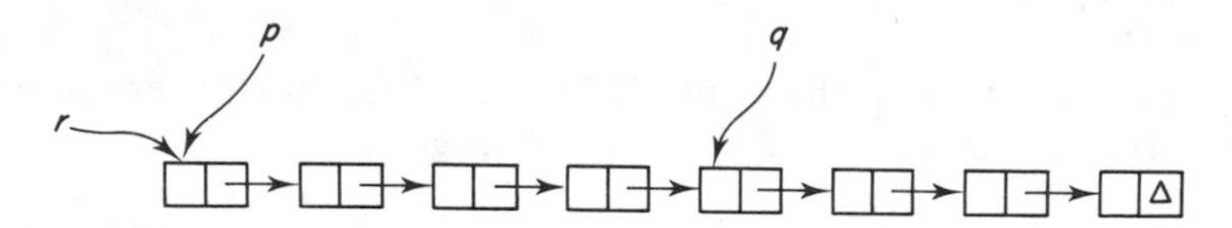

Figure 8.2.2 An unsorted input list N is pointed to by pointer r. The pointers p, and q point to each of the two unsorted halves of N

8.3 Merge in External Sorting

The two-way merge can in an obvious way be extended to a multiway (or P-way) merge when not 2 but P initially sorted runs are used simultaneously in the merging process to produce one sorted run. This idea is demonstrated in Fig. 8.3.1.

AA	AA	AA	AA	AG	GI	IN	NO	OR	RS	ST	T
			DE	EG	GG	GI	IM	MN	NN	NR	
		AD	AD	AE	GG	IG	NI	OM	RN	SN	TR
			DA	EE	GE	GL	IM	MP	NX		
	IN		SO	RT	IN	GA	ND	ME	RG		
			IN	NG	EX	AM	PL	LE			

Figure 8.3.1 A six-way merge; the smaller key AA is output, the first sub-file is updated and the merge step repeated

In this extended form merging is widely used in external sorting. We shall now turn to discuss some of the methods, in particular the methods associated with the use of magnetic (mag) tapes as the means for auxiliary storage.

Suppose that the data file is much too large to fit into internal memory of a computer and has to be stored on some peripheral memory device, such as mag tape. Suppose further that at the start of a sorting process a specified number of blank tapes is available which can be used to store intermediate runs. This number of tapes is an important parameter that affects performance of a sorting process. Normally the keys are stored on the tape sequentially, and so to access them in the same order as they have been read-in the tape needs first to be rewound. This operation takes time which has to be noted accordingly.

To avoid rewinding one can think of reading the tape backwards, but in this case the generated runs will have the reverse ordering, in other words, if the original run has been read onto the tape in ascending order then reading it from the tape backwards will produce the run in descending order. The features of sorting which are associated with the use of the tapes or any other storage device depend on the current computer technology, and the method's efficiency relies upon the 'systems' aspect of the problem as much as the 'algorithm' aspect. For example, 'good' sorting algorithms should require shorter rewinding time and a balanced distribution of the operations of reading the tapes forward and backward. The 'best' algorithms developed will remain such only as long as the present-day computer environment is valid. To illustrate some specific features of external sorting using the tapes we consider an example.

Example 8.3.1

This is a merge using four tapes.

A set of 7000 keys is to be sorted, $K(1), \ldots, K(7000)$. Only 1000 of the keys will fit into the core memory of a computer. An obvious way to deal with the problem is to sort each of the seven runs of 1000 keys internally and independently, and then merge the sorted runs together. For brevity we shall denote the runs as $K(1-1000), \ldots, K(6001-7000)$.

The initial sorted runs are placed on the tapes alternately to ensure a balanced as possible distribution of them. In general, at the start of an external sort it is not known how many initial runs there will be so we cannot store 'half' of them, say, on one tape and the other half on the other tape. The four tapes can be used to store the intermediate runs in various ways. Three such ways are shown, they differ by the total number of times that the keys are read before the sort is complete.

Version One

Phase	Tape	Runs
Initial (distribution) phase	$T1$	$K(1-1000)$ $K(2001-3000)$ $K(4001-5000)$ $K(6001-7000)$
	$T2$	$K(1001-2000)$ $K(3001-4000)$ $K(5001-6000)$ dummy
	$T3$	empty
	$T4$	empty

Rewind $T1$ and $T2$.

Phase	Tape	Runs
First merging phase	$T1$	empty
	$T2$	empty
	$T3$	$K(1-2000)$ $K(4001-6000)$
	$T4$	$K(2001-4000)$ $K(6001-7000)$

Rewind $T3$ and $T4$.

Phase	Tape	Runs
Second merging phase	$T1$	$K(1-4000)$
	$T2$	$K(4001-7000)$
	$T3$	empty
	$T4$	empty
Third merging phase	$T1$	empty
	$T2$	empty
	$T3$	$K(1-7000)$
	$T4$	empty

The sorting is now complete. We note that in this version, given a certain number of tapes, the initial runs are distributed onto one half of the tapes ($T1$ and $T2$), and the runs of the first merging phase onto the other half of the tapes ($T3$ and $T4$). The roles of the tapes are then reversed. The process is repeated until the entire file is sorted in order.

Reading (moving) the keys is one of the essential operations in external sorting. The number of these operations in Version One is $7000 + 7000 + 7000 + 7000 = 28\,000$.

Version Two

Phase	Tape	Runs
Initial (distribution) phase	$T1$	$K(1-1000)$ $K(3001-4000)$ $K(6001-7000)$
	$T2$	$K(1001-2000)$ $K(4001-5000)$
	$T3$	$K(2001-3000)$ $K(5001-6000)$
	$T4$	empty

Rewind $T1$ and $T2$.

First merging phase	$T1$	$K(6001-7000)$	
	$T2$	empty	
	$T3$	$K(2001-3000)$	$K(5001-6000)$
	$T4$	$K(1-2000)$	$K(3001-5000)$

Rewind $T1$, $T3$, and $T4$.

Second merging phase	$T1$	empty	
	$T2$	$K(1-3000)$	$K(6001-7000)$
	$T3$	$K(5001-6000)$	
	$T4$	$K(3001-5000)$	

Rewind $T2$ and $T4$.

Third merging phase	$T1$	$K(1-5000)$
	$T2$	$K(6001-7000)$
	$T3$	$K(5001-6000)$
	$T4$	empty

Rewind $T1$

Fourth merging phase	$T1$	$K(1-5000)$
	$T2$	empty
	$T3$	empty
	$T4$	$K(5001-7000)$

Rewind $T4$

Fifth merging phase	$T1$	empty
	$T2$	$K(1-7000)$
	$T3$	empty
	$T4$	empty

The sorting is now complete.

The total number of the keys read is given as

$$7000 + 4000 + 4000 + 5000 + 2000 + 7000 = 29\ 000.$$

In this version, aiming to have just one tape vacant at the end of each phase, we had to pay a 'fine' of an extra 1000 keys read.

In the next variant, starting with the same initial distribution as in Version Two, and ensuring just one tape vacant at the end of each phase, the three-way merge is used instead of the two-way merge.

Version Three

Initial (distribution) phase	$T1$	$K(1-1000)$	$K(3001-4000)$	$K(6001-7000)$
	$T2$	$K(1001-2000)$	$K(4001-5000)$	
	$T3$	$K(2001-3000)$	$K(5001-6000)$	
	$T4$	empty		

Rewind $T1$, $T2$, and $T3$.

First merging phase	$T1$ K(6001–7000) $T2$ empty $T3$ empty $T4$ K(1–3000) K(3001–6000)

Rewind $T4$.

Second copying phase	$T1$ K(6001–7000) $T2$ K(1–3000) $T3$ empty $T4$ K(3001–6000)

Rewind $T2$.

Third merging phase	$T1$ empty $T2$ empty $T3$ K(1–7000) $T4$ empty

The total number of keys read is equal to

$$7000 + 6000 + 3000 + 7000 = 23\,000.$$

Version Three is more 'economical' than both Version One and Two. Example 8.3.1 shows that a clever distribution of the initial runs on the tapes leads to an efficient external sort, that is the sort with a decreased or even a minimum required number of complete passes over the keys.

Another important factor of an efficient external sort is the way in which the initial runs are generated. Since the aim of the sort is to produce a single sorted file, it is clearly desirable to generate initial runs which are as long as possible; this will call for fewer complete passes needed by the process. It will be shown later that in fact the initial runs can be produced of length exceeding the capacity of the core memory. On the other hand increasing the length of an initial run requires additional processing operations during the internal sort phase. Since for internal sorts the processing cost grows as $n$long$n$ with the size of the file there is a limit to the 'size to which one may be willing to go'. An external sort with mag tapes is described by the following three parameters:

T is the number of tapes
S is the number of initial sorted runs
P is the merge order of a multiway merge.

Now assuming that the initial runs are all of the same approximate relative length unity consider further examples of distribution patterns on the tapes.

Example 8.3.2

This is a merge using three tapes.

Let $T = 3$; $P = 2$; $S = 13$. For these parameters, three versions of merging the initial runs are given in Table 8.3.1. Here '1, 1, 1, 1, 1, 1, 1' denotes seven initial runs; '2, 2, 2' denotes three runs of relative length 2; and so on.

Table 8.3.1 Three variants of the balanced multiway sorting for $P=2$, $T=3$ and $S=13$

	Contents of			
	$T1$	$T2$	$T3$	Remarks
Phase 1	1, 1, 1, 1, 1, 1, 1	1, 1, 1, 1, 1, 1	—	Initial distribution
Phase 2	—	—	2, 2, 2, 2, 2, 2, 1	Merge 7 runs to $T3$
Phase 3	2, 2, 2, 1	2, 2, 2	—	Copy the runs alternately to $T1$ and $T2$
Phase 4	—	—	4, 4, 4, 1	Merge 4 runs to $T3$
Phase 5	4, 4	4, 1	—	Copy the runs alternately to $T1$ and $T2$
Phase 6	—	—	8, 5	Merge 2 runs to $T3$
Phase 7	8	5	—	Copy the runs alternately to $T1$ and $T2$
Phase 8			13	Merge 1 run to $T3$
Phase 1	1, 1, 1, 1, 1, 1, 1	1, 1, 1, 1, 1, 1	—	Initial distribution
Phase 2	—	—	2, 2, 2, 2, 2, 2, 1	Merge 7 runs to $T3$
Phase 3	2, 2, 2, 1	—	2, 2, 2	Copy half of the runs onto $T1$ (Take every second run and copy it)
Phase 4	—	4, 4, 4, 1	—	Merge 4 runs to $T2$
Phase 5	4, 4	4, 1	—	Copy half of the runs onto $T1$
Phase 6	—	—	8, 5	Merge 2 runs to $T3$
Phase 7	—	8	5	Copy half of the runs onto $T2$
Phase 8	13	—	—	Merge 1 run to $T1$
Phase 1	1, 1, 1, 1, 1, 1, 1, 1	1, 1, 1, 1, 1		Initial distribution
Phase 2	1, 1, 1	—	2, 2, 2, 2, 2	Merge 5 runs onto $T3$
Phase 3	—	3, 3, 3	2, 2	Merge 3 runs onto $T2$
Phase 4	5, 5	3	—	Merge 2 runs onto $T1$
Phase 5	5	—	8	Merge 1 run onto $T3$
Phase 6	—	13	—	Merge 1 run onto $T2$

To complete sorting, the first two versions require eight phases each, and the last variant requires six phases and it has no copying phases as opposed to the first two variants. The copying phases are 'passive' phases and an algorithm which avoids them is bound to be more efficient.

Version One, where at each phase the runs are distributed as evenly as possible on the tapes, is an example of the balanced P-way merge (in the example, $P=2$). Version Two is a certain modification of the balanced two-way merge using three tapes. Version Three deserves a more careful analysis as it uses a very particular distribution of the initial runs, eight on $T1$ and five on $T2$. This specific partitioning of 13 initial subfiles into eight and five may be explained using the Fibonacci numbers. In fact, Version 3 is an example of the so-called polyphase sort, a sorting method which is based on the 'perfect Fibonacci distributions'.

To compare the variants in terms of the number of complete data file passes, we find

for Version One:

$$1+1+1+1+1+1+1=7$$

for Version Two:

$$1+1+\frac{7}{13}+1+\frac{8}{13}+1+\frac{8}{13}+1=6\frac{10}{13}$$

for Version Three:

$$1+\frac{10}{13}+\frac{9}{13}+\frac{10}{13}+\frac{8}{13}+1=4\frac{11}{13}.$$

In Version Three only phases 1 and 6 are complete passes over all data; Phase 2 processes $\frac{10}{13}$ of the initial runs, Phase 3 only $\frac{9}{13}$, and so on.

The Number of Passes Required by Balanced Multiway Merge Sort

Consider once again Version One of Example 8.3.2. It is an example of the balanced multiway merge sort for $P=2$. We wish to express the number of complete passes required by the method as a function of parameter S, the number of the initial runs. Specifically, for Version One of Example 8.3.2 we have:

in Phase 2 $(=2^1)$ $\left\lceil \frac{S}{2} \right\rceil$ runs are merged

in Phase 4 $(=2^2)$ $\left\lceil \frac{S}{2^2} \right\rceil$ runs are merged

in Phase 2^k $\left\lceil \frac{S}{2^k} \right\rceil = 1$ run is merged.

It follows that

$$k = \lceil \log_2 S \rceil \tag{8.3.1}$$

is the number of merging phases.

We also note that the sort illustrated by Version One would typically have $k-1$ copying phases.

Since each phase (the merging as well as the copying) traverses the complete file once, the total number of passes required is given by

$$k+(k-1)+1=2k=2\lceil \log_2 S \rceil \tag{8.3.2}$$

(where '1' stands for the initial distribution pass).

When $T > 3$, we have the generalized balanced $(P, T-P)$-way merge sort. Consider Example 8.3.3 for $P = 3$ and $T = 5$.

Example 8.3.3

This is a balanced three-way merge.

	$P=3$			$T-P=2$		
	$T1$	$T2$	$T3$	$T4$	$T5$	
Phase 1	1, 1, 1, 1, 1	1, 1, 1, 1	1, 1, 1, 1	—	—	Initial distribution
Phase 2	—	—	—	3, 3, 1	3, 3	Three-way merge is used to merge the runs onto $T4$ and $T5$
Phasc 3	6	6	1	—	—	Two-way merge is used to merge the runs onto $T1$, $T2$, and $T3$
Phase 4	—	—	—	13	—	Three-way merge is used to merge the runs onto $T4$ and $T5$

At the initial distribution phase the runs are distributed onto $T1$, $T2$, and $T3$. Then, in Phase 2 these are merged onto $T4$ and $T5$ using the three-way merge. In Phase 3 a two-way merge is used and the generated runs are placed onto $T1$, $T2$, and $T3$, and so on. We note that in the general case of S initial runs, there are k P-way and $k-1$ $(T-P)$-way merge phases, respectively, and k is determined from the relation

$$\left\lceil \frac{S}{P^k (T-P)^{k-1}} \right\rceil = 1, \qquad (8.3.3)$$

which gives

$$k = \log_{P(T-P)} S. \qquad (8.3.4)$$

Since each phase of the process requires one complete pass over the data the number of total passes is given as

$$k + (k-1) + 1 = 2k = 2\log_{P(T-P)} S \qquad (8.3.5)$$

In (8.3.5) the quantity $\log_{P(T-P)} S$ is called the effect power of the $(P, T-P)$-way merge. The greater the logarithm base the fewer is the total number of required passcs, and for a fixed T the base achieves its maximum when $P = T - P$, that is when $P = T/2$. The number of complete passes in this case is equal to $\log_{T/2} S$.

8.4 Polyphase Merge

In balanced multiway merging for tape sorting the process of copying all of the data file between merging phases uses up as many passes as the merging process, thus effectively doubling the total number of passes required. Several ingenious tape-sorting algorithms have been contrived which eliminate nearly all of this copying by means of changes in the ways in which intermediate sorted subfiles are merged together. The most distinguished of these methods is polyphase merge.

The polyphase merge was discovered in the late 1950s (the three-tape case is due to Betz (1956), and the general pattern has been developed by Gilstad (1960), who also gave the method its present name). Analysis of the method has also been given (see, for example, MacCallum, 1972) but it was Knuth (1973) who first enhanced the theory of the method by observing its relation to the Fibonacci distributions.

The Fibonacci distributions are based on the Fibonacci numbers which are defined by the recurrence:

$$F_n = F_{n-1} + F_{n-2}, \quad n \geqslant 2, \quad F_0 = 0, \quad F_1 = 1. \tag{8.4.1}$$

For example, the first ten Fibonacci numbers are:

$$F_0 = 0, \quad F_1 = 1, \quad F_2 = 1, \quad F_3 = 2, \quad F_4 = 3, \quad F_5 = 5,$$
$$F_6 = 8, \quad F_7 = 13, \quad F_8 = 21, \quad F_9 = 34, \quad F_{10} = 55,$$

Let us now recall Version Three in Example 8.3.2. There for $T = 3$ and $P = 2$, we have

$$\begin{aligned} S &= F_n = F_{n-1} + F_{n-2} \\ &= F_7 = F_6 + F_5 \\ &= 13 = 8 + 5. \end{aligned} \tag{8.4.2}$$

The initial distribution of the runs onto $T1$ and $T2$, in the way defined by (8.4.2), ensures that the copying stage is completely avoided, and the method requires exactly $(n - 1)$ phases to complete the sort. The distribution of initial runs defined by (8.4.2) and (8.4.1) is called the perfect Fibonacci distribution. If a given number of initial runs does not match the perfect Fibonacci distribution the dummy runs can be added to make the match perfect.

The Fibonacci numbers given by (8.4.1) can be explicitly related to the parameter P. For example, for $P = 2$ the following notation may be used:

$$F_n^{(2)} = F_{n-1}^{(2)} + F_{n-2}^{(2)}, \quad n \geqslant 2, \quad F_0^{(2)} = 0, \quad F_1^{(2)} = 1. \tag{8.4.3}$$

This notation may seem somewhat artificial but it is convenient for the analysis of the general T-tape polyphase sort since the recurrence (8.4.3) is easily generalized to define the Fibonacci numbers for any p; in other words,

$$\begin{aligned} &F_n^{(p)} = F_{n-1}^{(p)} + F_{n-2}^{(p)} + \ldots F_{n-p}^{(p)}, \quad n \geqslant p, \\ &F_0^{(p)} = 0, \quad F_1^{(p)} = 0, \ldots, \quad F_{p-2}^{(p)} = 0, \quad F_{p-2}^{(p)} = 1. \end{aligned} \tag{8.4.4}$$

Sequence (8.4.4) is called the pth order Fibonacci numbers. The first 11 fourth-order Fibonacci numbers are:

$$F_0^{(4)} = 0 \quad F_1^{(4)} = 0 \quad F_2^{(4)} = 0 \quad F_3^{(4)} = 1 \quad F_4^{(4)} = 1 \quad F_5^{(4)} = 2$$
$$F_6^{(4)} = 4 \quad F_7^{(4)} = 8 \quad F_8^{(4)} = 15 \quad F_9^{(4)} = 29 \quad F_{10}^{(4)} = 56 \quad F_{11}^{(4)} = 108. \tag{8.4.5}$$

Example 8.4.1 illustrates how sequence (8.4.4) is used to produce the perfect Fibonacci distributions in external sorting when $P > 2$. An important detail of the sorts based on the perfect Fibonacci distributions is an assumption that $P = T - 1$ for any given T, that is in these sorts one tape only is made vacant at the end of each phase.

Example 8.4.1

This is a polyphase merge using five tapes.

Let $T = 5$ and $P = T - 1 = 4$. The perfect Fibonacci distribution for this case is shown in Table 8.4.1. Here '108' denotes 108 initial runs of relative length 1 placed on $T1$, '4×56' denotes 56 runs of relative length 4 placed on $T5$, etc. Phase 0 is the initial distribution phase and Phases 1–8 are merging phases.

The distribution of the runs shown in Table 8.4.1 has a direct relation to table 8.4.2, based on the fourth order Fibonacci numbers.

Here $a_n = F_{n+3}^{(4)}$, the fourth order Fibonacci number, and b_n, c_n, and d_n are obtained from this number by straightforward procedures. The t_n will be useful in calculation of the total number of passes over the data.

Table 8.4.1 A polyphase merge using five tapes

	$T1$	$T2$	$T3$	$T4$	$T5$	The number of the initial runs processed by the current phase
Phase 0	108	100	85	56		108 + 100 + 85 + 56 = 349
Phase 1	52	44	29		(56)	4 x 56 = 224
Phase 2	23	15		(29)	27	7 x 29 = 203
Phase 3	8		(15)	14	12	13 x 15 = 195
Phase 4		(8)	7	6	4	25 x 8 = 200
Phase 5	(4)	4	3	2		49 x 4 = 196
Phase 6	2	2	1		(2)	94 x 2 = 188
Phase 7	1	1		(1)	1	181 x 1 = 181
Phase 8			(1)			349 x 1 = 349

Table 8.4.2

Level					Total
0	1	0	0	0	1
1	1	1	1	1	4
2	2	2	2	1	7
3	4	4	3	2	13
4	8	7	6	4	25
5	15	14	12	8	49
6	29	27	23	15	94
7	56	52	44	29	181
8	108	100	85	56	349
n	a_n	b_n	c_n	d_n	t_n

where

$$a_n = a_{n-1} + a_{n-2} + a_{n-3} + a_{n-4},$$

$$b_n = a_{n-1} + a_{n-2} + a_{n-3},$$

$$c_n = a_{n-1} + a_{n-2},$$

$$d_n = a_{n-1},$$

$$t_n = t_{n-1} + 3a_{n-1}.$$

In the general case of $P = T - 1$ we have

$$a_n = F^{(P)}_{n+(P-1)} \tag{8.4.6}$$

and

$$t_n = t_{n-1} + (P-1)a_{n-1}. \tag{8.4.7}$$

The total number of initial runs in the sort that requires n merging phases is given by

$$S = t_n = PF_{n+P-2} + (P-1)F^{(P)}_{n+P-3} + \ldots + F^{(P)}_{n-1}. \tag{8.4.8}$$

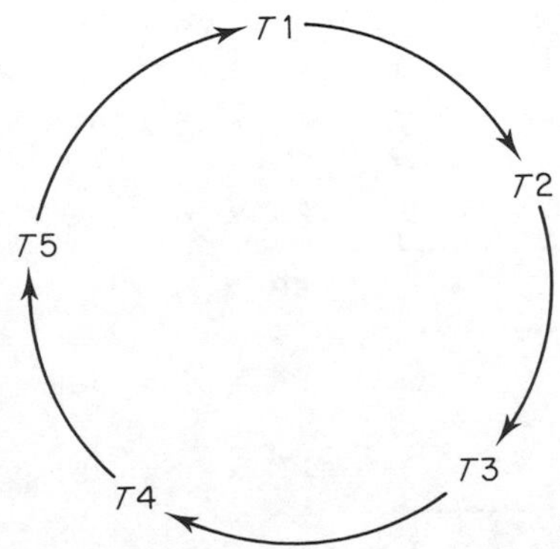

Figure 8.4.1 A 'cyclic' presentation of five tapes

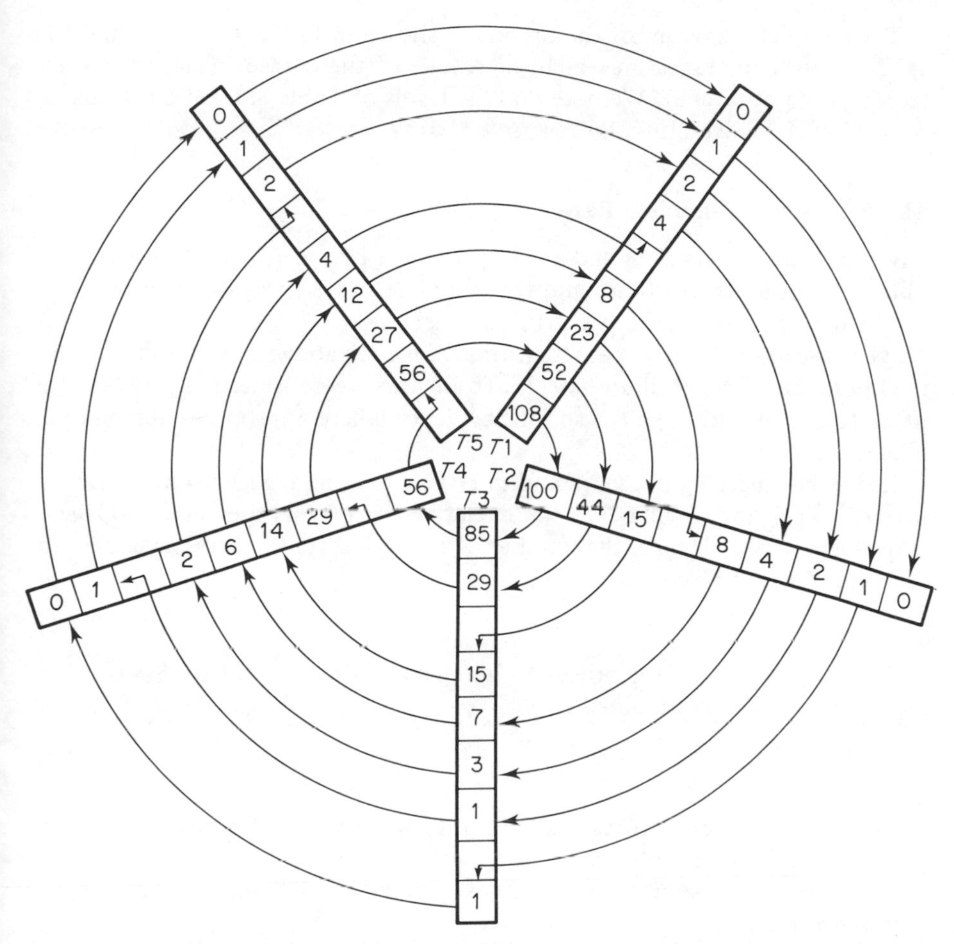

Figure 8.4.2 The distribution of runs corresponds to Tables 8.4.1 and 8.4.2. The levels of Table 8.4.2 and phases of Table 8.4.1 have mutually reverse ordering. So, for example, the results of Phase 1 of Table 8.4.1 are stored on $T5$ and shown at level 7 ($=n-1$), the results of Phase 2 are stored on $T4$ and shown at level 6 ($=n-2$), etc. N is the total number of phases (levels). The sequence of numbers on $T1$ is the sequence of the fourth order Fibonacci numbers. This sequence controls the distribution on other tapes

In order to see the relation between Tables 8.4.1 and 8.4.2 consider Figs 8.4.1 and 8.4.2.

The distribution illustrated in Table 8.4.1 can be obtained using Table 8.4.2. We start at the last level of Table 8.4.2 (i.e. the 8th level) and distribute the sequence {108, 100, 85, 56} on the tapes shown in Fig. 8.4.1, starting from $T1$ and proceeding in a clockwise direction. Before moving to the next level in Table 8.4.2 (i.c. the 7th level) we make one (cyclic) step forward and one step 'upwards' to mark the beginning of a new phase. We then 'place' the integer on the tape and proceed as before.

The complete diagram of the process is shown in Fig. 8.4.2. The 'upward' shifts identify the tapes on which the results of the current phase are stored (e.g. tape $T5$ at level 8). We note that the levels of Table 8.4.2 and the phases of Table 8.4.1 have mutually reverse ordering.

The Number of Complete Passes

Given the parameters S, T and $P = T - 1$, for a polyphase merge, we wish to obtain the number of phases and the complete passes over the data required by the method, as functions of the parameters.

We first obtain the approximate formulae for the above quantities following the approach of MacCallum (1972). To do this we consider the general case of S, T, and P, although for specific evidence where appropriate reference is made to Example 8.4.1.

At the kth merging phase, t_k and a_{n-k} give the relative length of the newly merged runs (they are all of the same relative length) and the number of these runs, respectively. Hence, the number of the initial runs processed at the kth merging phase is given as

$$t_k a_{n-k}. \tag{8.4.1}$$

The total number of the processed initial runs when $S = t_n$ (e.g. $S = t_8 = 349$ in Example 8.4.1) is obtained as

$$\sum_{k=1}^{n} t_k a_{n-k} + S, \tag{8.4.2}$$

where the second term represents the complete pass over the data required by the initial distribution phase.

In order to express relation (8.4.2) as a function of S and P, consider the recurrence relation

$$t_k = t_{k-1} + (P - 1)a_{n-k+1}. \tag{8.4.3}$$

This recurrence can be expressed in the form which enables one readily to obtain its solution. We write for (8.4.1):

$$\begin{aligned} &t_k = t_{k-1} + t_{k-2} + \ldots + t_{k-P}, \quad k = 1, \ldots, n, \\ &t_0 = t_{-1} = \ldots = t_{-P+1} = 1, \end{aligned} \tag{8.4.4}$$

where n is the total number of merging phases required.

In example 8.4.1, the equation corresponding to (8.4.4) is

$$\begin{aligned} &t_k = t_{k-1} + t_{k-2} + t_{k-3} + t_{k-4}, \quad k = 1, \ldots, 8 \\ &t_0 = t_{-1} = t_{-2} = t_{-3} = 1. \end{aligned} \tag{8.4.5}$$

The general solution of the recurrence (8.4.4) has the form

$$t_k = \sum_{i=1}^{P} A_i \alpha_i{}^k, \tag{8.4.6}$$

where the α's are roots of the characteristic polynomial equation

$$Q(x) = x^P - x^{P-1} - \ldots - x - 1 = 0. \tag{8.4.7}$$

The polynomial equation (8.4.7) has one real root α in the interval $[1, 2]$ since values of the polynomial at $x = 1$ and $x = 2$ have opposite signs and since it can be shown that all the roots except one are less than unity, in modulus.

Thus for reasonably large P we have

$$t_k \approx A\alpha^k \quad \text{for some constant } A, \tag{8.4.8}$$

where α is the dominant real zero of the polynomial $Q(x)$.

Next, since $S = t_n \approx A\alpha^n$ and $1 = t_0 \approx A$, we have

$$n \approx \log_\alpha S, \tag{8.4.9}$$

that is the number of merging phases required by the polyphase sort is approximately equal to $\log_\alpha S$.

It can further be shown that the root α is given by

$$\alpha = 2 - \frac{1}{2^{-P} - 1} + O(2^{-2P}) \tag{8.4.10}$$

which indicates that $\alpha \to 2$ when $P \to \infty$.

Now, the recurrence relation involving a's is given as

$$a_j = a_{j-1} + a_{j-2} + \ldots + a_{j-P}, \quad j = 1, \ldots, n,$$

$$a_{-P} = a_{-P+1} = \ldots = a_{-1} = 0, \qquad a_0 = 1. \tag{8.4.11}$$

It can be solved in a way similar to that of relation (8.4.4) and its general solution is given by the approximate expression

$$a_j \approx B\alpha^j \quad \text{for some constant } B, \tag{8.4.12}$$

where α is the dominant real root of the polynomial equation (8.4.7).

Using (8.4.12) and (8.4.8) we can now estimate the number given by (8.4.12), i.e.

$$t_k a_{n-k} \approx A\alpha^k B\alpha^{n-k} = AB\alpha^n, \tag{8.4.13}$$

which shows that at each merging phase the number of the processed initial runs is approximately the same. A study of Table 8.4.1 shows that the above result holds quite well for all the merging phases except possibly the last one, which naturally corresponds to the complete pass over the data.

To estimate the quantities given by (8.4.13) we turn to the first merging phase of the polyphase sort. At this phase we have

$$t_1 a_{n-1} = P a_{n-1} = P\,\frac{t_n - t_{n-1}}{P - 1}, \tag{8.4.14}$$

where the use is made of the recurrence:

$$t_n = t_{n-1} + (P - 1)a_{n-1}.$$

Since $t_n = S$, and by virtue of (8.4.14) and (8.4.8), the processed fraction of the initial subfiles at any given merging phase is given as

$$\frac{P}{P-1}\frac{t_n - t_{n-1}}{t_n} \approx \frac{P}{P-1}\frac{\alpha^n - \alpha^{n-1}}{\alpha^n} = \frac{P}{P-1}\frac{\alpha - 1}{\alpha}. \tag{8.4.15}$$

As it was shown earlier, the polyphase sort using S initial runs requires

$$n \approx \log_\alpha S = (\ln \alpha)^{-1} \ln S \tag{8.4.16}$$

merging phases. This together with (8.4.16) gives the total number of complete passes required as

$$\rho \approx \frac{P}{P-1}\frac{\alpha - 1}{\alpha}\log_\alpha S = \frac{P}{P-1}\frac{\alpha - 1}{\alpha}(\ln \alpha)^{-1} \ln S, \tag{8.4.17}$$

and

$$\frac{\rho}{n} \approx \frac{P}{P-1}\frac{\alpha - 1}{\alpha} \approx 0.5, \quad \text{for large } P.$$

With the initial distribution pass taken into account we have

$$\rho + 1 \approx \frac{P}{P-1}\frac{\alpha - 1}{\alpha}\log_\alpha S + 1. \tag{8.4.18}$$

By virtue of (8.4.10), for reasonably large P formula (8.4.18) can be expressed as

$$\rho + 1 \approx \frac{1}{2}\frac{P}{P-1}\log_2 S + 1 \tag{8.4.19}$$

or asymptotically

$$\rho + 1 \approx \tfrac{1}{2}\log_2 S + 1. \tag{8.4.20}$$

The estimates (8.4.19) and (8.4.20) can be used for asymptotic comparison of the polyphase merge with other external merge sorts.

A more precise analysis of the number of complete passes over the data required by the polyphase merge is given in Knuth (1973).

To estimate the quantity given by (8.4.2), the sequences

$$\{a_k = F^{(P)}_{k+P-1}, \quad k = 0, 1, \ldots n\}$$

and

$$\{t_k = t_{k-1} + (P-1)a_{k-1} = t_{k-1} + (P-1)F^{(P)}_{k+P-2}, \quad k = 1, \ldots, n, \ t_0 = 1\}$$

are first expressed in concise form using the concept of the generating function.

In particular, it is shown that for the two sequences the corresponding generating functions are given as

$$a(z) = \sum_{k \geqslant 0} a_k z^k = \frac{1}{1 - z - \ldots - z^P}, \tag{8.4.21}$$

and

$$t(z) = \sum_{k \geq 1} t_k z^k = \frac{Pz + (P-1)z^2 + \ldots + z^P}{1 - z - \ldots - z^P}. \tag{8.4.22}$$

From (8.4.21) and (8.4.22) it follows that for the general case of $S = t_n$, the number of the processed initial runs is given as the coefficient of z^n in $a(z)t(z)$, plus t_n, to include the initial distribution pass. Accordingly, the following problem is formulated.

Assume the merging pattern with the properties which are characterized by the polynomials

$$H(z) = Pz + (P-1)z^2 + \ldots + z^P \quad \text{and} \quad Q(z) = 1 - z - \ldots - z^P$$

in the following way:

(i) The number of initial runs present in a 'perfect distribution' requiring n merging phases is the coefficient of z^n in $H(z)/Q(z)$,
(ii) The number of initial runs processed during these merging phases is the coefficient of z^n in $H(z)/Q(z)^2$,
(iii) There is a 'dominant root' α of $Q(z^{-1})$ such that $Q(\alpha^{-1}) = 0$, $Q'(\alpha^{-1}) \neq 0$, $H(\alpha^{-1}) \neq 0$ and $Q(\beta^{-1}) = 0$ implies that $\beta = \alpha$ or $|\beta| < |\alpha|$.

Prove that there exists $\varepsilon > 0$ such that, if S is the number of the initial runs in a perfect distribution requiring n merging phases, and if S initial runs are processed during those phases, then

$$n = \ln S + b + O(S^{-\varepsilon}), \tag{8.4.23}$$

and

$$\rho = c \ln S + d + O(S^{-\varepsilon}), \tag{8.4.24}$$

where

$$a = (\ln \alpha)^{-1},$$

$$b = a \ln\left(\frac{H(\alpha^{-1})}{-Q'(\alpha^{-1})}\right) - 1,$$

$$c = a \frac{\alpha}{-Q'(\alpha^{-1})},$$

$$d = \frac{(b+1)\alpha - H'(\alpha^{-1})/H(\alpha^{-1}) + Q''(\alpha^{-1})/Q'(\alpha^{-1})}{-Q'(\alpha^{-1})}.$$

The values of n and ρ, computed for various P using the expressions (8.4.16) and (8.4.23), and (8.4.17) and (8.4.24), respectively, are compared in Table 8.4.3.

When the number of initial subfiles in the general T-tape polyphase merge does not match a 'perfect Fibonacci distribution', dummy runs are added to

Table 8.4.3 Approximate number of merging phases and passes in the polyphase merge using S initial runs

	Phases		Phases		Pass/phase ratio	
Tapes T	$n_1 = \log_\alpha S$	$n_s = a \ln S + b$	$\rho_1 = \frac{P}{P-1}\frac{\alpha-1}{\alpha} \times \log_\alpha S + d$	$\rho_2 = c \ln S + d$	$\frac{\rho_1}{n_1}$	$\frac{\rho_2}{n_2}$
3	$1.958 \ln S$	$2.078 \ln S + 0.672$	$1.566 \ln S$	$1.504 \ln S + 0.992$	0.80	0.72
4	$1.615 \ln S$	$1.641 \ln S + 0.364$	$1.117 \ln S$	$1.015 \ln S + 0.965$	0.69	0.62
5	$1.517 \ln S$	$1.524 \ln S + 0.078$	$0.977 \ln S$	$0.863 \ln S + 0.921$	0.64	0.57
6	$1.477 \ln S$	$1.479 \ln S - 0.185$	$0.908 \ln S$	$0.795 \ln S + 0.864$	0.61	0.54
8	$1.451 \ln S$	$1.451 \ln S - 0.642$	$0.843 \ln S$	$0.744 \ln S + 0.723$	0.58	0.51
10	$1.445 \ln S$	$1.445 \ln S - 1.017$	$0.811 \ln S$	$0.728 \ln S + 0.568$	0.56	0.50
20	$1.443 \ln S$	$1.443 \ln S - 2.170$	$0.762 \ln S$	$0.721 \ln S - 0.030$	0.53	0.50

make the match perfect but the way in which these dummies are dispersed on the tapes affects the overall performance of the polyphase merge. For example, it turns out that the very best method for distributing dummies among the tapes involves using extra phases and more dummies than would seem to be needed. The reason for this is that some subfiles are used in merges much more often than others.

Various ways of distributing the dummy runs have been proposed (see Knuth (1973) and Wirth (1976)), although the optimal solution of this problem still awaits its discovery.

Asymptotic efficiency of the polyphase sort may be compared with that of the balanced P-way merge using an approximate estimate of the number of passes given by (8.4.20). We have

$$\frac{1}{2}\log_2 S + 1 < \log_{T/2} S = \frac{\log_2 S}{\log_2 T - 1}\,. \tag{8.4.25}$$

8.5 Cascade Merge

Another important method which also makes use of a 'perfect distribution' is known as the cascade merge. It was proposed in 1959 by Betz and Carter, The method is illustrated in Example 8.5.1, for $T = 5$ and $S = 2037$. In this method the merging stages are distinguished by the complete passes over the data. Pass 2, for example, is obtained by doing a four-way merge from $T1$, $T2$, $T3$ and $T4$ onto $T5$ until $T4$ is empty; then a three-way merge from $T1$, $T2$ and $T3$ onto $T4$ until $T3$ is empty; finally, a two-way merge from $T1$ and $T2$ onto $T3$ leaves $T2$ empty; for uniformity, the content of $T1$ is then copied onto $T2$. The pass is now complete. The copying phases are not necessary for the process; they are included for the purpose of giving the process a fully uniform

Example 8.5.1

This is a cascade merge using five tapes.

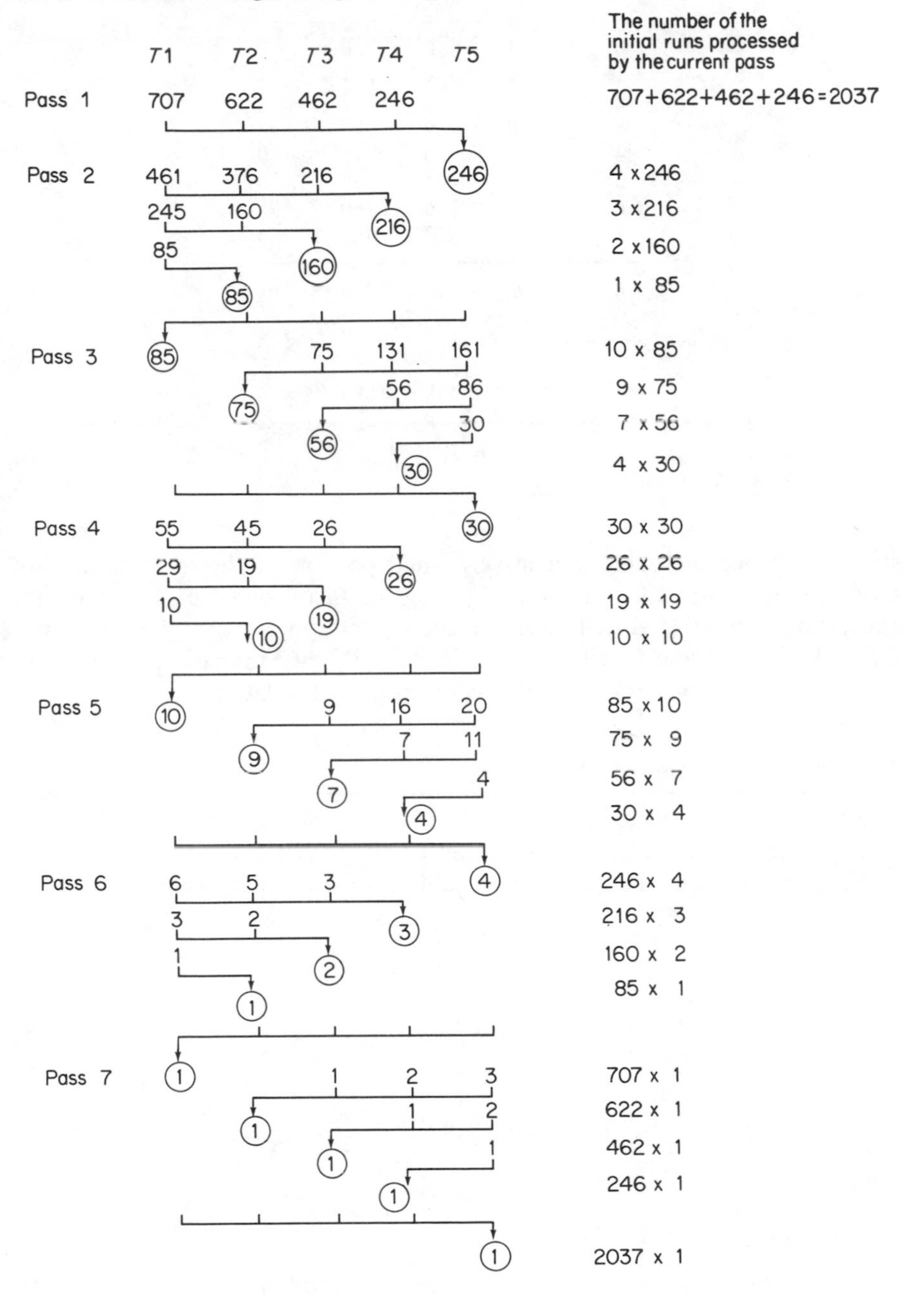

The initial runs are read in total $2037 \times 8 = 16{,}296$ times, out of which 756 reads are for copying purposes.

Table 8.5.1

Level				
0	1	0	0	0
1	1	1	1	1
2	4	3	2	1
3	10	9	7	4
4	30	26	19	10
5	85	75	56	30
6	246	216	160	85
7	707	622	462	246
n	a_n	b_n	c_n	d_n

where

$$a_n = a_{n-1} + b_{n-1} + c_{n-1} + d_{n-1}$$
$$b_n = a_{n-1} + b_{n-1} + c_{n-1}$$
$$c_n = a_{n-1} + b_{n-1}$$
$$d_n = a_{n-1}$$

structure. Since they take a relatively small portion of the total number of moves of the data, they are not of any great inconvenience. The rules for generating a 'perfect distribution' in the cascade merge are given in Table 8.5.1. As is seen these distributions are not exactly the Fibonacci distributions, although they are generated in the same spirit as the latter.

The Number of Complete Passes

The cascade merge has been analysed by many people. The main results of the analysis for the general T-tape case can be summarized as follows.

The parameters in the distribution table (Table 8.5.1) are given as

$$a_n = \frac{4}{2T-1} \sum_{-T/2 < k < \lfloor T/2 \rfloor} \cos^2 \theta_k \left(\frac{1}{2 \sin \theta_k}\right)^n,$$

$$b_n = \frac{4}{2T-1} \sum_{-T/2 < k < \lfloor T/2 \rfloor} \cos \theta_k \cos 3\theta_k \left(\frac{1}{2 \sin\theta_k}\right)^n, \tag{8.5.1}$$

$$c_n = \frac{4}{2T-1} \sum_{-T/2 < k < \lfloor T/2 \rfloor} \cos \theta_k \cos 5\theta_k \left(\frac{1}{2 \sin \theta_k}\right)^n,$$

where

$$\theta_k = (4k+1)\pi/(4T-2), \quad k = 0, 1, \ldots, n.$$

The formulae are obtained using the generating functions which happen, for this problem, to be related to the Chebyshev polynomials of the second kind. Hence, the form of the solution.

Table 8.5.2 Approximate number of passes required to merge 2037 runs by polyphase and cascade merges

Tapes	Passes in		
T	Polyphase merge	Cascade merge (without copying)	Cascade merge (with copying)
3	12.451	12.451	16.504
4	8.698	9.216	10.163
5	7.496	7.634	8.004
6	6.921	6.697	6.886
8	6.392	5.649	5.721
10	6.115	5.060	5.098
20	5.463	3.926	3.930

In Table 8.5.2 are given the approximate number of passes required to merge $S = 2037$ runs by each, the polyphase and cascade merges, the latter for both, with and without copying. We note that the cascade merge without copying is asymptotically better than polyphase, on six or more tapes.

In a way similar to the polyphase merge, if the given number of the runs does not match the 'perfect distribution', some dummy subfiles are added to make the match perfect. Since in the cascade merge, the merging operates by complete passes, the way in which the dummies are dispersed on the tapes is unimportant.

8.6 Oscillating Sort

This sort uses a somewhat different approach to merge sorting. It was proposed by Sobel in 1962, and instead of assuming the initial distribution phase where all the initial runs are disseminated on the tapes, the algorithm uses alternatively distribution and merging, so that much of the sorting is carried out before the input has been completely inspected.

The method uses both backward and forward tape-reading, thus producing the intermediate runs of ascending and descending orderings alternately.

It is therefore important after every merging phase explicitly to emphasize the ordering of the current merged runs. We shall denote an ascending run by A and a descending run by D. We assume that all initial runs are of relative length 1 (the relative lengths of intermediate runs can be easily deduced at every particular phase).

Example 8.6.1 illustrates the oscillating sort that uses five tapes.

The Number of Complete Passes

Example 8.6.1 shows that the key which, say, was read in the first distribution phase, is then involved in the processing only in the first and the last merging

Example 8.6.1

	$T1$	$T2$	$T3$	$T4$	$T5$	The number of initial runs processed
Distribution phase	A	A	A	A		4
Merging phase					D	4
Distribution phase		A	A	A	D A	4
Merging phase	D				D	4
Distribution phase	D A	A		A	D A	4
Merging phase	D		D		D	4
Distribution phase	D A		D A	A	D A	4
Merging phase	D	D	D		D	4
Merging phase				A		4 x 4 = 16

phases. In fact, it is easily observed that if the number of initial runs S is $(T-1)^n = 4^n$ then the oscillating sort processes each key exactly $n+1$ times (once during the distribution phase and n times during a merge). Now, assuming that the dummy runs are present when S is between 4^{n-1} and 4^n, bringing it up to 4^n, the total sorting time amounts to $\lceil \log_4 S \rceil + 1$ passes over all the data, and hence in the general T-tape case this time given by

$$\lceil \log_{T-1} S \rceil + 1. \qquad (8.6.1)$$

Comparison with other merge sorts shows that the oscillating sort is the best method when S is a power of $T-1$.

Moreover, when S is a power of $T-1$ the oscillating sort is the 'best possible' method under an optimizing sorting model that has been developed by Karp. The model establishes the lower bound on the sorting time for any T-tape external sort as

$$\log_{T-1} S + O(1),$$

and as is seen from (8.6.1) the oscillating sort achieves this lower bound when $S = (T-1)^n$.

8.7 Replacement Selection

Up to now in the analysis of different external sorting techniques it has been assumed that all initial runs have the same relative length which is determined by the size of the computer internal memory. However, one would generally expect that a sorting process which starts with 'longer' initial runs would require each key processed fewer times in merging phases. In particular, it is desirable to remove the constraint on the run size induced by considerations of the computer internal memory. One method which produces initial runs of length that is independent of the size of internal memory is the replacement selection.

Replacement selection uses a multiway merge of order P. In a P-way merge the basic operation is to repeatedly output the smallest of the smallest keys not yet output from each of the P runs to be merged. That key should be replaced with the next key from the run from which it came. Continuing in this way the sorted file is produced. When a run is exhausted, a sentinel key, larger than all other keys, is placed so as to block out the exhausted run from subsequent examining. When all runs are blocked out by sentinels, the merge is complete.

The same idea of doing P-way merge but on unsorted runs is used in the replacement merge with an additional proviso: if the new key is smaller than the last one put out then it should be marked as a member of the next 'initial run' and treated as greater than all keys in the current 'initial run'; this is because the new key could not possibly become part of the current sorted initial run. Replacement selection method is illustrated in Example 8.7.1. A four-way merge is used on the runs S1, S2, S3, and S4. When the key is output it is replaced by the next key from the data file. If the new key is smalle than the one just output it is not considered for inclusion in the current initial run. Instead it is 'suspended' on line 4. As soon as all four blocks S1, S2, S3 and S4 have 'suspended' keys as their first keys, the generating of a new initial run is started.

Example 8.7.1

This is the replacement selection using four-way merge.

Memory content				Output
*S*1	*S*2	*S*3	*S*4	
03	87	12	61	03
75	87	12	61	12
75	87	53	61	53
75	87	(26)	61	61
75	87	(26)	(08)	75
70	87	(26)	(08)	70
97	87	(26)	(08)	87
97	(54)	(26)	(08)	97
(09)	(54)	(26)	(08)	end subfile
09	54	26	08	08
09	54	26	12	09
etc.				

The runs produced in this way can contain more than four keys, each even though in the P-way selection there are never more than P keys at any time. In practice P is set fairly large so that the internal memory is essentially filled and, what is important, the unsorted data file is in fact never partitioned explicitly, since each time simply the next in-line key of the data file is placed in the internal location vacated by the key put out.

It can be shown that if the keys are random then the runs produced using replacement selection are about $2P$ keys long, on average. The practical effect of this is a saving of one merge pass, since rather than starting with sorted runs about the size of the internal memory and then taking a merging pass to produce runs about twice the size of the internal memory, one starts off with runs which are already about twice the size of the internal memory. If there is some partial order in the keys then the runs will be significantly longer. It is also obvious that while the minimum length of the runs created by replacement selection equals P, their maximum length equals the complete data file.

8.8 Merge Trees and Optimum Merge Sorting

The replacement selection method produces initial runs of varying size, the average size being, with small variance, twice the capacity of the internal computer memory. Consider now a set of runs $S = \{S_1, \ldots, S_k\}$ which have to be merged into a sorted file using P-way merge. The merge can be performed in a number of ways. The process of a P-way merge of S runs can be set forth as a P-way merge tree with S terminal nodes where the terminal nodes represent the initial runs, and the internal nodes, the merge phases. In Fig. 8.8.1(a) a merge tree is shown which depicts the instance of the polyphase merge of Example 8.3.2, Version Three.

Here each internal node is labelled with the reverse order of the step at which the run was generated; for example, the last step (in Version Three) is numbered 1st node in the tree, the penultimate step, 2nd in the tree, etc. The branch immediately above a node is labelled with the name of the tape on which the generated run is stored. Similarly, Fig. 8.8.1(b) shows a merge tree which represents a balanced merge for $T = 5$, $P = 3$ and $S = 13$.

Let key K be a member of an initial run R which like any other initial run is represented by a terminal node in a merge tree. The number of passes that involve key K is given as the length of the path from the terminal node of run R to the root of the merge tree. If we allow runs of unequal sizes then the total number of passes over the data is given as the sum of the path lengths (where each path is weighted by the run size) divided by the size of the data file.

Given a set of runs S and a merge order P the way in which a P-way merge can be done best depends heavily on the type of backing store available. For example, in tape sorting its sequential structure of tape storage restricts the set of runs to which one may economically obtain access at any given time. However, in order to derive some lower bounds on an optimal merge performance we assume a merge model which allows at any time convenient access to

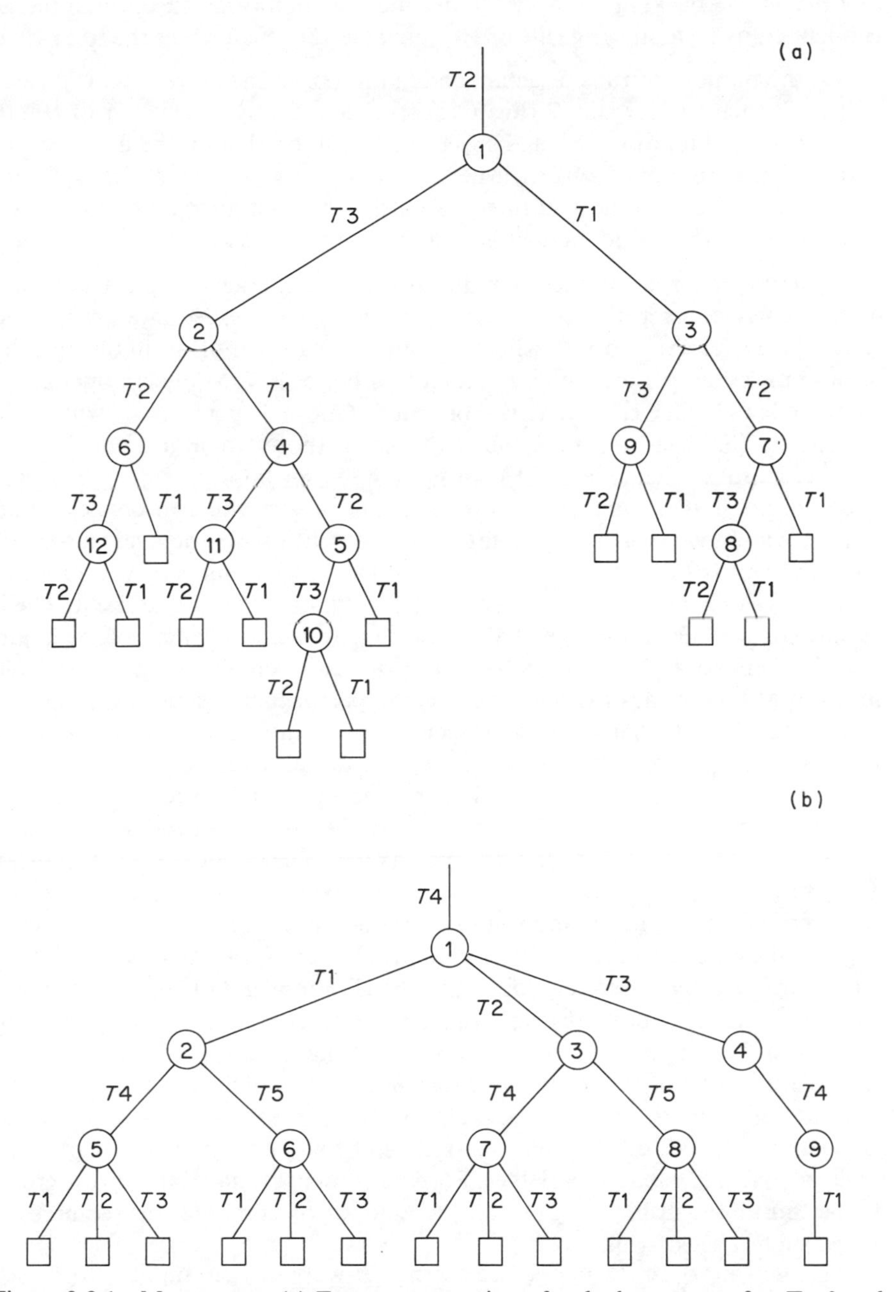

Figure 8.8.1 Merge trees. (a) Tree representation of polyphase merge for $T = 3$ and $S = 13$; (b) tree representation of balanced merge for $T = 5$, $P = 3$ and $S = 13$

any run in the backing memory. Under these assumptions an optimal merge is obtained using a merging rule of Huffman (1952) which states the following:

> If the number of runs S, actual and dummies, in the current set is greater than P then replace the P runs of shorter lengths by a new run of length equal to the sum of their lengths with the condition that on the first application of this rule only m runs, where $S - m = 0 \bmod (P-1), 2 \leqslant m \leqslant P$, are merged. (The condition ensure that after the first merge operation the number of runs to be merged is a multiple of $(P-1)$.)

The Huffman rule implies that in an optimal merge the short runs are merged early and repeatedly while the long runs are merged late and passed fewer times. Of the two sets with the same average size of an initial run the set with many short runs requires greater number of key movement operations compared with the set with few short runs. Optimal merge trees which are asymptotic (or 'distorted') are thus superior to those which are not.

Now, suppose that one must choose between an internal sort which produces initial runs of almost constant size, and a sort which produces initial runs of widely varying size. Assuming that for both sorts the average size of an initial run is the same the sort with the runs of varying size will produce some very long and some very short intermediate runs. As a result it will permit a more efficient merge. This conjecture was conceived and later quantified by Bennett and Frazer (1971) and Frazer and Bennett (1972). They have shown that the bounds on achievable merge performance depend on internal sort strategy. A stochastic model to generate the initial runs which was used in the analysis assumes that all output key permutations are equally likely and the size of a run does not depend on its neighbours. Under these assumptions whatever method is used in the initial phase the size of a produced run is a random variable. This size may be taken as a dimensionless rational number if the minimum possible run size, $S_{\min}$, is used as a normalizing unit. Averaging over an ensemble of applications of the same initial phase the merge method allows one to regard the size of the ith run S_i as having been chosen according to some discrete probability distribution, $p_i(x)$, on the rational numbers. The probability distributions $p_i(x)$ normally converge quite rapidly with i to a limiting distribution $p(x)$ which is characteristic of the algorithm used (Knuth, 1963). Apart from some final runs, which are almost always atypical and must be regarded as exceptions, for most of the duration of the internal sort the size of the produced runs can be viewed as a sequence of identically distributed random variables. Frazer and Bennett have studied the effect of the limiting distribution $p(x)$ on the amount of computation required to merge such a set of runs.

Consider an optimal merge tree of a P-way merge applied to a set $S = \{S_1, \ldots, S_k\}$ of runs of varying size. If all internal nodes of the tree are at level k or less then an upper bound on the average number of key movement per key is given as

$$1 + \lceil \log_p k \rceil . \qquad (8.8.1)$$

However, this bound does not take into account the size distribution function and thus cannot be sharp. Sharper bounds can be obtained using a merge model where a set of k independent samples is given from some run size distribution $p(x)$. It has been shown that under this model the average number of merge passes for a key in an optimal P-way merge is bound by

$$\mu[\log_P kx] - \frac{k}{k-1}\mu[x\log_P x](\mu[x])^{k/(k-1)} < M_{\text{opt}}$$
$$\leqslant 1 + \log_P(k\mu[x]) - \frac{k}{(k-1)\mu[x]}\mu\left[\frac{x\log_P x}{1+\dfrac{x}{(k-1)\mu[x]}}\right] \qquad (8.8.2)$$

where all averages, $\mu[y]$, are taken with respect to $p(x)$.

The optimal size distribution function has the form

$$p_{\text{opt}}(x) = \begin{cases} \dfrac{\bar{x}(\alpha-1)}{\alpha\bar{x}-1}, & x = 1, \\ 0, & 1 < x < \alpha\bar{x}, \\ \dfrac{\bar{x}-1}{\alpha\bar{x}-1}, & x = \alpha\bar{x}, \qquad \text{where} \quad \bar{x} = \mu[x]. \end{cases} \qquad (8.8.3)$$

8.9 Comments

The wide variety of external storage device types and costs of accessing an item make the development of external sorting methods very dependent on current technology. Many parameters affect performance of these methods, and many a time an elegant method might remain unused and unappreciated as a result of the latest change in the technology. We have seen that one of the major factors which needs to be considered in any tape-sorting method is the time it takes to rewind a tape. Sophisticated methods, like polyphase merge, cascade merge and oscillating sort, have been specially designed to reduce the required tape rewind time. However, as analyses show the savings achievable by such methods over the simple multiway balanced merge are quite limited.

It is therefore prudent to check the configuration of the local computer system environment when a sorting of very large files needs to be undertaken. For example, many modern computer systems provide a large virtual memory capability. A good virtual memory system allows one to address a large amount of data while the system takes responsibility to make sure that addressed data is transferred from external to internal storage when need. Pertinent to this approach are the programs with a relatively small 'locality of reference', that is the likelihood that each reference by the program to memory is to an area of memory which is relatively close to other recently referenced areas. An internal sorting method with a small locality of reference could be ideally implemented for sorting very large files on a virtual memory system.

Exercises

1. Algorithm *simple_two_way_merge* can be easily modified to produce a merge sort algorithm if '$m+n$' is replaced by 'n' and '$N(j)$' by $M(n+1-j)$. Explain what should be done in the new algorithm about the sentinel keys. What is the smallest number of steps the new algorithm could use?
2. Assuming that the average and the maximum numbers of comparisons required to merge two files with n and m keys are

$$C_{av}(n, m) = m + n - \frac{m}{n+1} - \frac{n}{m+1}$$

and

$$C_{max}(n, m) = m + n - 1,$$

respectively, show that in the straight two-way merge sort of a file of $N = 2^k$ keys the average and the maximum numbers of comparisons required are

$$C_{av} = N\left(-1 + \log_2 N - \sum_{s=0}^{k-1} \frac{1}{2^s + 1}\right)$$

and

$$C_{max} = N(-1 + \log_2 N) + 1.$$

3. How much additional storage is required by the simple two-way merge sort?
4. Consider the straight two-way merge sort to sort in order a file $K_1 \ldots K_N$, $N \neq 2^k$. Derive the formula for the maximum number of comparisons assuming the following:
   (i) random conditions on the file,
   (ii) if $n-1 = (b_k \ldots b_0)_2$ in binary notation, then there are $(b_k \ldots b_{j+1})_2$ merges of the ordered subfiles of equal length 2^j, and at every pass, whenever $b_j = 1$, there is one 'special' merge of the two subfiles of the lengths 2^j and $1 + (b_{j-1} \ldots b_0)_2$, respectively.

   Use your formula to show that for $N = 12$ the maximum number of comparisons required is equal to 33.
5. Show that the time complexity of the recursive algorithm *recmergesort* when applied to sort an input of size n is of $0(n \log n)$.
   (*Hint*. Derive the recurrence $T(n) = 2T(n/2) + n$ then solve it.)
6. Sort in order the following sequence of 17 keys using the algorithm recmergesort:

   FO AN SO OR TO IN NO GO EX XI AM MA PI LE EN DO

   Give the total number of comparisons taken to do the sort.
7. Show the contents of the linked lists passed as arguments to each call when the *recemergesort* is used in Exercise 6.

8. Outline how an external selection can be done: find the kth largest in a file of n keys, where n is too large for the file to fit in main memory?
9. Compare the multiway balanced merge with four tapes and with six tapes to polyphase merge with the same number of tapes, for 63 initial runs.
10. What happens if the replacement selection is applied again to an output produced by replacement selection?
11. What is the output of the replacement selection method when the input keys are in decreasing order $K(1) \leqslant \ldots \leqslant K(n)$?
12. Using ordinary alphanumeric order and treating each word as one key, apply the four-way replacement selection method to successive words of the following sentence:
    'There was an old fellow of Trinity who solved the square root of infinity but it gave him such fidgets to count up the digits that he chucked maths and took up divinity.'
13. Illustrate the polyphase sort process, given that the data file consists of 193 initially sorted subfiles, each of relative length unity, and that four tapes are available for the purpose.
14. Show that $(P-1)n \log_2 n$ comparisons are required by the balanced P-way merge.
15. Assume that a sequence of 912 sorted runs, each of the same relative size unity, forms a file which has to be sorted 'externally' using six tapes. Carry out the sorting using (a) balanced multiway merge, (b) polyphase merge, (c) cascade merge, and (d) oscillating sort. Dummy runs should be added where necessary. In each case, how many complete passes over the data are made?
16. How many phases does five-tape polyphase merge use when started up with four tapes containing 56, 52, 44, 29 runs?
17. Explain how small files should be handled in a Quicksort implementation to be run on a very large file within a virtual memory environment.

9

Searching

9.1 Introduction

The searching problem is that of locating a specific item in a given sequence of n items. As in the case of sorting we assume that each item contains within itself a piece of information called key. We then wish to store the sequence which initially arrives in an arbitrary order, and upon demand quickly retrieve those items whose keys match the given key K, called the search key. The purpose of the search is typically to access information within the item (not merely the key) for processing. Normally one thinks of the items in the sequence as records, and the collection of all stored records is called a (symbol) table, or a dictionary.

Technically speaking, in the information processing known as the search two distinct tasks can be recognized: the process of storing a key, where a key is entered into the table; and the process of retrieval of a key, where a key is accessed in the table. In the storage process an empty table and an arbitrarily ordered sequence of records is assumed at the outset. Using a specified storage algorithm the records from the sequence are then entered into the table. (Thereafter the same algorithm would be used whenever a record from the table needs to be retrieved.) In the retrieval process a table already filled up, partially or fully, is assumed, and one searches for a match for the search key, again using a specified retrieval algorithm. In general setting the search algorithms are called storage and retrieval algorithms.

In our discussion we assume that a complete table is stored in a random-access core memory of the computer. The table may be either full or only partially filled in. The table may also be either static, where collection of the records is given once and for all, or dynamic, where periodically one may wish to delete previously sorted record or to include further new records. In this context a search is 'successful' if a record whose key is identical to the search K has been located in the table, and a search is 'unsuccessful' if no match for the search key is found in the table. Other common operations of interest are insertion of a new record, deletion of a record, joining two tables to make a large table, sorting the dictionary (table) in order and outputting all the records in sorted order.

Like sorting, searching is one of the most time-consuming processes of many data processing systems, so there is a great need for efficient search algorithms. We shall look at some basic ways in which a given record sequence is structured in the table and a specified record is searched for. In a large group of searching techniques the search mechanism is an operation of comparison between the search key K and a key $K(j)$ in the table. For these techniques the number of comparisons required by a search serves as a measure of the method's efficiency. As was noted earlier, in practice the key $K(j)$ is normally a part of a larger record of information, $R(j)$, which is being retrieved via its key. However, for the purpose of our discussion it suffices to concentrate solely on the keys themselves, since they are the only things which significantly enter into the search algorithms.

9.2 Search Algorithms

Linearly Ordered Tables

A large group of search methods is based on the use of linearly ordered tables. That is, the table corresponding to the given sequence is obtained by sorting the sequence into a linear order, for example,

$$K(1) \leqslant K(2) \leqslant K(3) \leqslant \ldots \leqslant K(n).$$

For the table so obtained, one rather obvious way to search is to start at the beginning of the table and to search in turn through the keys until either the key required is located or the end of the table is reached. Wood (1973) has shown that for this method the average number of comparisons required for a successful or an unsuccessful search varies between n and $\frac{1}{2}n$, depending on the degree to which the table is filled.

The degree of 'fullness' of the table is quantitatively expressed by the load factor, $\alpha = m/n$, which is the ratio of the number of filled locations, m, to the size of the table, n. The 'last' location of the table is normally reserved for the purpose of signalling the end of the table.

Another way to search through a table of linearly ordered keys is to start by comparing the search key K with a particular key $K(j)$ in the table. As a result one of these cases would follow: $K > K(j)$, $K = K(j)$, or $K < K(j)$. The next step is then decided strictly in accordance with the outcome case instance. The search methods of this group use binary tree structures and are known as the tree search methods. The specifics of how to decide on the next key to use in the comparison process bring about different binary tree structures and hence different tree search algorithms. Another factor to affect the tree search algorithm is the key access frequency. Static search trees with known access probabilities for the keys have been studied in depth (see Nievergelt, 1974). Here one distinguishes between the use of balanced and Fibonacci trees for uniformly distributed keys, and unbalanced trees for various non-uniform distributions of keys. Algorithms for developing 'optimal binary search trees' under assumption of different distributions on the keys have also been developed. These algorithms are of significant theoretical interest though, perhaps, less so for practical applications, with few exceptions, such as the Huffman code/decode trees.

Hashing: Open Addressing and Chaining Algorithms

A completely different group of search methods uses an alternative approach which avoids sorting the given sequence altogether. Instead, for every key K of the sequence, a transformation function $h(K)$ is computed and then taken as the location of K in the table. The goal is to define the function $h(K)$ such that for any two different keys, $K(i)$ and $K(j)$, we shall get $h(K(i)) = h(K(j))$. To obtain such a function is extremely difficult. What is usually done instead is that the different keys are allowed to yield the same value $h(K)$ and then a special mechanism is used to resolve the conflict situation.

A prerequisite of a good transformation function is that it distributes the keys as evenly as possible over the table. With this exception, the distribution is not bound by any pattern, and it is actually desirable if it gives the impression that it is entirely at random. This process of computing the location address, $h(K)$, for a given key has got the name hashing, i.e. 'making mess' or 'chopping the argument up', and $h(K)$ is called the hash function.

An occurrence that more than one key is hashed to the same location, i.e. that

$$h(K(i)) = h(K(j)) \quad \text{when } K(i) \leqslant K(j),$$

is called a collision or a conflict.

Several algorithms have been developed for resolving collisions. These can be grouped into two major sets. One, open addressing hashing, is based on the approach where a hash table for keys is built up directly and whenever a collision occurs it is resolved by simply finding an unoccupied space in the table for those keys whose natural 'home' location is already occupied. The collision resolution mechanism ensures that the 'misplaced' key can be later retrieved in a usual way. The other approach, hashing with chaining, resolves the problem of collisions by using indirect addressing which allows all keys that collide to maintain a claim to their 'home' location.

In search methods many special factors can be encountered such as the possibility of variable length records, retrieval on secondary keys, and the possibility of duplicate keys in distinct records. These problems will not be explicitly dealt with since many special circumstances can often be accommodated within the framework of a basic algorithm.

9.3 Ordered Tables

We assume a sequence of keys that are linearly ordered by a relation $'<'$, $K(1) < \ldots < K(N)$. We then want to know whether this sequence contains a key equivalent to the search key, K. From time to time we may want to insert a new key into the sequence or delete a key from it.

Algorithm Sequential Search

This is the simplest method for searching when the keys are stored as an array and then looked up through the array sequentially for every search key K. If $K(k)$ is found such that $K = K(k)$, the search is successful and k is returned; if no key in the array matches the key K, the search is unsuccessful. The following algorithmic code is a straightforward implementation of the method.

Algorithm seqsearch

```
if n > 0 then
        k := 0
        repeat k := k + 1 until (K = K[k] or k = n)
endif
if k = K[k]
then return (k)      //K[k] is found to be equal to K//
else return (nil)
endif
```

The algorithm takes n steps for an unsuccessful search since every key must be examined for equality with K before deciding that no key in the array equals the search key. For a successful search about $n/2$ steps, on average, are required since in a random search the average number of steps equals the sum of the numbers of steps, required by every possible successful outcome, divided by the number of such outcomes:

$$(1 + 2 + \ldots + n)/n = (n - 1)/2. \tag{9.3.1}$$

The algorithm *seqsearch* uses purely sequential access to the keys and thus can be naturally adapted for a linked list representation for the keys. We assume as usual that the linked list is pointed to by pointer r and the link field of a key is denoted by *'.next'*. This would yicld thc following implementation of a search on a linked list of keys.

Algorithm seqlistsearch

```
if r < > null then
    t := r
    while K > t.key do t := t.next enddo
    if K = t.key
    then return (t)
    else return (nil)
    endif
endif
```

With a sorted list an unsuccessful search ensues whenever a key larger than the search key is found. It follows that the argument similar to that of formula (9.3.1) holds here, and thus only about half keys on average need to be examined for an unsuccessful search.

In searching techniques substantial savings can often be achieved if some information is available about the relative frequency of access of different keys. For example, an optimal arrangement of the keys in a linked list representation would be to put the most frequently accessed keys at the beginning of the list and the last frequently accessed keys at the end of the list. In other words, the sequence of keys would be 'ordered' not on the key values but on their frequency of access, and in reverse order for that.

Algorithm Binary Search

The most efficient search of an ordered set of keys uses the 'divide-and-conquer' approach: compare the search key with the key closest to the middle of the set. If the search key equals the key with which it was compared, the search terminates successfully, but if non-equality between the two keys is encountered, then one half of the keys is simply discarded and the search concentrates on the half to which the key being sought belongs. The method is applied recursively. In the implementation which follows, the method is expressed iteratively rather then recursively, since the algorithm would use just

Algorithm binsearch

```
i := 1
j := n
repeat
    index := (i + j) div2
    if K < K[index] then j := index - 1 else i := index + 1 endif
until (K = K[index] or i > j)
if K = K[index]
then return (index)
else return (nil)
endif
```

one recursive call and an iteration in this case is a simpler implementation.

In similarity with Quicksort, the pointers i and j indicate the end elements of the array on which the algorithm is currently working. During one loop step, the variable index is set to point to the midpoint of the current interval. The loop either terminates successfully or the right pointer is changed to '*index – 1*' or the left pointer to '*index + 1*' depending on whether the search key is equal, less than or greater than the key in the array, with which the search key is compared.

A simple analysis shows that the total number of steps that the repeat loop would take is about $\log_2 n$, since the size of the search interval is halved at each step. Overholt (1973a) has shown that the binary search method is one of a class of optimal binary search methods, and the class itself may be rather large for certain values of n. In particular, he has shown that for $n = 2^m + 2^{m-1} - 1$ the number of optimal methods is of order $2^{2^m} \approx 2^{2^n/3}$. An alternative to binary search is provided by the method which uses the Fibonacci numbers.

Algorithm Fibonaccian Search

The 'divide-and-conquer' paradigm is again used in this interesting search method, except that now after every comparison the set of keys is divided into two parts, the sizes of which are determined by the sequence of the Fibonacci numbers. Let n be such that $n = F_k - 1$ for some Fibonacci number F_k, $k > 1$. The search key is first compared with the array key which occupies position given by the Fibonacci number F_{k-1}. If the two keys are equal, then the search terminates successfully; if the search key is less than the array key, the new search interval is given as the left part of the array of size $F_{k-1} - 1$, and if the search key is greater than the array key, the new search interval is given as the right part of the array with the end elements given as $F_{k-1} + 1$ and $F_k - 1$. The algorithm is used recursively. This brings us to the following implementation which is again given in an iterative form.

Algorithm fibsearch

```
    if n > 2 then
    //compute the initial values of the pointers p and q//
        k := 1; F[0] := 1; F[1] := 1
        repeat
            q := F[k - 1]; p := F[k]; k := k + 1
            F[k] := F[k - 1] + F[k - 2]
            index := F[k]
        until F[k] = n + 1
        //search for the match for the search key K//
        j := 0
        notdecided := true
```

```
        while notdecided do
            case K of
                : K = K[index] :    begin j := index
                                          notdecided := false
                                    end
                : K = K[index] :    if q > 0 then
                                    begin index := index − q
                                          temp := p; p := q
                                          q := temp − q
                                    end
                                    else notdecided := false
                                    endif
                : K > K[index] :    if p > 1 then
                                    begin index := index + q
                                          p := p − q; q := q − p
                                    end
                                    else notdecided := false
                                    endif
            endcase
        enddo
        if K = K[index]
        then return (index)
        else return (nil)
        endif
    endif
```

The following example shows the parts of the array examined by *fibsearch* when searching for NP in a table built by inserting the keys AC AE CE EG EN GH GI IL IP LM MX MN NP PR PS RS SX SY XZ.

Example 9.2.1

This is a Fibonaccian search for the key NP on twenty keys.

```
AC AE CE EG EN GH GI HI IL IP LM LX (MN) NP  PR  PS  RS  SX  SY XZ
                                     NP  PR  PS  RS (SX) SY XZ
                                     NP  PR (PS) RS
                                     NP (PR)
                                    (NP)
```

Encircled are the keys in the sequence which are compared with the search key NP at the first, second, third, fourth and fifth steps, after which the search terminates successfully.

Efficiency of the Fibonaccian search method has been studied by Overholt (1973b) who has shown that the method is not optimal in terms of the number of comparisons required. Specifically, its search time is some 4% greater than that of the binary search and it has much greater maximum search time and standard deviation.

9.4 Search Trees

The sequence of comparisons made by the binary and Fibonaccian search algorithms are predetermined: in each case the specific sequence is based on the value of the search key and on the array size n. The comparison structure of the algorithms can be simply described by a binary tree structure. In Fig. 9.4.1 three examples of the comparison structures are shown: for the binary comparison search on seven keys, for the sequential search on seven keys, and for the Fibonaccian search on 12 keys. The search key in such a structure is first compared with the root value of the tree on level 1, then either with the left child's value or with the right child's value depending on whether the search key is less than or greater than the root value, respectively (of course, if on the first comparison the search key is found to be equal to the root value, we are complete), and so on.

The binary search tree is in fact a data structure which is extremely well suited for many data processing problems including search processes. It is very convenient for representing tables and lists because it offers an easy access to the elements of the structure, easy inserting of new elements and easy deleting of the keys from the structure. It is one of the most flexible and best understood techniques for organizing the search of large data files.

We shall specify a binary search tree for an array **A** as a labelled binary tree in which each node i is labelled by an element $K(i) \in \mathbf{A}$ such that

(a) for each node j in the left subtree of i, $K(j) < K(i)$,
(b) for each node j in the right subtree of i, $K(j) > K(I)$, and
(c) for each key K which is in **A**, there is exactly one node i such that $K(i) = K$.

The levels in a binary search tree are numbered from one (at the root level) upwards, where k is the last level.

We have noted earlier that the common metric for search techniques studied here is the number of comparisons used to find a match for the search key. In terms of the tree structure concept it implies path length measures. We define the depth of a tree as its last or maximum level, the length of the longest path from the root to a terminal node. Furthermore, a binary tree is called height-balanced if the depth of the left subtree of every node never differs by more than ± 1 from the depth of its right subtree. Examples of the balanced trees are given in Fig 9.4.1(a) and (c). An example of an unbalanced tree is given in Fig. 9.5.1. (This tree fails the height restriction on subtrees at both the 'November' and 'February' nodes.) When the access frequencies of the keys are known, it is more convenient to use the so-called weighted path length which is defined by taking into account the frequencies of the keys.

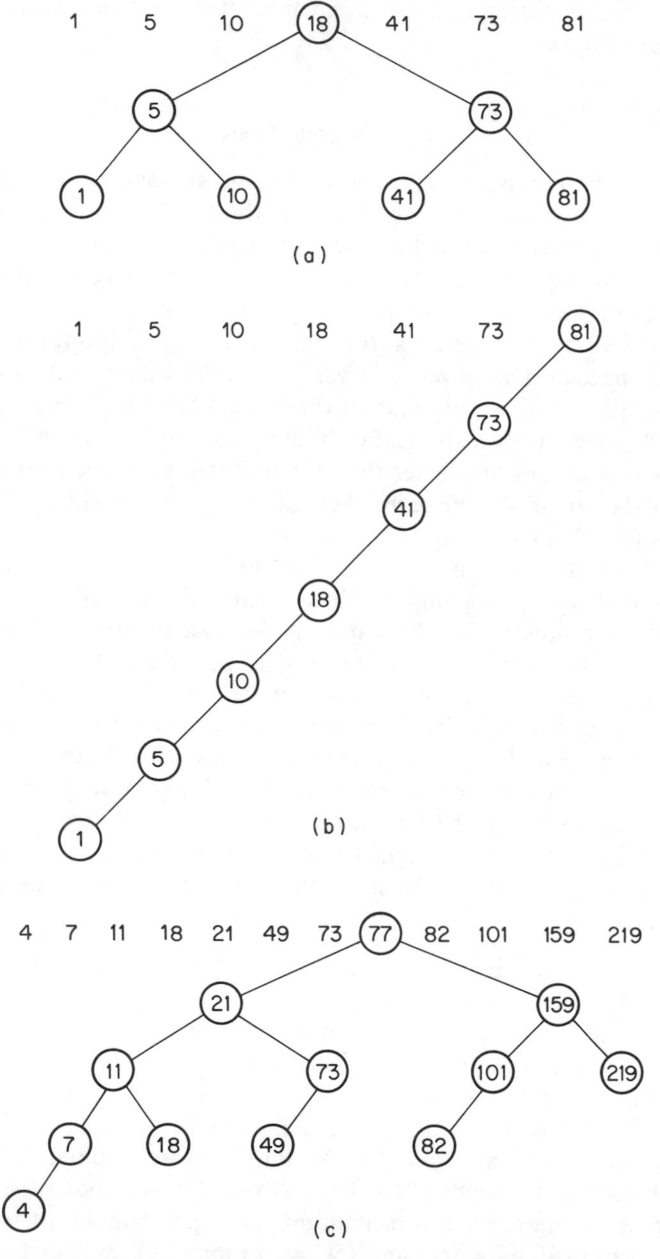

Figure 9.4.1 Examples of binary trees. (a) A balanced tree for seven nodes representing standard binary search; (b) a degenerate tree for seven nodes representing sequential search; (c) a search tree for twelve nodes representing Fibonaccian search

Frequencies are then referred to as the node-weights. The weighted paths are discussed in Section 9.5.

Binary Tree Search

In search-by-comparison methods the binary tree structure plays two different and both very important roles. These two roles can be specified by the terms 'a search tree' and 'a tree search'. Examples of the use of the binary tree in the first sense are given in Fig. 9.4.1 where the binary tree is used as a representation form to illustrate the comparison structures for binary search, sequential and Fibonaccian search. Here, the idea is that given an array of keys in linear order, a particular search method is developed and its comparison structure is then presented in the form of a binary tree. The binary tree in this situation represents the search method. As the next stage we generalize the set of comparison search methods by considering various (or even all possible) binary tree structures which can be associated with a given array of sorted keys. The group of these methods is collectively called binary tree search. Binary tree search is one of the most essential algorithms in computing; it is simple, fast and most versatile. Its basic structure is as follows.

Compare the search key with the root key value and one of the four cases arise:

(a) there is no root (the binary tree is empty): the search key is not in the tree, and the search terminates unsuccessfully;
(b) the search key matches the key at the root: the search terminates successfully;
(c) the search key is less than the key at the root: the search continues by examining the left subtree of the root in the same way;
(d) the search key is greater than the key at the root: the search continues by examining the right subtree of the root in the same way.

The following implementation is of the recursive algorithm which searches for the search key K in a tree defined by treenode; if K is not found, the algorithm inserts it. Each node in the tree has three fields: value which contains the key value, and left and right which are links to the left and right subtrees. The algorithm returns the pointer to the node which contains the key value equal to the search key.

Algorithm treesearchinsert (treenode)

```
if treenode = nil            //the tree is empty//
then //insert new node with key K//
begin
    create newnode (t)
    t.key := K
    t.left := nil; t.right := nil
```

```
        return (t)
    end
    else
        case K of
            :K < treenode.key : treesearchinsert (treenode.left)
            :K > treenode.key : treesearchinsert (treenode.right)
            :K = treenode.key : return (treenode)         //found//
        endcase
    endif
```

The algorithm *treesearchinsert* can be naturally adapted to build a search tree on a random sequence of n keys. For this we can simply assume that n key values are read in random sequence and inserted into the tree which is initially empty.

The time efficiency of algorithms on binary search tree are fully dependent on the shapes of the trees. In the best case, the tree could take the form given in Fig. 9.4.1(a) for describing the comparison structure for binary search; it then has about $\log_2 n$ nodes between the root and each external node. In the worst case, the tree could turn out to be of a degenerate form given in Fig. 9.4.1(b) for sequential search; then it obviously takes between $n/2$ to n comparisons to complete the search for a given key. For a random shape of a search tree which may be constructed by the algorithm treesearchinsert, we might roughly expect logarithmic search times on the average; the analysis in this case is facilitated by the following argument.

Assume that the keys are the integers 1 to n and that all permutations of the keys are equally likely. Let some key, i, arrive first and so to become the root of the tree. When the rest of the keys have been inserted, there will be $i-1$ nodes to the left of the root and $n-i$ nodes to the right. If i happen to be the midpoint of the range than it would divide the keys in half; if i is equal to either 1 or n then the first branches of the tree will be completely unbalanced. Continuing the argument recursively we note that if the second key to arrive is j and it is less than i then when the tree is filled, there will be $j-1$ nodes in the branch to the left of j and $i-j$ nodes in the branch to the right.

To determine the average number of comparisons in this case, we make use of the notion of the tree path length. Let us recall that the path length of the tree is the sum of the distances of all nodes from the root; the distance of a node from the root is one less than its level and the root is on level 1. Noting that a node in the binary search tree represents the operation of comparison of two keys, we now consider the average path length in a tree constructed by a random insertion.

At the root the path length is equal to 1 (for the root node itself) plus the length of a subtree with $i-1$ nodes plus the length of a subtree with $n-i$ nodes. These lengths are not known but can be calculated by applying the same procedure at the next level in the tree. Ultimately when the end of each branch is reached every node would have a path length of either 0 or 1. It remains to average the recursive definition of the path length over all possible values of i from 1 to n. We have

$$P_n(i) = [(i-1)(P_{i-1}+1) + 1 + (n-i)(P_{n-i}+1)]/n, \quad (9.4.1)$$

and then

$$P_n = \sum_1^n P_n(i) = \sum [(i-1)(P_{i-1}+1) + (n-i)(P_{n-i}+1)]$$

$$= 2(1+1/n)H_n - 3 = O(\log n,), \text{ for large } n. \quad (9.4.2)$$

Regretfully binary tree search suffers from the same worst-case problems as Quicksort; in other words, when processing the sequence of keys which are in order (or in reverse order) the binary tree search is no better than the sequential search. However, while eliminating the worst-case for Quicksort we have to resort to a random choice of a partitioning element so that the laws of probability could save the situation, for binary tree searching we can do much better. A general technique, called balancing, enables us to guarantee that the worst-case will never happen. A brief excursion into the ideas for the construction of 'balanced trees' is given in a later section. As it will be shown there the implementation of balanced tree algorithms is often a complex task, in spite of an easy general concept behind the algorithm. Therefore it is useful to remember that on average the performance of a completely unbalanced search tree, which is constructed by inserting the keys, read at random, into an initially empty tree, is worse off by no more than 38% compared with the performance of an optimal (that is perfectly balanced) binary search tree. This amazing result follows from the observation that while the perfectly balanced binary search tree requires no more than $\log_2 n$ comparisons for any key input, an unbalanced search tree, such as constructed by the algorithm treesearch-insert, requires on average

$2(1+1/n)H_n - 3 = 2\log_e n - 1.845$ comparisons for large n.

9.5 Search Methods for Unequal Distribution Tables

The search problem under the assumption of unequal probabilities of occurrence of keys has been studied by Knuth (1971, 1973), Hu and Tucker (1971) and others. In this case the best binary tree will not necessarily be balanced. Let $K(1), \ldots, K(n)$ as usual be a linearly ordered table of keys and let $\alpha_0, \alpha_1, \ldots, \alpha_n$ and $\beta_1, \ldots, \beta_n$ be two sets of frequencies, where α_i is the probability that the search key K lies between $K(i)$ and $K(i+1)$, α_0 and α_n correspond to the case when the search key is less than $K(1)$ and greater than $K(n)$, respectively; and β_i is the probability that K equals $K(I)$. In the extended binary tree shown in Fig. 9.5.1(a) the square nodes denote fictitious terminal nodes where no keys are stored and where a new key is inserted in the event of unsuccessful search. These nodes are denoted as 0, 1, 2, 3, 4, 5.

Our aim is to develop an algorithm for building an optimal binary search

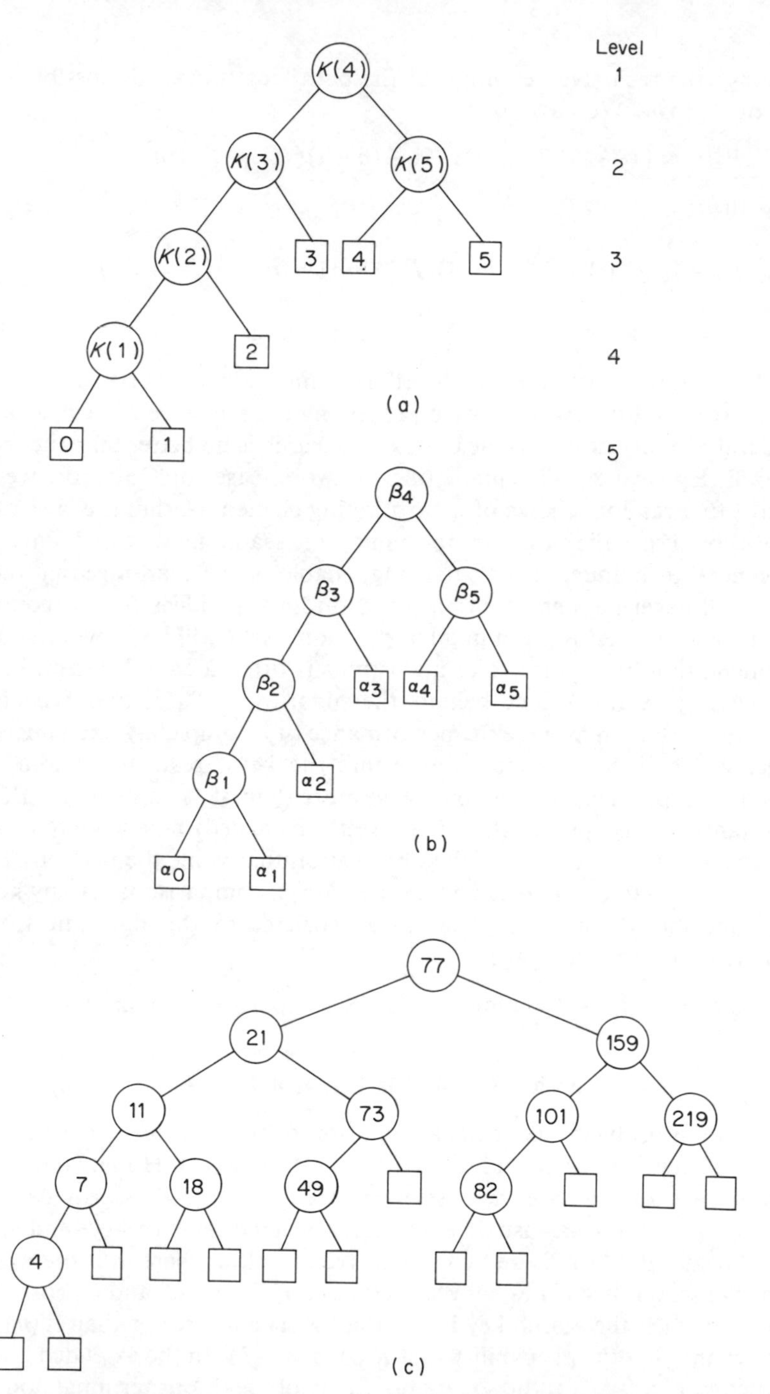

Figure 9.5.1 Comparison structures for three different search methods. (a) A binary search tree for five keys and six added (fictitious) terminal nodes; (b) the binary tree of (a) with explicitly shown probabilities attached to each node; (c) Fibonaccian search tree of Fig. 9.4.1(c) with added fictitious external nodes

tree for the keys with unequal probabilities of access. For this purpose, we define the cost of the binary search tree for n keys as the number of comparisons required by the tree search. Noting that if the search key K is present in some node i of the tree, then to establish this fact the number of examined nodes equals to the path length of node i, and if K is not in the tree and such that $K(i) < K < K(i+1)$ then the number of examined nodes is one less than the path length of the fictitious terminal node i, we can write

$$\mathrm{cost}(0, n) = \sum_{i=1}^{n} \beta_i \,\mathrm{level}(i) + \sum_{i=1}^{n} \alpha_i(\mathrm{level}(i) - 1). \tag{9.5.1}$$

For example, the score for the tree shown in Fig. 9.5.1(b)is

$$\begin{aligned}\mathrm{cost}(0, 5) = {} & 5\alpha_0 + 5\alpha_1 + 4\beta_1 + 4\alpha_2 + 3\beta_2 \\ & + 3\alpha_3 + 3\alpha_4 + 3\alpha_5 + 2\beta_3 + 2\beta_5 + \beta_4\end{aligned}$$

We further denote by weight $(0, n)$ the sum of all probabilities associated with the tree:

$$\mathrm{weight}(0, n) = \sum_{i=1}^{n} \beta_i + \sum_{i=1}^{n} \alpha_i. \tag{9.5.2}$$

Now, if the root of the tree contains the key $K(i)$ then its left subtree will contain all keys between $K(1)$ and $K(i-1)$ and its right subtree, all keys between $K(i+1)$ and $K(n)$. It is then easy to see that the formula (9.5.1) can be expressed as

$$\mathrm{cost}(0, n) = \mathrm{cost}(0, i-1) + \mathrm{weight}(0, n) + \mathrm{cost}(i, n). \tag{9.5.3}$$

An optimum binary search tree in the search tree of the least cost and to determine such a tree the following problem must be solved:

$$\begin{aligned}\min \mathrm{cost}(0, n) = {} & \min_i \{\mathrm{cost}(0, i-1) + \mathrm{cost}(i, n)\} \\ & + \mathrm{weight}(0, n).\end{aligned} \tag{9.5.4}$$

This problem is solved recursively.

An Optimum Cost Tree

Recursive solution of the problem (9.5.4) means that whichever key is an optimal choice for the root of the tree, succeeding choices in the construction of the left and right subtrees must themselves be optimal, that is all subtrees of an optimal tree are optimum for their own set of keys. This is known as the *principle of optimality*. The difficulty in applying this principle lies in the simple fact that the only way we know how to choose the root is to try many possibilities and for each one of them constructing all optimal left and right subtrees. This approach can be sketched as follows.

$$
\begin{array}{c}
\text{An optimal} \\
\text{binary search tree,} \\
\text{Obst}(0,\ n)
\end{array}
$$

$$
= \min \left\{ \begin{array}{l}
\text{Set } K(1) \text{ as the} \\
\text{root and solve} \\
\text{two problems:} \\
\text{find Obst}(0,\ 0) \\
\text{and} \\
\text{find Obst}(1,\ n)
\end{array},
\begin{array}{l}
\text{Set } K(2) \text{ as the} \\
\text{root and solve} \\
\text{two problems:} \\
\text{find Obst}(0,\ 1) \\
\text{and} \\
\text{find}(2,\ n)
\end{array},
\ldots,
\begin{array}{l}
\text{Set } K(n) \text{ as the} \\
\text{root and solve} \\
\text{two problems:} \\
\text{find Obst}(0,\ n-1) \\
\text{and} \\
\text{find}(n,\ n)
\end{array} \right\}
$$

The solution (or solutions) with the lowest cost will be an optimal solution. One algorithm which solves the problem is based on the approach of dynamic programming where a sequence of interrelated decisions yields an optimum solution at the end of the day. The algorithm first selects the key which becomes the root of the tree; it then 'improves' subtrees systematically, each such improvement leading to an improvement to the whole tree. The method is explained in the example below.

Example 9.5.1

This is a construction of an optimal binary search tree for five keys with unequal probabilities.

$K(1) < K(2) < K(3) < K(4) < K(5)$

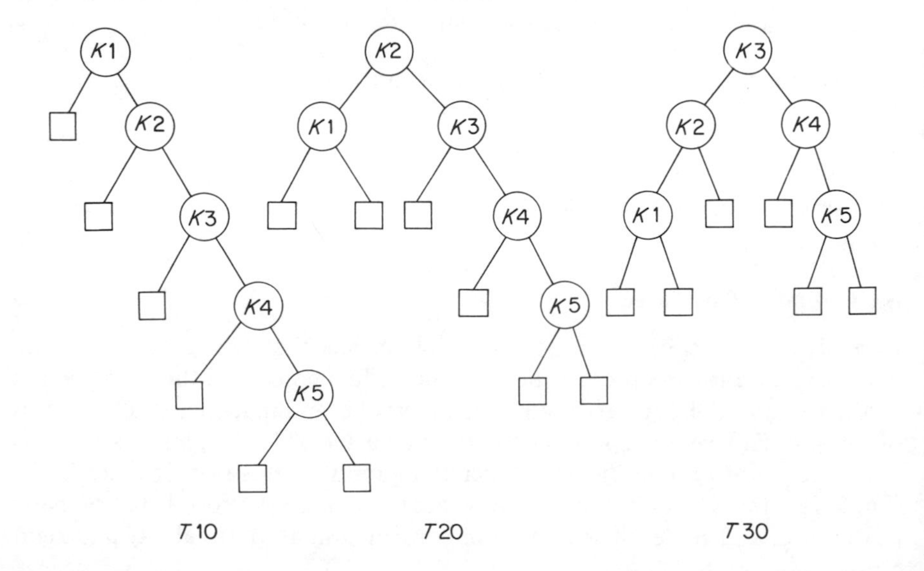

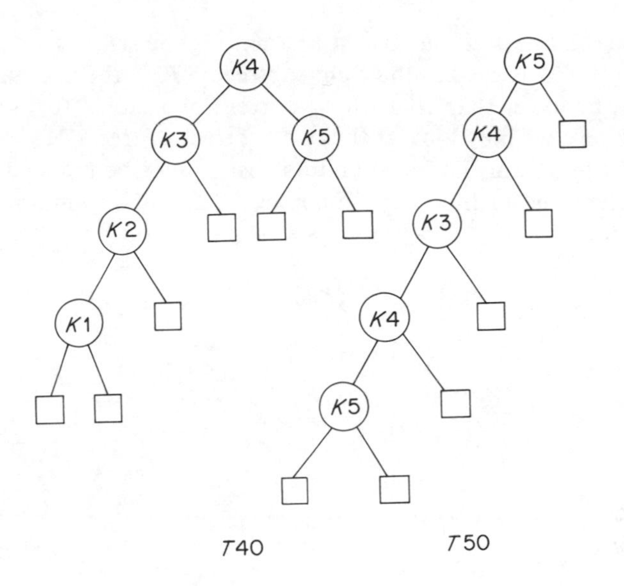

Let $T20$ be the tree with the lowest cost among the current trees. We can show it diagrammatically as follows.

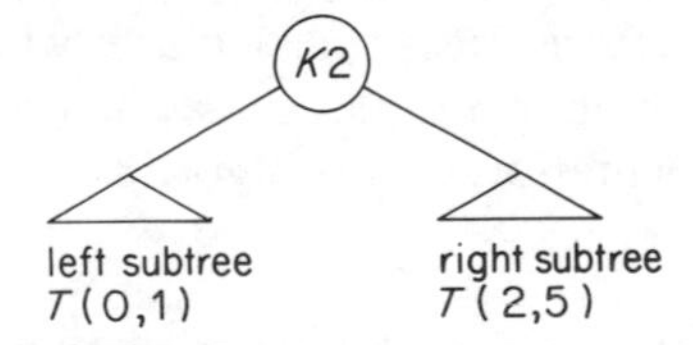

Apply the method to each of the two subtrees. As it happens in this example the left subtree $T(0, 1)$ is elementary (that is, it contains one key only) and thus no further improvement is possible. Let us next consider the right subtree. It contains the keys: $K(3) < K(4) < K(5)$, and hence we have the following possible trees to consider

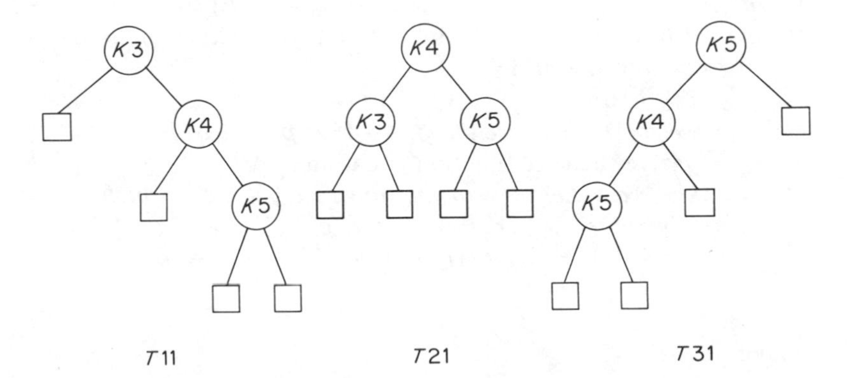

Let $T31$ be the tree with the lowest cost among the set $T11$, $T21$ and $T31$. It means that if $T31$ is used as the right subtree of $T20$, the overall cost of the new $T20$ will be lower than that of the current one. For a further reduction in the overall cost we may look at the two subtrees of tree $T31$: the right subtree is trivial (it contains no keys) and its cost cannot be reduced, but to the left subtree the same method can be applied. The final optimum binary tree is shown below.

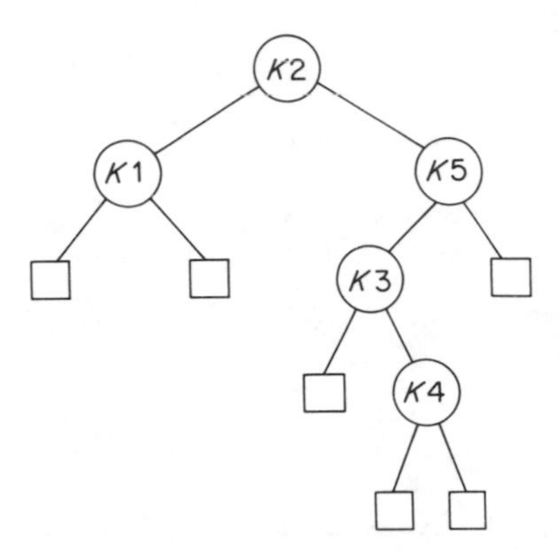

What we have demonstrated is how the algorithm works. It also needs to be proven that the tree which is produced by the algorithm is indeed optimal. Proving the algorithm's correctness, particularly where the algorithm is not 'obviously correct', is an important part of the algorithm design. We are now ready to give an implementation of the algorithm.

Procedure obst (p, q, n)

```
for i := 0 to n − 1 do              //initialize//
    weight [i,i] := q[i]; root[i, i] := 0; cost[i, i] := 0
    weight[i, i + 1] := q[i] + p[i + 1] + q[i + 1]
    weight[i, i + 1] := i + 1; cost[i, i + 1] := weight[i, i + 1]
enddo
weight[n, n] := q[n]; root[n, n] := 0; cost[n, n] := 0
for m := 2 to n do                  //find optimal trees with m nodes//
    for i := 0 to n − m do
        j := i + m
        weight[i, j] := weight[i, j − 1] + p[j] + q[j]
        //solve the cost function equation//
        k := {a value of g in the range i + 1 ≤ g ≤ j which
        minimizes the function cost[i, g − 1] + cost[g, j]}
        cost[i, j] := weight[i, j] + cost[i, k − 1] + cost[k, j]
        root[i, j] := k
    enddo
enddo
```

Procedure tree(i, j)

```
    if i = j then return (nil)          //no subtrees//
    else
        create newnode (t)
        t.value := root[i, j]
        t.left := tree(i, t.value − 1)
        t.right := tree(t.value, j)
        return (t)
    endif
```

Algorithm optimaltree (p, q, n)

```
    begin
    obst(p, q, n)
    tree(0, n)
    end
```

In the implementation the first algorithm, Procedure *obst* (p, q, n), computes the cost(i, j) of optimal binary search trees $t(i, j)$ on the keys $K(i+1) < \ldots < K(j)$ with $p(i), \ldots, p(j)$ and $q(i), \ldots, q(j)$, given n distinct keys $K(1) < \ldots < K(n)$ with the probabilities of occurrence $p(i)$, $1 \leqslant i \leqslant n$, and the unsuccessful search probabilities $q(i)$, $0 \leqslant i \leqslant n$. It also finds the $root[i, j]$ and $weight[i, j]$ of $t(i, j)$, so that $root[i, j]$ contains the index of key in the (sorted) data array. The second algorithm, Procedure *tree* (i, j), recursively constructs an optimal binary search tree given the global variable where $root[i, j]$ is the index of the root key of the tree containing keys $K[i+1]$ to $K[j]$. Each node of the tree has three fields: value which is equal to the order of key in original input, and left and right which are links to the left and right subtrees. Procedure tree returns the newly created node. The principal algorithm *optimaltree* consists of just two calls, to *obst* and to *tree*.

In order to estimate the time performance of algorithm *optimaltree* we note that in Procedure *obst* the first (simple) **for loop** requires $0(n)$ steps. This is followed sequentially by triply nested loop, where the outer **for loop** takes $0(n)$ steps, the middle **for loop** requires $0(n-m)$ steps, and the innermost **for loop** when computing the value for the variable k needs $0(m)$ steps. The time complexity of the algorithm is thus given as $0(n) + 0(n^2m) = 0(n^3)$ since $2 \leqslant m \leqslant n$.

Knuth (1971) improved the algorithm *obst* to run in time proportional to n^2 by proving that the root of an optimal tree for $K(i) < \ldots < K(j)$ need never lie outside the interval bracketed by the two roots of an optimal tree for $K(i) < \ldots < K(j-1)$ and one for $K(i+1) < \ldots < K(j)$. To implement the Knuth result we simply need to replace the range '$i+1 \leqslant g \leqslant j$' in the innermost **for loop** by the range '$root[i, j-1] \leqslant g \leqslant root[i+1, j]$'. Next, the recursive procedure tree when used on n keys will use $2n$ calls of itself since each time a node is created via *create newnode*(t), *tree* is called twice. Hence the time of algorithm *tree* is of $0(n)$. Assuming that the Knuth result is implemented in procedure *obst* we obtain the time complexity of algorithm optimal tree as $0(n^2)$. The algorithm also uses the memory space proportional to n^2, in order to store a triangular $n \times n$ matrix of intermediate results in procedure *obst*. In Example 9.5.2 an optimal binary search tree is constructed for five keys.

Example 9.5.2

This is a construction of an optimal binary search tree for five keys.
Given

$$K(1) < K(2) < K(3) < K(4) < K(5).$$

with $\beta_1 = 0.1$, $\beta_2 = 0.3$, $\beta_3 = 0.05$, $\beta_4 = 0.2$, $\beta_5 = 0.01$, and

$$\alpha_0 = 0.08,\ \alpha_1 = 0.04,\ \alpha_2 = 0.1,\ \alpha_3 = 0.03,\ \alpha_4 = 0.06,\ \alpha_5 = 0.03.$$

Results of the computation are shown in Table 9.5.1. Here the values given in column $m = 4$, for example, are obtained as follows:

For $m = 4$ consider $i = 0$ which gives $j = i + t = 4$.

$$w_{04} = w_{03} + \beta_4 + \alpha_4 = 0.96.$$

Table 9.5.1 The value of variables weight(i_j), cost(i_j) and root(i_j) computed for five keys with unequal probabilities of occurrence

$t = j - i \rightarrow$ $i \downarrow$	0	1	2	3	4	5
0	$w_{00} = 0.08$ $c_{00} = 0$	$w_{01} = 0.22$ $c_{01} = 0.22$ $r_{01} = K(1)$	$w_{02} = 0.62$ $c_{02} = 0.84$ $r_{02} = K(2)$	$w_{03} = 0.70$ $c_{03} = 1.10$ $r_{03} = K(2)$	$w_{03} = 0.96$ $c_{04} = 1.80$ $r_{04} = K(2)$	$w_{05} = 1.00$ $c_{05} = 1.98$ $r_{05} = K(2)$
1	$w_{11} = 0.04$ $c_{11} = 0$	$w_{12} = 0.44$ $c_{12} = 0.44$ $r_{12} = K(2)$	$w_{13} = 0.52$ $c_{13} = 0.70$ $r_{13} = K(2)$	$w_{14} = 0.78$ $c_{14} = 1.40$ $r_{14} = K(2)$	$w_{15} = 0.82$ $c_{15} = 1.58$ $r_{15} = K(2)$	
2	$w_{22} = 0.1$ $c_{22} = 0$	$w_{23} = 0.18$ $c_{23} = 0.18$ $r_{23} = K(3)$	$w_{24} = 0.44$ $c_{24} = 0.62$ $r_{24} = K(4)$	$w_{25} = 0.48$ $c_{25} = 0.76$ $r_{25} = K(4)$		
3	$w_{33} = 0.03$ $c_{33} = 0$	$w_{34} = 0.28$ $c_{34} = 0.28$ $r_{34} = K(4)$	$w_{35} = 0.32$ $c_{35} = 0.42$ $r_{35} = K(4)$			
4	$w_{44} = 0.06$ $c_{44} = 0$	$w_{45} = 0.1$ $c_{45} = 0.1$ $r_{45} = K(5)$				
5	$w_{55} = 0.03$ $c_{55} = 0$					

For $1 \leqslant g \leqslant 4$ we have

$$c_{00} + c_{14} = 1.40,$$
$$c_{01} + c_{24} = 0.84,$$
$$c_{02} + c_{34} = 1.12,$$
$$c_{03} + c_{44} = 1.10,$$

and the minimum sum is clearly achieved for $k = 2$. This gives

$$c_{04} = w_{04} + c_{01} + c_{24} = 1.80 \quad \text{and} \quad r_{04} = K2.$$

Next consider $i = 1$, which gives $j = i + t = 5$.

$$w_{15} = w_{14} + \alpha_5 + \beta_5 = 0.82.$$

For $2 \leqslant g \leqslant 5$ we have

$$c_{11} + c_{25} = 0.76,$$
$$c_{12} + c_{35} = 0.86,$$
$$c_{13} + c_{45} = 0.80,$$
$$c_{14} + c_{55} = 1.40,$$

and the minimum is achieved for $k = 2$. This gives

$$c_{15} = w_{15} + c_{11} + c_{25} = 1.58 \quad \text{and} \quad r_{15} = K2.$$

The results of the table are summarized in the two-dimensional root array $R(ij) = K(r_{ij}) = K(k)$.

$$\begin{bmatrix} \text{null} & K(1) & K(2) & K(2) & K(2) & K(2) \\ & \text{null} & K(2) & K(2) & K(2) & K(2) \\ & & \text{null} & K(3) & K(4) & K(4) \\ & & & \text{null} & K(4) & K(4) \\ & & & & \text{null} & K(5) \\ & & & & & \text{null} \end{bmatrix}.$$

An optimum tree $t(0, 5)$ is then obtained as shown in Fig. 9.5.2.

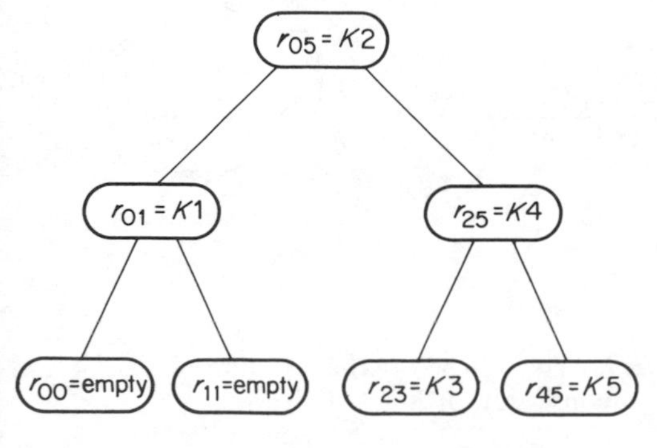

Figure 9.5.2 A minimum-cost binary search tree for five keys with unequal probabilities of occurrence

9.6 A Tree Search Followed by Insertion or Deletion of the Key

We shall now consider the search process where the operation of search is normally followed by insertion or deletion of the key. The corresponding binary trees in these cases are called dynamic binary trees. Consider the following example. Given a binary tree as shown in Fig. 9.6.1(a), we wish to insert in it the key 'July'.

To do this, we start at the root of the tree, comparing 'July' to 'May', then move along the tree to the left and compare 'July' to 'January'; the search is again unsuccessful and we move further to the right and compare 'June' to 'July'; finally we move to the left and the search key is now inserted into the empty external node 5. The new tree is shown in Fig. 9.6.1(b). The operation of insertion is straightforward and its implementation was given in Section 9.4, algorithm *treesearch*.

Apparently, if the right subtree of the node 'January' were empty, we could have simply deleted the node itself and joined the branch from 'February' to 'May'. But this cannot be done in our example because the right subtree in question is not empty. What we can do, though, is to find the next node for

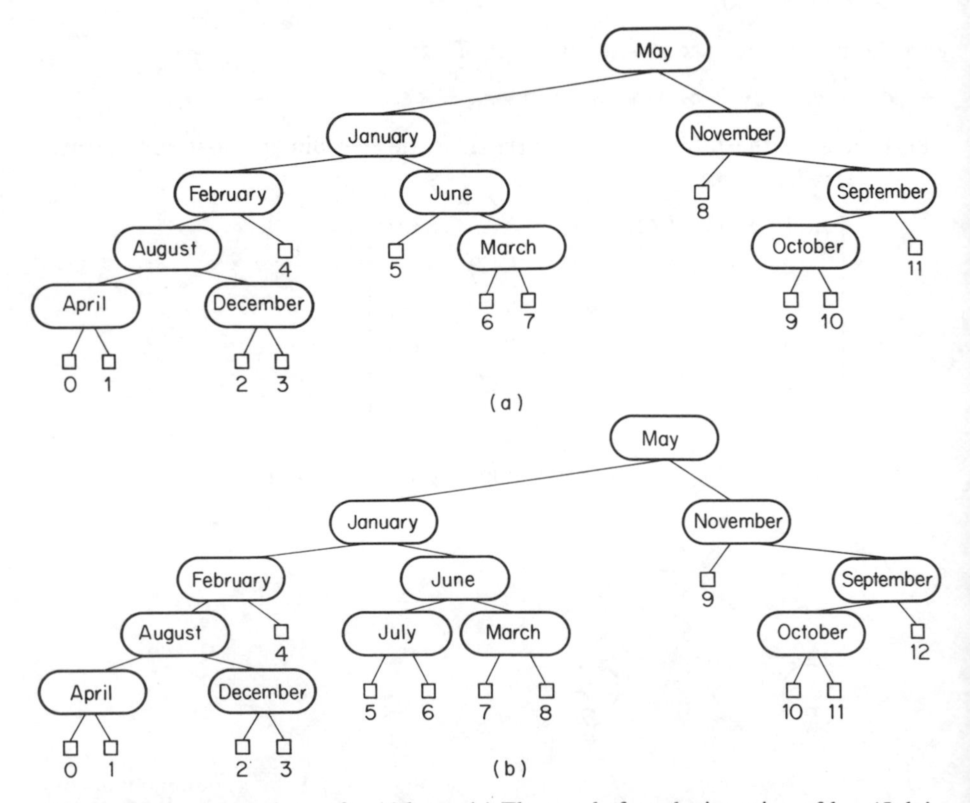

Figure 9.6.1 A search tree for 11 keys. (a) The tree before the insertion of key 'July'; (b) the tree after the insertion of key 'July'

which the argument introduced holds. We then can 'delete' the latter node and reinsert it in place of the node that we really wanted to delete.

In our example we may decide to replace 'January' by its left subtree 'February'. Since 'February' has no right subtree, we delete 'February' first, then join up 'August' to 'January' and then replace 'January' by 'February'. The tree obtained is shown in Fig. 9.6.2(a). As an alternative we can also decide to look for a replacement for 'January' in its right subtree. One such

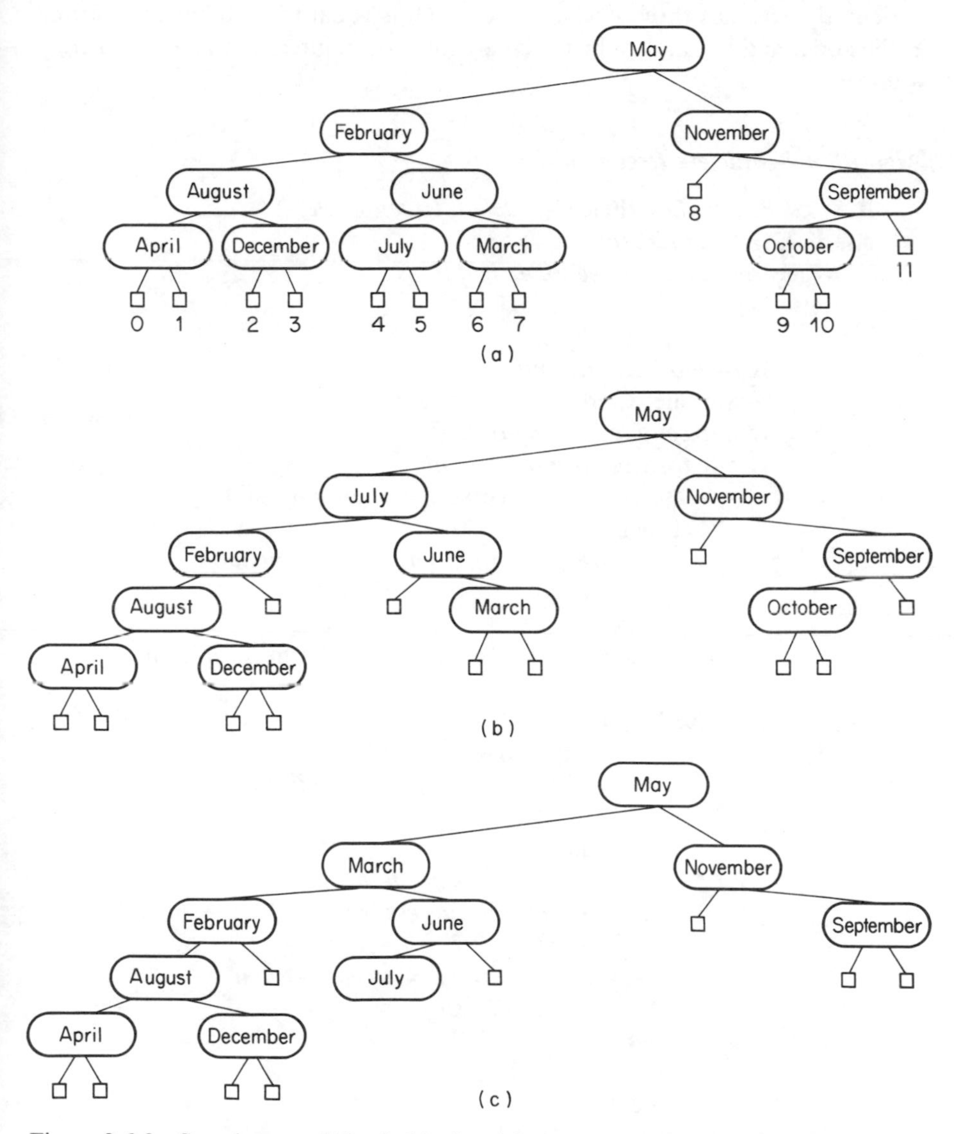

Figure 9.6.2 Search trees of Fig. 9.6.1 after deletion of node 'January' using three different sets of arrangements. (a) 'January' is deleted via 'February'; (b) January is deleted via 'July'; (c) 'January' is deleted via 'March'

node is 'July' since it has no right subtree. We delete 'July' by making 'June's' left subtree equal to null, and then replace 'January' by 'July'. Similarly, we could use the node 'March' to replace 'January'. The two trees which are obtained as a result of such operations are shown in Fig. 9.6.2(b) and (c).

The example shown in Fig. 9.6.2 demonstrates that the operation of deletion in a tree search is a complex task, and requires a suitable technique. We shall now specify the algorithm just used for deletion. As usual we assume that each node in the tree has three fields: value which is equal to the key value stored in the node, and left and right which are links or pointers to the left and right subtrees.

Algorithm rightdelete (treenode)

```
if K < treenode.key then rightdelete (treenode.left)
else if K > treenode.key
     then rightdelete (treenode.right)
     else begin
          copy := treenode
          if treenode.right = nil
          //case one: there is no successor//
          then treenode := treenode.left
          else if treenode.right.left = nil
               //case two: the successor is the right child//
               then begin
                         treenode.right.left := treenode.left
                         treenode := treenode.right
                    end
               //case three: the successor is the leftmost child in the
               right subtree//
               else begin
                         succ := treenode.right
                         while succ.left.left <> nil do
                              succ := succ.left
                         enddo
                         successor := succ.left
                         succ.left := successor.right
                         successor.left := treenode.left
                         successor.right := treenode.right
                         treenode := successor
                    end
               endif
               dispose (copy)
          endif
          end
     endif
endif
```

In the given implementation of the algorithm a node's successor is defined to be the smallest node in its right subtree. (Similarly a node's predecessor can be defined to be the largest node in its left subtree.) To delete a node from a binary tree, replace the node with its successor, that is, the node that contains the next larger key. This implementation is called asymmetric because it specifically uses the right and left subtrees of the node to define the node's successor.

The algorithm can be made symmetric if the node is allowed to be deleted by replacing it with either its successor or its predecessor. Alternatively choose the successor and predecessor, so that half the time the rightdelete procedure is called and half the time a suitably modified version of this procedure, left-delete, is called.

9.7 Balanced Search Trees

Suppose that an initial set of data is structured as a fairly balanced tree, then its search time is close to $0(\log n)$. Now, if insertion and deletion operations are repeatedly applied to this tree, the tree may grow very unbalanced. For example, if a sequence of deletion operations removes nodes from only the right side of the tree, we shall finish up with a tree that is unbalanced to the left. As a result the search time of the tree will now be closer to $0(n)$. To prevent these worst-case situations a provision has to be made for some specific mechanism which rebalances the tree after each insertion or deletion operation.

The oldest and most well-known mechanism for balancing the search trees was developed by Adelson-Velski and Landis (1962). It is based on the use of the balanced tree data structure, known as the AVL tree. This tree has the property that the heights of the two subtrees of each node differ by at most one. The tree is preserved balanced throughout the complete process of insertions and deletions. When the property is violated because of an insertion or a deletion, it is reinstated using rotations. A rotation is a transformation operation on trees that changes the shape of the tree without altering its lexicographic order. The basic algorithm is to search for the value being inserted, then proceed up the tree along the path just travelled adjusting the heights of nodes using rotations. The algorithm must know whether each node has a height that is one less than, the same, or one greater than the height of its sibling. What actually happens when a node is inserted into a balanced tree is illustrated in Figs 9.7.1 and 9.7.2.

In Fig. 9.7.1(a) is shown a subtree of an AVL tree in which node N is about to be inserted. Before the insertion, the heights of the two subtrees of P are equal, and the left subtree of Q is of greater height than the right subtree of Q. After the insertion of node N the height constant is violated at node Q. Consequently the tree is rearranged using a single rotation. The subtree of Fig. 9.7.1(a) is rotated to the right and T^2 becomes attached to Q instead of P. The new tree is again an AVL tree.

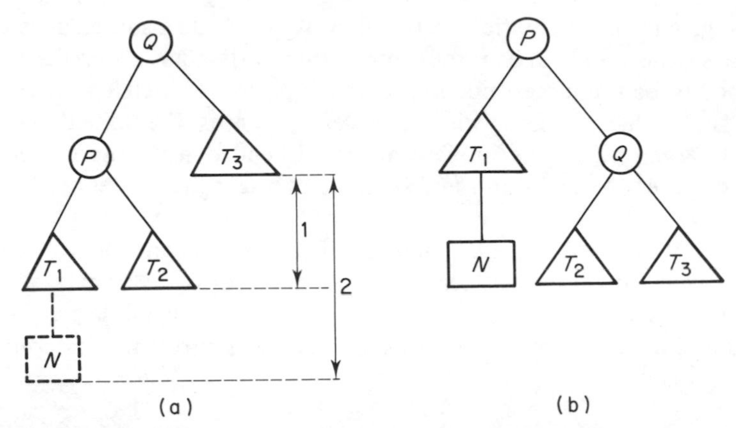

Figure 9.7.1 An AVL tree, by means of single rotation, is rearranged into a height balanced tree, i.e. into an AVL tree again, after insertion of node N. (a) Node N is about to be inserted in the AVL tree. (b) The tree of (a) transformed into a new AVL tree.

The only other case which may arise in rearranging a height-balanced tree is shown in Fig. 9.7.2. It is handled by a double rotation. Here at first the subtree of R and Q is rotated to the right and T^3 becomes attached to P. We can see that the AVL algorithm is simply a rule that requires a single or a double rotation whenever this rotation can reduce the height of a subtree. A straightforward implementation of the algorithm for double rotation is given below.

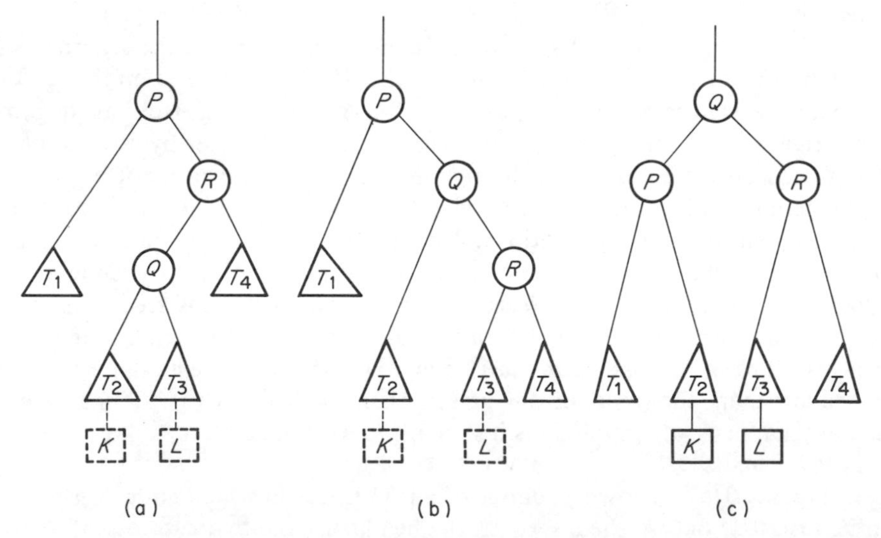

Figure 9.7.2 An AVL tree, by means of double rotation, is rearranged into a height balanced tree, i.e. a new AVL tree. (a) Nodes K and L are about to be inserted in an AVL tree. (b) First step of the 'double rotation', a device used to transform the tree (a) into an AVL tree. (c) The tree transformed into a new AVL tree insertion of nodes K and L

Algorithm doublerotate (treenode)

```
if K < treenode.key then child := treenode.left
else child := treenode.right
endif
if K < child.key then
begin
    grandchild := child.left
    child.left := grandchild.right
    grandchild.right := child
end
else begin
        grandchild := child.right
        child.right := grandchild.left
        grandchild.left := child
    end
endif
if K < treenode.key then treenode.left := grandchild
else treenode.right := grandchild
endif
if K < treenode.key then
begin
    treenode.left := grandchild.right
    grandchild.right := treenode
end
else begin
    treenode.right := grandchild.left
    grandchild.left := treenode
    end
endif
return (grandchild)
```

The steps of algorithm *doublerotate* can be easily traced in Fig. 9.7.2. In Fig. 9.7.2(a) an initial pointer *treenode* points to node P, then the pointer's *child* (pointing to node R) and *grandchild* (pointing to node Q) are created. Subsequent steps of the algorithm follow Fig. 9.7.2(b) and (c). At the end of the process a pointer (*grandchild*) to the new root is returned.

Insertion in height-balanced trees requires at most one single or one double rotation. Deletion, on the other hand, may require as many as $h/2$ transformations where h is the height of the tree, though on average the number of required transformations is a small constant.

The AVL trees represent a compromise between optimum binary trees for which as we know all external nodes required to be on two adjacent levels, and arbitrary binary trees. In view of such a placing of the balanced tree it is interesting to know how far from optimum a balanced tree can be. The answer to this question is given by a theorem of Adelson-Velski and Landis. It proves that the depth of a balanced tree with n internal nodes always lies between $\log_2(n+1)$ and $1.4404 \log_2(n+2) - 0.328$, and, hence, its search paths will never be more than 45% longer than the optimum.

The AVL trees have been generalized by Foster (1973) and Bayer (1972). The Bayer trees are weight-balanced (as opposed to height-balanced AVL trees) in the sense that the number of nodes of each subtree is balanced; these trees are known as BB trees. Balanced trees have been studied intensively (for a list of references see, for example, Gonnet, 1983). One of the more interesting recent additions are the self-adjusting binary search trees or the splay trees of Sleator and Tarjan (1985). Splaying is a restructuring heuristic which is applied whenever the tree is accessed with the view to restructuring the tree so as to keep its access time short. Suppose a sequence of access operations on a binary search tree is to be carried out. For the total access time to be short, frequently accessed keys should be near the root of the tree. The goal is then to devise a simple way of restructuring the tree after each access that moves the accessed key closer to the root, on the plausible assumption that this key is likely to be accessed again soon. Splaying is a rotation-like transformation which cleverly restructures the tree so that the accessed key becomes the root or a node close to the root of the tree. Sleater and Tarjan (1985) have shown that splay trees are as efficient as balanced trees and, for sufficiently long access sequences, as efficient as static optimum search trees.

Another well-known balanced tree structure is the *two–three tree*, where only *two-nodes* and *three-nodes* are allowed.

More recent developments in the design of balanced tree structures are known as the top-down *two–three–four tree* and the *red–black tree*. These structures have very high degree of flexibility and allow particularly efficient ways in rebalancing the search trees. A top-down *two–three–four tree* consists of *two-nodes*, *three-nodes* and *four-nodes*, which hold one key and two links, two keys and three links, and three keys and four links, respectively. For example, a *three-node* has three links coming out of it, one for all keys smaller than both of its keys, one for all keys with the keys in between its two keys, and one for all keys larger than both its keys. Similarly, a *four-node* has four links coming out of it, one for each of the intervals defined by its three keys.

In our example below we have a *two–three–four tree*

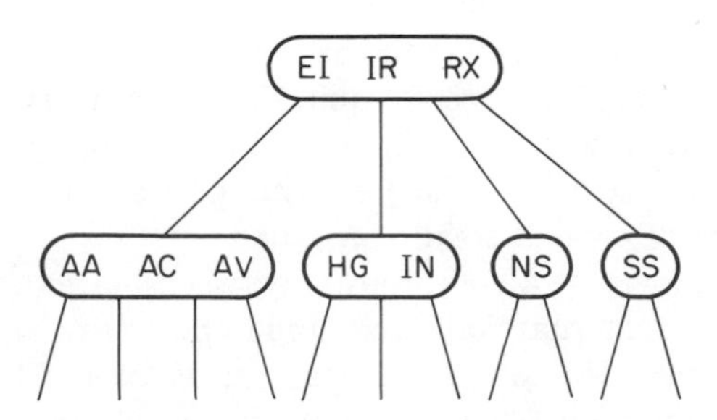

To search for OP in this tree, one would follow the third link from the root,

OP is between IR and RX and terminate the unsuccessful search at the right link from the *two-node* containing NS.

To insert a new node in a *two–three–four tree* after an unsuccessful search, simply hook the node on. If the node at which the search terminates is a *two-node* or a *three-node*: just turn it into a *three-node*, or a *four-node*, respectively. If the new node needs to be inserted into a *four-node*, then the *four-node* is first split into two *two-nodes* and one further key is passed on up in the tree, and finally the new node is inserted in a usual way. The process is demonstrated in Example 9.7.1 where the keys from DI IS ST TR RI IB BU UT TI IO ON are inserted into an initially empty tree.

Example 9.7.1

This is insertion of eleven keys into an initially empty *two–three–four tree*.

Start out with a *two-node*, then a *three-node*, then a *four-node*:

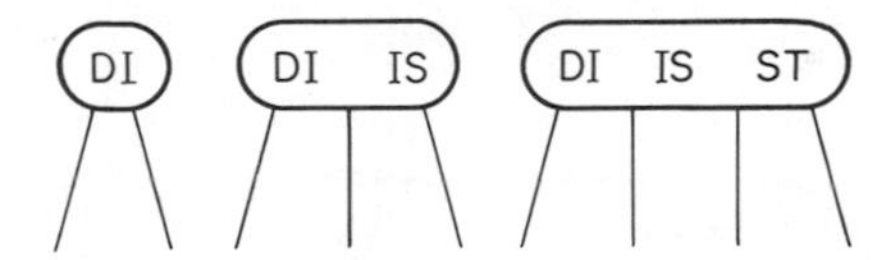

To insert *TR* into the *four-node* we cannot and thus first split it as shown

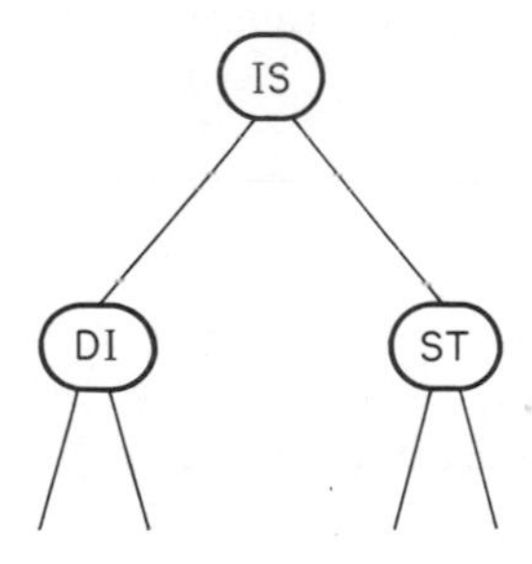

and then insert *TR* into the *two-node* at which the unsuccessful search terminates:

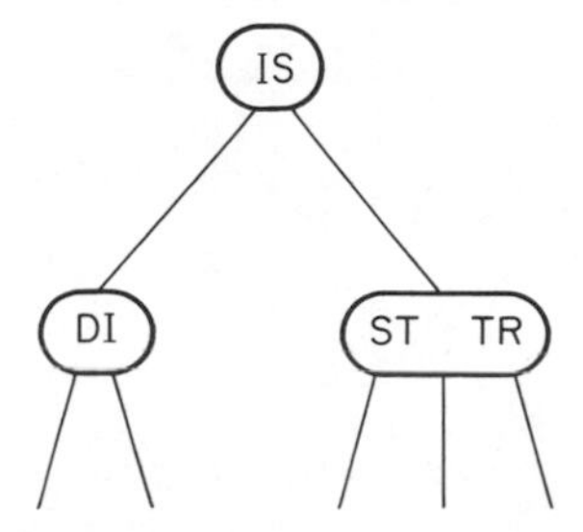

RI, *IB* and *BU* are inserted without difficulty:

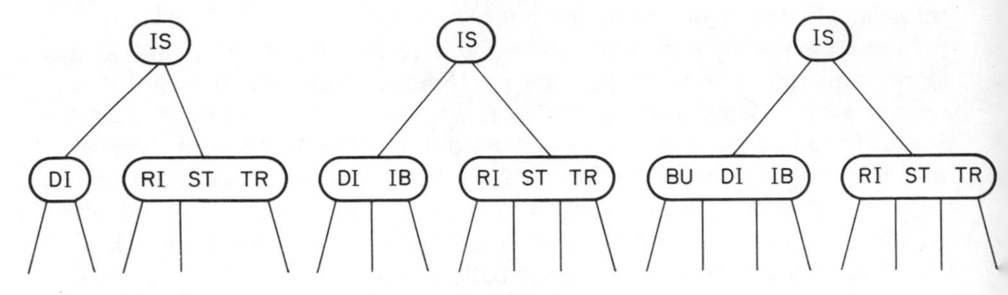

but when *UT* arrives for insertion, the *four-node* on the right is full. Again, this *four-node* is first split into two *two-nodes* and a further key is inserted into the parent, changing the parent from a *two-node* to a *three-node*, then *UT* is inserted:

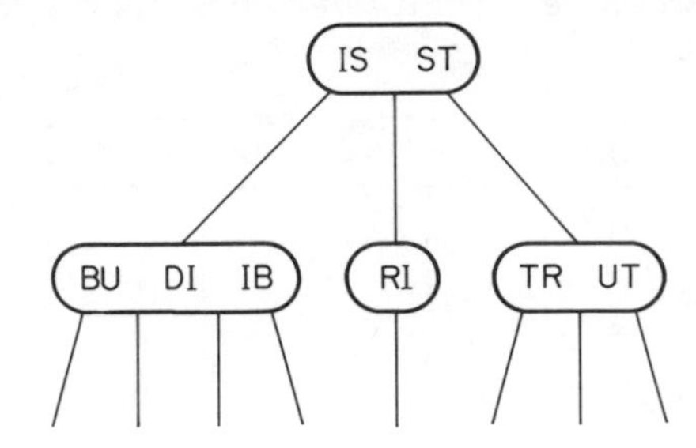

TI is next with no problem, and *IO* causes another split and is then inserted:

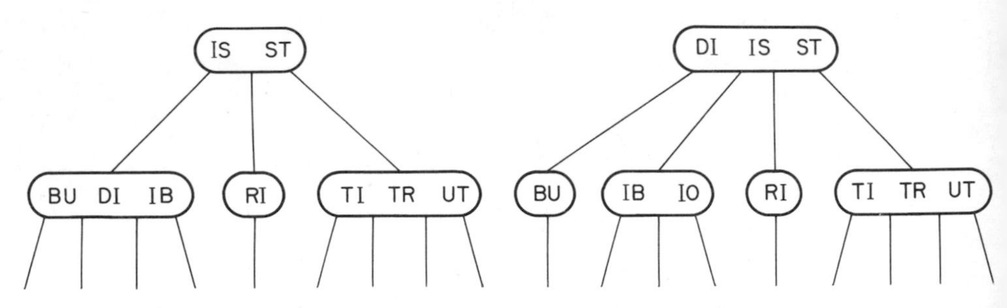

Finally *ON* is inserted in a straightforward way:

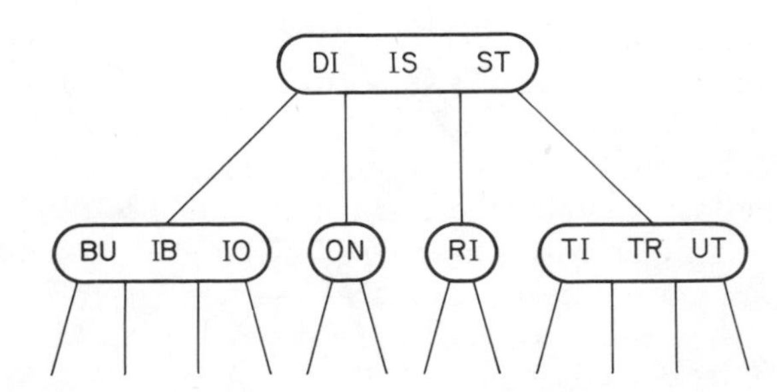

One complex case that may arise in this algorithm is where a *four-node* needs to be split and whose parent is also a *four-node* and thus it cannot absorb the extra node resulting from the split. If the parent is also split, the situation could persist all the way up the tree. What is done instead is to make sure that the parent of any node considered will not be a *four-node*, this is achieved by splitting any *four-node* that is encountered on the way down the tree. This process gave the name to the trees produced by the algorithm as top-down *two–three–four trees*, because the *four-nodes* are split up on the way from the top down. The most remarkable fact about the top-down *two–three–four trees* is that they are perfectly balanced, though no direct action about rebalancing the tree after every insertion is ever taken. The search and the insertion times required by the algorithm which searches or inserts a new node into a top-down *two–three–four tree* is always proportional to log$n$. The proof of this fact is left to the reader as an exercise.

The actual implementation of the described algorithm is done by first representing top-down *two–three–four trees* as standard binary trees with two different kinds of links: 'red' links which bind together *three-nodes* and *four-nodes* as small binary trees, and 'black' links which bind the *two–three–four tree* together. These representations are referred to as *red–black trees*. The rotations and splittings, required by the algorithm for structuring a *two–three–four tree* are then carried out on standard binary trees with a proper attention to a distinct use of two types of links.

9.8 Hashing

We shall now examine another important group of searching methods known as hashing techniques. Hashing is a completely different approach to searching from that of the comparison-based tree structures. In hashing, the keys in the table are referenced directly by doing arithmetic transformations on the keys into table addresses. If we assume that the table to be searched is of size n, and its addresses are numbered from 0 to $n-1$, then if one were to know that the keys are distinct integers from 0 to $n-1$, they could be stored in table positions directly, ready for immediate access by their value. In general, in the absence of any special knowledge about the key values, the first step in a search using hashing is to compute a hash function, $h(K)$, which transforms the search key into a table address, that is $0 \leqslant h(K) < n-1$. Ideally one would wish to produce a hash function which gives a unique address value for each K but this is simply too difficult to achieve and so from time to time two or more different keys hash to the same table address and create a collision. The second part of a hashing search is to resolve a collision, so that either the key is located or the fact is established that the key is not in the table.

Prime concerns of any hashing method is an efficient use of available memory and fast access to the memory. The two extreme cases of these resources—no memory limitation (then the search could always be done with only one memory access by simply using the key as a memory address), or no time limitation (then a sequential search method would always do with only a minimum amount of memory)—are of no interest in practical applications. In fact, various hashing algorithms compete in terms of simplicity with which the hash function is computed and collision is resolved.

In the next section we shall consider several methods for selecting the initial hash function and subsequent collision resolution.

Computing Hash Function

The computation of the hash function is an arithmetic computation with the aim of producing a perfectly random distribution of the hash function values in the range 0 to $n-1$—that is such that for each input, K, every output, a number between 0 and $n-1$, should be equally likely. The key K can be, of course, either integer or a character string (usually short) and since the hashing involves arithmetic calculations, the first step is to transform the key into a number which can be operated on. From now on we assume that the keys are integers which fit into a machine word.

Distribution-dependent Hash Functions

Consider a collection, G, of k integer keys which are included in a universe of possible integers, U. Define a random variable X on G as

$$X(G) = G,$$

and let F_X be its (cumulative) distribution function. We then say that the keys in G are distributed according to a discrete distribution F_X.

We wish to find a hash function h which is defined on a domain U and with the value equal to $\{0, 1, \ldots, n-1\}$, such that the random variable $h(X)$ defined on G has a discrete uniform distribution function on $\{0, 1, \ldots, n-1\}$. In other words, we wish to find a has function h which maps our k given keys to the addresses $0, 1, \ldots, n-1$ so that the keys are uniformly distributed over the address locations. The reason for a uniform distribution of the hash values is to minimize the number of collisions.

By definition we have

$$F_X(t) = P(X \leqslant t) \tag{9.8.1}$$

where $P(Y)$ denotes the probability of Y. We wish to find h such that

$$P(h(X) = i) = \frac{1}{n} \quad \text{for} \quad 0 \leqslant i \leqslant n. \tag{9.8.2}$$

In order to do this, consider the random variable $F_X(X)$ and note that

$$P\left(F_X(X) \leqslant \frac{r}{k}\right) = \frac{r}{k} \quad \text{for} \quad 0 \leqslant r \leqslant k. \tag{9.8.3}$$

Relation (9.8.3) is strictly true only when all keys are distinct. (The assumption that the keys are distinct has been made at the beginning of this chapter.) From (9.8.3) it follows that $F_X(X)$ has a discrete uniform distribution on

$$\left\{\frac{1}{k}, \ldots, \frac{k-1}{k}, 1\right\}, \tag{9.8.4}$$

and so $nF_X(X)$ has a discrete uniform distribution on

$$\left\{\frac{n}{k}, \frac{2n}{k}, \ldots, n\right\}. \tag{9.8.5}$$

Therefore $nF_X(X) - 1$ is approximately uniform on $\{0, 1, \ldots, n-1\}$, especially when $n \leqslant k$. Thus a good choice for h is

$$h(K) = \lceil nF_X(K) \rceil - 1. \tag{9.8.6}$$

When F_X is known and is easily computable, it can be used to obtain a hash function which is tailored to the collection of keys to be stored.

When F_X is itself a uniform distribution on $\{1, 2, \ldots, s\}$, our hash function becomes

$$h(K) = \left\lceil \frac{nK}{s} \right\rceil - 1.$$

where s is the largest key occurring in G.

One may take s as the largest key in U if the largest key occurring among the actual keys is not known.

Other variants of defining F_X can also be used, for example, an approximation of the F_X by a piecewise linear function or an approximation by a polynomial of degree higher than one. These techniques assume the use of a partial knowledge of F_X, etc.

Cluster-separating Hash Functions

If two or more keys in G are close to each other in some sense, they are said to form a *cluster*. If it is known that a given collection tends to have such clusters among its keys, than a good hash function is expected to destroy these clusters, that is the hash values of the 'clustered' keys should be not equal. In practice, however, no matter what hash function we choose, there are notions of what constitutes a cluster for which the hash function fails to achieve reasonable results. In this sense, there is no such thing as a general purpose hash function.

The best we can select is a hash function which achieves a reasonably uniform distribution of hash values given certain general asusmptions about the distribution of the keys to be processed.

One common assumption is that the set of keys of collection G is a random sample from the set of all possible keys, U.

A hash function which is suitable under this assumption is the function with range $\{0, 1, \ldots, n-1)$ which partitions the universe of possible keys, U, as nearly as possible into n equinumerous sets of synonymous keys corresponding to the addresses $0, 1, \ldots, n-1$. Knott (1975) calls such a hashing function a *balanced hash function.*

A more special assumption which can be made is that the keys tend to occur in arithmetic progression, i.e. if K is an occurring key, then $K+s$ for some fixed s is relatively likely to occur also. One notion for the choice of s (or the distance between two keys) is adapted from Hamming (1950). It is known as the *Hamming metric* and may be defined as follows:

> Two keys, Q and R, considered to be t-tuples whose components are symbols of some alphabet, are separated by a distance s under the Hamming metric, $D(Q, R)$ when Q and R have different symbols in exactly s different component positions. A hashing function, h, is called d-separating with respect to the metric $D(Q, R)$ if $0 < D(Q, R) \leqslant d$ implies $h(Q) \neq h(R)$.

Based on the notion of the Hamming metric, a common clustering assumption is that if K is a key actually occurring then keys which are close to K with respect to the Hamming metric are relatively likely to actually occur also.

For example, if keys are English words of t or fewer letters then the keys are distributed in a certain manner in the set of all t-component strings of letters, including blank. This distribution is such that our intuitive notion of clusters is indeed approximated by using the Hamming metric as a measure of closeness. If w is an English word then w changed in one or two positions is also likely to be an English word.

Methods of constructing 'cluster-destroying' hash functions under the notion of cluster in terms of the Hamming metric, have been successfully studied using the results of algebraic coding theory. Ths was first suggested by Muroga (unpublished report, IBM, 1961). One approach for the choice of the hash function which has emerged from these studies is the use of polynomials.

Let $g(x)$ be a polynomial of degree s. Also, let $y(x)$ by a polynomial of degree $t-1$ or less, where

$$y(x) = y_0x^{t-1} + y_1x^{t-2} + \ldots + y_{t-1}.$$

Let K be the set of coefficients of $y(x)$, i.e.

$$K = (y_0, y_1, \ldots, y_{t-1}),$$

then the hash function $h(K)$ is obtained as the set of coefficients of the polynomial $y(x) \bmod g(x)$.

The coefficients of $g(x)$ are defined and a suitable polynomial division algorithm is used to compute $y(x) \bmod g(x)$, where all the arithmetic among coefficients is done in the symbol field.

A care has to be exercised to ensure an appropriate choice of polynomial $g(x)$ with which one can construct d-separating functions. Such polynomials can be found but these constructions require some knowledge of the theory of finite fields (Schay and Raver, 1963).

Distribution-independent Hash Functions

When no special information on the initial distribution of keys is known, the hash function can be computed using methods similar to those of generating pseudo-random numbers. If one considers a key as a bit-string, then some simple hash functions are as follows.

(i) *Division method.*
$h(K) = K \bmod n$, where n is the table size.

(ii) *Linear congruential method.*
$h(K) = k$ leading bits of the number $bK \bmod w$ where the table size $n = 2^k$, w is the computer word size and b is an arbitrary constant (normally chosen to be a number smaller than w, eg. a number with one digit less than w).

The above are most commonly used techniques but many other hash function computing formulae have been proposed as well and in some cases a special techique may be preferable to a general-purpose method. We can further list some hash function computing methods.

(iii) *Addition method.*
$h(K) =$ a k-bit value is extracted from the value obtained by adding together segments of the key K.

(iv) *Multiplication method.*
$h(K) =$ a k-bit value is extracted from the value K^2, or from the value cK for some constant c.

(v) *Extraction method.*
$h(K) =$ a k-bit value composed by extracting k bits from K and concatenating them in some order. This method is convenient for the table size $n = 2^k$ and produces the range of hash addresses $\{0, \ldots, n = 2^k - 1\}$.

(vi) *Radix conversion method.*
$h(K) =$ a k-bit value extracted from the value obtained by treating K as a binary-coded sequence of base p digits and converting this coded value into a similarly coded base value. This method again assumes the table size $n = 2^k$.

Of the methods listed the division method is perhaps the most attractive. It involves very simple arithmetic and produces a hash function with a number of useful properties. For example, the division hash function destroys clusters under the Hamming metric. Knuth has noted that for a better distribution the division method should be used on the table sizes, n, which are not divisible by integers 2 or 3. The choice of n to be prime may be recommended but is not a necessity (Buchholz, 1963). Also the byproduct of the division method arithmetic, the quotient, is available 'without cost' and can be put to good use as various hash methods show; see for example, the quadratic quotient method for collision resolution of Bell (1970) which is discussed below (page 324).

The linear congruential method can on some computers be more efficiently computed than the division method. It is also advantageous in that it can spread out clusters. The multiplication method is sensitive to the distribution of keys and to the choice of a constant multiplier. Below we discuss one interesting multiplication scheme developed by Floyd (1970).

A Multiplicative Hash Function

In this section we describe the construction of a particular multiplicative hash function which was proposed by Floyd, the function is given as

$$h(K) = n(cK \bmod 1), \tag{9.8.7}$$

where K is as usual a non-negative integer key value, c is a constant defined in the interval $0 \leqslant c < 1$, and $cK \bmod 1$ produces a number in the unit interval. The crucial point of the method is the choice of c.

Let us assume that keys have a tendency to occur in arithmetic progression so that we may have keys

$$K,\ K+s,\ K+2s,\ \ldots \tag{9.8.8}$$

In this case it is desirable that

$$h(K), \qquad h(K+s), \qquad h(K+2s),\ \ldots$$

be a sequence of distinct addresses, in so far as this is possible. Now, this goal is achieved if the set of values $\{cK \bmod 1,\ c(K+s) \bmod 1,\ \ldots\}$ is 'well-spread' in the unit interval, for then the addresses achieved by scaling will be well-spread among $\{0,\ 1,\ \ldots,\ n-1\}$.

A set of points is considered to be 'well-spread' in the unit interval when the ratio of the minimum of the lengths of the subintervals defined by the given points placed in the unit interval to the maximum subinterval length is large, that is,

$$D = \frac{\text{minimum subinterval}}{\text{maximum subinterval}} \quad \text{is large.} \tag{9.8.9}$$

Floyd then proves that for $c = p/q$ with p and q integers, the general sequence

$$\{c(K + as^{i}) \bmod 1,\ a=0;\ a=1,\ i=i_1;\ a=2,\ i=i_2;\ a=3,\ i=i_3;\ \ldots\} \tag{9.8.10}$$

where the integer $s \equiv 1 \bmod q$ is the increment in arithmetic progression (9.8.8) corresponds to the sequence

$$\{(cK + \mathrm{ac}) \bmod 1,\ a = 0,\ 1,\ \ldots\} \tag{9.8.11}$$

regardless of the values of i_1, i_2, ... This implies that for q and s such that

$s \equiv 1 \bmod q$ the sequence of keys

$$K, \quad K + s^{i_1}, \quad K + 2s^{i_2}, \ldots \tag{9.8.12}$$

corresponds to the values

$$cK \bmod 1, \quad (cK + c) \bmod 1, \quad (cK + 2c) \bmod 1, \ldots \tag{9.8.13}$$

regardless of the values of $i_1, i_2, \ldots$.

With a further assumption (which can be made without loss of generality) that K is such that

$$cK \bmod 1 = 0 \text{ (that is, } cK \text{ is an integer)},$$

the problem is reduced to that of obtaining c such that the points

$$0, c \bmod 1, 2c \bmod 1, 3c \bmod 1, \ldots \tag{9.8.14}$$

are well spread in the unit interval.

The distribution (9.8.14) at both the early and asymptotic stages has been studied by Halton (1965) and Zaremba (1966). In particular, it has been shown that in the process of successively placing the points from (9.8.8) in the unit interval, the current largest subinterval is split into two smaller subintervals by every new point placed, that is relation (9.8.9) holds throughout the process.

The analysis further shows that an appropriate choice for the integer p is a value such that the constant $c = p/q$ would be a good rational approximation to the golden ratio

$$\rho = \frac{-1 + \sqrt{5}}{2} = 0.618. \tag{9.8.15}$$

For example, we may take $s = 1024 = 2^{10}$, having in mind the storage of items with 10-bit character string keys, and then we may take $q = 1023$ and $p = \lfloor q\rho \rfloor = 632$, whence the Floyd hash function becomes

$$h(K) = \left\lfloor n\left(\frac{632}{1023} K \bmod 1\right) \right\rfloor . \tag{9.8.16}$$

Another scheme for computing (9.8.7) which involves no division is given by Knuth (1973). The parameter c is suggested as a rational fraction p/q with $q = 2^m$ where one is dealing with m-bit values in the arithmetic operations. The parameter p is a prime to q. If n is chosen to be a power of 2, say 2^k, then $\lfloor n(cK \bmod 1) \rfloor$ is just the high k bits of the low order word of the double length product pK.

We have discussed four major groups of the methods for computing the initial hash function, $h(K)$. Other methods some even more elaborate than those discussed, cater for various specific forms of data arrays, like band matrices, triangular arrays, etc. We shall not discuss these methods trusting that the interested reader can turn to appropriate sources which are available in the literature on the subject.

We finish the section by noting that, in the opinion of Knuth (1973), none of the methods suggested for computing the hash functions has proved to be superior to the simple divison and multiplication methods described above.

9.8 Collision Resolution by Open Addressing

The hash function will convert a key into a table address. We next need to decide how to resolve the case when two keys hash to the same address. The simplest method is to rely on empty locations in the table to help with collision resolution. This is particularly suitable if the number of keys to be put in the hash table can be estimated in advance. A number of methods have been devised using this approach, called open-addressing hash methods.

Suppose that given a table with a number of keys in it we wish to find out whether the key K is in the table. The open addressing method looks at various entries in the table one by one until either the key K or an empty location is found. The sequence of locations inspected during the search is called the probe sequence, and the idea is to formulate some rule by which every key K determines a probe sequence. If using such a probe sequence, an empty location is encountered, one concludes that K is not in the table. The same sequence of probes will be made every time K is processed.

Let $\{h_i(K), i=0, 1, \ldots\}$ be a sequence of relative addresses through which the search proceeds until either the key is found or an empty location is encountered. The first address, $h_0(K)$, is called the initial hash address of K. The simplest way in which the subsequent addresses, $h_i(K)$, can be calculated is to add, modulo n, an increment d_i to the initial address

$$h_i(K) = (h_0(K) + d_i) \bmod n, \ i = 1, 2, \ldots \tag{9.8.1}$$

in order to obtain a new address in the table. The process is illustrated in Fig. 9.9.1.

Open addressing hash techniques are distinguished by different algorithms used to define the increment d_i. Efficiency of a particular method is measured by the average number of probes required for a successful or unsuccessful search. To know how an open addressing hash method performs on average is especially important because the worst case of hash algorithms is usually bad, so one needs to have reassurance that on average the performance is very good. The average number of required probes is a function of the load factor

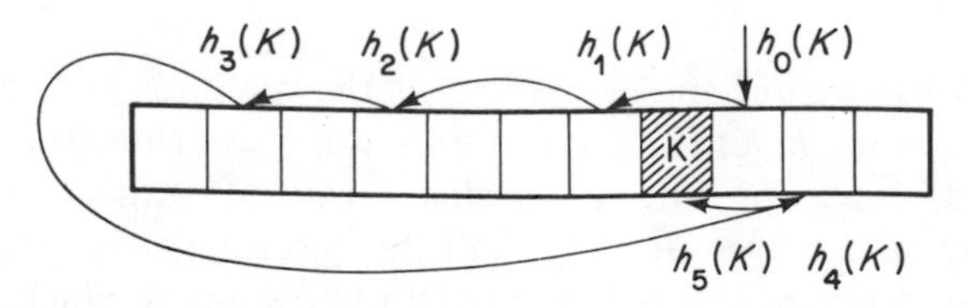

Figure 9.9.1 A probe sequence in an open addressing hashing

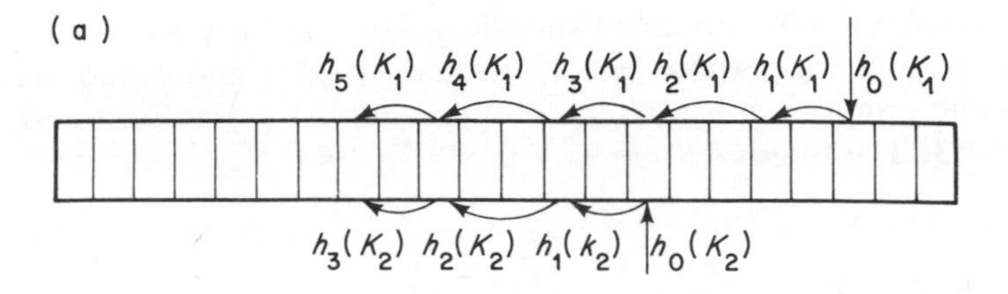

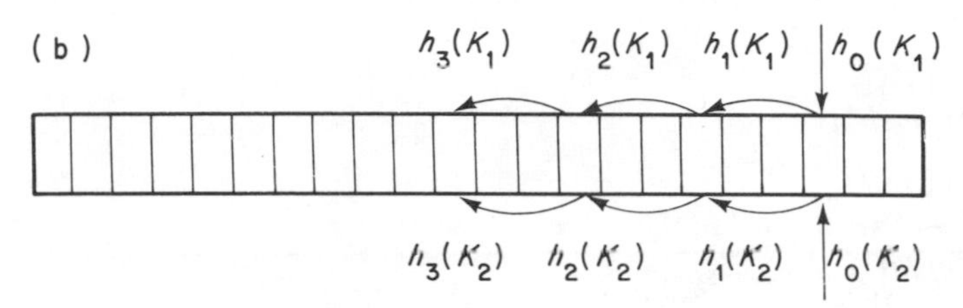

Figure 9.9.2 Clustering phenomena in open addressing hash: (a) a pile-up; (b) a secondary clustering

α, which is defined as the number of occupied locations in the table to the size of the table ($0 \leqslant \alpha \leqslant 1$). We denote the average number of probes for a successful search by $C_s(\alpha)$ and for an unsuccessful search by $C_u(\alpha)$. The functions $C_s(\alpha)$ and $C_u(\alpha)$ are affected by specific phenomena occurring in searching when for two distinct keys, a partially or fully identical sequence is traced through the table. These phenomena are called the pile-up (primary clustering) and secondary clustering.

By the pile-up one describes the situation which occurs in hashing when for two distinct keys, K_1 and K_2, the hash functions $h_i(K_1)$ and $h_j(K_2)$, where $j \bmod n = i$, designate the same location and thereafter $h_{i+r}(K_1) = h_{j+r}(K_2)$ for $r = 1, 2, \ldots$. The pile-up case is illustrated in Fig. 9.9.2(a). The secondary clustering occurs when for two distinct keys K_1 and K_2 the hash functions $h_i(K_1)$ and $h_i(K_2)$ designate the same location for some i and thereafter $h_{i+r}(K_1) = h_{i+r}(K_2)$ for $r = 1, 2, \ldots$. Secondary clustering phenomenon is illustrated in Fig. 9.9.2(b).

The clustering phenomena slow down the table search. To eliminate the clusters the increment d_i in (9.9.2) should be computed so that a unique probing sequence would result for every key. A number of algorithms are known for computing d_i. Each algorithm copes with the clustering phenomena in its own way. Below we consider some of the well-known schemes.

Linear Search

This is the oldest and simplest open addressing hash method that just probes the next position in the table whenever the search key is hashed to a place in the table which is already occupied and whose key is not the same as the search key. The method was expounded by Peterson (1957) and Ershov (1958b), though some primitive form of linear open addressing was discussed as far back as 1953. The rule is

$$h_i(K) = [h_0(K) + i] \bmod n, \; i = 1, 2, \ldots \qquad (9.9.2)$$

which is the same as

$$h_i(K) = [h_{i-1}(K) + 1] \bmod n, \; i = 1, 2, \ldots \qquad (9.9.3)$$

The linear search is implemented as follows.

Algorithm hashinsert

```
//initialize//
for i := 1 to n do
        table[i].key := specvalue
enddo
//hash process//
index := h(K)
while table[index].key <> specvalue do
index := (index + 1) mod n
enddo
table[index].key := K
return(index)
```

The algorithm uses specvalue to convey an empty location in the table. One important point is not implemented in the algorithm: it does not test for the table being full.

Linear search simply places colliding entries as near as possible to their nominally allocated position. A more general method follows if the next probe is made at a distance q, $q > 1$, from the place of collision:

$$h_i(K) := [H_0(K) + qi] \bmod n, \; i = 1, 2, \ldots, \qquad (9.9.4)$$

where q is a constant.

Linear search is affected by both the pile-up and secondary clustering.

Random Search

The method which virtually eliminates the clustering problem was presented by Morris (1968) and independently by de Balbine (1968) who named it double hashing. Instead of examining each successive entry following a collided position the method uses two hash functions, $h_0(K)$ and $g_0(K)$, to compute the probing sequence

$$h_i(K) = [h_0(K) + ig_0(K)] \bmod n. \qquad (9.9.6)$$

As usual $h_0(K)$ should produce a value between 0 and $n-1$, inclusive. The function $g_0(K)$ must be chosen with some care otherwise the method might fail altogether: $g_0(K)$ must produce a value between 1 and $n-1$, and relatively prime to n. Say, if n is prime then $g_0(K)$ can be any value between 1 and $n-1$, and if $n=2^r$ then $g_0(K)$ can be any odd value between 1 and 2^r-1. The fact that $g_0(K)$ is relatively prime to n ensures that the probe sequence is not too short so that no part of the table is examined twice until all n locations have been probed. Obviously $g_0(K)$ can be computed in many different ways. What we need, however, is a fast simple algorithm for obtaining $g_0(K)$, yet with the property that distinct keys would tend to have different hash addresses. We consider five algorithms for computing hash functions $h_0(K)$ and $g_0(K)$.

$$\text{Let } n \text{ be prime. Compute } h_0(K) = K \bmod n. \tag{9.9.7}$$

Knuth's algorithm for computing the second function is to let

$$g_0(K) = 1 + K \bmod (n-2). \tag{9.9.8}$$

This suggests choosing n so that n and $n-2$ are 'twin primes', like, for example, 513 and 511.

The Bell–Kaman algorithm (Bell and Kaman, 1970) uses the quotient of the division K by n so that

$$g_0(K) = 1 + (K \operatorname{div} n) \bmod (n-2). \tag{9.9.9}$$

Computation of increment in this scheme does not require any extra computer time as the quotient is a byproduct of the division method used to determine $h_0(K)$.

Let $n=2^r$. Compute $h_0(K)$ using the multiplication method. Morris's algorithm for computing increment $d_i = ig_0(K)$ is as follows.

Algorithm morrishash

```
//initialize//
for i := 1 to n do
    table[i].key := specvalue
enddo
//hash process//
index := h(K)
while table[index].key <> specvalue do                    (9.9.10)
    index := (index + (n − 1)) mod n
enddo
table[index].key := K
return(index)
```

The Luccio algorithm (Luccio, 1972) is for the function $g_0(K)$ to take the form

$$g_0(K) = (2p(K) + 1) \bmod n, \tag{9.9.11}$$

where $p(K)$ is an integer constant that can be extracted from the key K by any hash method. In particular, the hash function $h_0(K)$ can be substituted for $p(K)$, and the following probe sequence results:

$$h_i(K) = [h_0(K) + ((2h_0(K) + 1) \bmod n)i] \bmod n \tag{9.9.12}$$

The Luccio variant yields an easy analysis of the clustering phenomena. Indeed, from (9.9.12) it follows that for two distinct keys, K_1 and K_2, the same increment d_i in their respective probing sequences is used if

$$[2h_0(K_1) + 1] \bmod n = [2h_0(K_2) + 1] \bmod n. \tag{9.9.13}$$

Equation (9.9.13) is satisfied if either

$$h_0(K_1) = h_0(K_2) \tag{9.9.14}$$

or

$$h_0(K_1) + \frac{n}{2} = h_0(K_2). \tag{9.9.15}$$

Relation (9.9.14) indicates that scheme (9.9.12)is affected by the secondary clustering while relation (9.9.15) shows that the same increment is also used for the keys with different initial hash addresses when such addresses differ by half of the table size. In other words, the probing sequence for any K_2 traces the same locations as the probing sequence for K_1, where K_1 and K_2 are distinct and satisfy the relation

$$h_0(K_1) + \frac{n}{2} = h_0(K_2),$$

but only after one half of the table locations have been inspected. It may be concluded that for sufficiently large tables the Luccio algorithm is practically free from the pile-up.

The Luccio algorithm requires the operation of division when computing the increment. This division may be avoided if we simply set

$$h_1(K) = \begin{cases} 1 & \text{if } h_0(K) = 0 \\ n - h_0(K) & \text{otherwise.} \end{cases} \tag{9.9.16}$$

This version, which is due to Knott (1975), is more economical than Luccio's but is still affected by secondary clustering.

Quadratic Increment Search

Another method was presented by Maurer (1968) and further elaborated by Bell (1970). Maurer suggested forming the d_i by using a quadratic polynomial

$$d_i = ai + bi^2 = pi + q\,\frac{i(i-1)}{2}, \qquad i = 1, 2, \ldots \tag{9.9.17}$$

with the parameters

$$p = a + b \quad \text{and} \quad q = 2b.$$

The probing sequence is then given by

$$h_i(K) = \left[h_0(K) + pi + q\,\frac{i(i-1)}{2}\right] \bmod n. \qquad (9.9.18)$$

To see whether the method is affected by the pile-up consider the following problem. Assuming that

$$\left[h_0(K_1) + pi + q\,\frac{i(i-1)}{2}\right] \bmod n = h_i(K_1)$$

$$= h_j(K_2) = \left[h_0(K_2) + pj + q\,\frac{j(j-1)}{2}\right] \bmod n, \qquad (9.9.19)$$

find the conditions (if any) for the following to hold:

$$\left[h_0(K_1) + p(i+1) + q\,\frac{(i+1)i}{2}\right] \bmod n = h_{i+1}(K_1)$$

$$= h_{j+1}(K_2) = \left[h_0(K_2) + p(j+1) + q\,\frac{(j+1)j}{2}\right] \bmod n \qquad (9.9.20)$$

Condition (9.9.20) is equivalent to

$$(qi) \bmod n = (qj) \bmod n, \qquad (9.9.21)$$

and the latter is satisfied only if

$$q \bmod n = 1. \qquad (9.9.22)$$

Hence the Maurer algorithm provides the distribution of the hash addresses free from the pile-up provided that the value of q is chosen so that the condition (9.9.22) is not satisfied. The algorithm is, however, affected by the secondary clustering.

In (9.9.18) the initial hash address can be computed using the division method. In this case, the size of the table, n, must be a prime number with the implication that the algorithm (9.9.18) covers exactly half the table after which the table is declared 'full'. This situation is created by the fact that for any prime number, Q, the quadratic remainder R,

$$i^2 \bmod Q = R, \qquad i = 1, 2, \ldots,$$

has $(Q-1)/2$ distinct values only (see Radke, 1970).

These values, however, are fairly evenly distributed in the interval $[0, Q]$. The even distribution of the probed locations is a positive factor, since in hashing one always strives to obtain as even distribution as possible of keys in the table.

Quadratic increment search was further studied by Ecker (1973) who has shown that under assumptions that n is not a prime number and that $q \bmod n > 1$ the whole table can be searched. The first assumption, however, precludes using the division method for computing the initial hash address.

Use of the quadratic search for the cases when the table size is 2^r and p^r with p prime, was studied by Hopgood and Davenport (1972) and Ackerman (1974). Bell has further suggested a variant of the quadratic search which is said to completely eliminate the secondary clustering. The variant is known as the *quadratic quotient search* and is of the form:

$$h_i(K) = [h_0(K) + pi + q(K)i^2] \bmod n \tag{9.9.22}$$

where p is a constant and $q(K) = K \ div \ n$.

Like the quadratic search of Maurer, the probing function (9.9.22) will search exactly half the table before declaring the table 'full'.

9.10 Performance of the Open Addressing Algorithms

For the purpose of comparison, the hashing methods discussed above may be grouped under three distinct computational models:

(i) *linear probing*

$$h_i(K) = [h_0(K) + i] \bmod n, \tag{9.10.1}$$

(ii) *double hashing with secondary clustering*

$$h_i(K) = [h_0(K) + f(h_0(K))] \bmod n, \tag{9.10.2}$$

where $f(h_0(K))$ is a sufficiently random function, and

(iii) *independent double hashing*

$$h_i(K) = [h_0(K) + f(h_0(K))] \bmod n, \tag{9.10.3}$$

where each of the possible values of the pair $(h_0(K), f(h_0(K))$ is equally likely.

In Table 9.10.1 the average numbers of probes for the three hash models are given for both successful and unsuccessful hash table searches. The numbers are expressed as functions of the load factor α.

Table 9.10.1 The average efficiencies of the three hashing processes

	The average number of probes required for	
Hashing method	a successful search, $C_S(\alpha)$	an unsuccessful search, $C_U(\alpha)$
Linear probing	$\frac{1}{2}\left(1 + \frac{1}{1-\alpha}\right)$	$\frac{1}{2}\left(1 + \frac{1}{(1-\alpha)^2}\right)$
Double hashing with secondary clustering	$1 - \ln(1-\alpha) - \frac{1}{2}\alpha$	$\frac{1}{1-\alpha} - \ln(1-\alpha) - \alpha$
Independent double hashing	$-\frac{1}{\alpha}\ln(1-\alpha)$	$\frac{1}{1-\alpha}$

The formulae indicate how badly performance slows down for all three methods when α gets close to 1. For large n, with the table's density increasing form 80% to 90%, for an unsuccessful search linear probing takes from about 14 to about 50 probes, double hashing with secondary clustering, from about six to about 12 probes, and independent double hash, from about five to about ten probes.

Hash techniques have been analysed in depth by many authors, and below we present some details of the analysis. Our aim is to derive an estimate for an average number of probes required by a particular search model. In a successful search using hashing one starts with an empty table, which is then filled with the keys from a data file. For each key K the probing is continued until an empty location is encountered, the key is then inserted in the location. In an unsuccessful search one starts with a table which is filled to a prescribed density. Given the search key K, the locations are probed until match or no match is found for K. If no match is found, the search is unsuccessful. Obviously other models can be used. The analysis simply assumes a reasonable model and follows the development of the process.

If $C_S(\alpha)$ and $C_U(\alpha)$ denote the average number of probes required for both successful and unsuccessful searches, respectively, then in the open addressing hashing these quantities are related as

$$C_S\left(\frac{m}{n}\right) = \sum_{k=0}^{m-1} C_U\left(\frac{k}{n}\right) .$$

Analysis of Linear Probing

Linear search was first analysed by Schay and Spruth (1962). They gave an approximate estimate of the averge number of probes required by a successful search. Since then, Knuth (1973) has given the exact formulae for both successful and unsuccessful searches by linear probing. We shall follow the Schay and Spruth approach, while using an enhancing notation contributed to the analysis by Knuth. The approach is based on a remarkable property of linear probing which was first noticed by Peterson (1957):

> in a successful search by linear probing, the average number of probes does not depend on the order in which the keys were entered but only on the number of keys which hash to each location.

For example, given a table of size $n = 13$, i.e.

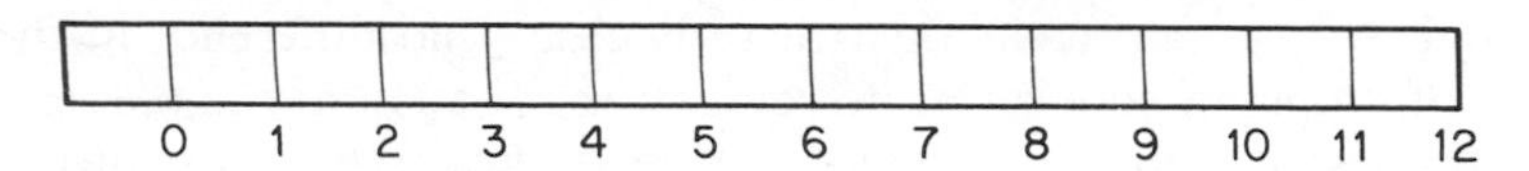

and eight keys, say,

8, 1, 5, 21, 33, 14, 2, 8

we obtain the following set of hash addresses:

$$8, 1, 5, 8, 7, 1, 2, 8. \qquad (9.10.4)$$

(9.10.4) is a hashing sequence. From this sequence it follows that

no keys hash to location 0,
two keys hash to location 1, (9.10.5)
etc.

(Note that the last location is reserved for signalling the end of the table.) The resulting table looks as follows

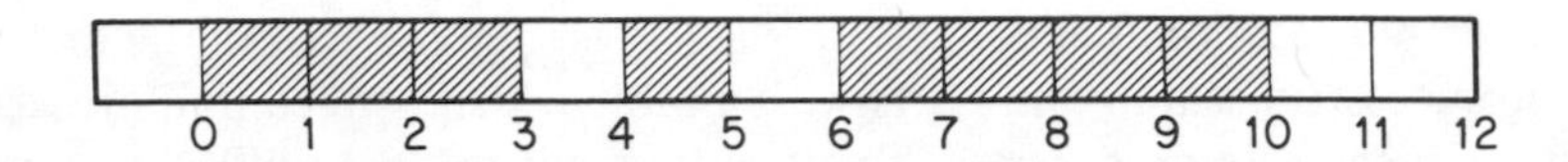

Sequence (9.10.5), given here as

0, 2, 1, 0, 0, 1, 0, 1, 3, 0, 0, 0, 0,

is the only factor that determines the number of probes required.

We assume that all keys in the table have the same access frequency, though the Peterson theorem can be extended to the case when some keys are more frequently accessed than others. Let

$$A_1 = h_0(K_1), \qquad A_2 = h_0(K_2), \ \ldots \qquad A_m = h_0(K_m) \qquad (9.10.6)$$

be a sequence of hash addresses computed for the given set of m keys, where $m < n$, and n is the table size. In (9.10.6), whenever a collision occurs it is resolved using the linear probing method that is the location immediately next to the hashed one is probed; we say in this case that the key is displaced by one from its hash address. The Peterson theorem states that any rearrangement of a hash sequence $A_1, A_2, \ldots, A_m$ results in a new sequence with the same average number of displacements of keys from their hash addresses. This statement will be asserted if we show that the total number of probes required to enter keys is the same for two hash sequences:

$$A_1, A_2, \ldots, A_{i-1}, A_i, A_{i+1}, A_{i+2}, \ldots, A_m$$

and (9.10.7)

$$A_1, A_2, \ldots, A_{i-1}, A_{i+1}, A_i, A_{i+2}, \ldots, A_m,$$
$$1 \leqslant i \leqslant m.$$

If A_i and A_{i+1} are different then there is clearly no difference in the total number of probes in both cases; however, if the $(i+1)$st key in the second sequence hashes to the location occupied by the ith key in the first sequence, then the ith and $(i+1)$st merely exchange places, so the number of probes for the $(i+1)$st is decreased by the same amount that the number for the ith is increased. This completes the assertion.

Next, we introduce a sequence of numbers $B_0, B_1, \ldots, B_{n-1}$ associated with the hash sequence $A_1, A_2, \ldots, A_m$ and defined so that B_j is the number of A's that equal j. Clearly, $B_0 + B_1 + \ldots + B_{n-1} = m$. The value $B_j > 1$ indicates that one or more collisions have occurred at some location A_i. It means that all the keys hashed to A_i, except one, need to be 'carried over' or displaced from their hashed location.

From sequence (B_j, $j = 0, \ldots, n-1$) we can determine the 'carry sequence' $C_0, C_1, \ldots, C_{n-1}$ where C_j is the number of keys for which both locations j and $j+1$ probed as the key is entered into the table. In other words, the value of C_j of location j is determined by the value of C_{j-1} of the previous location $j-1$ and the value of B_j of keys which have j as their hashed location.

Taking into account a 'circular' nature of the problem specified by the table size n, we deduce the relation:

$$C_{j \bmod n} = \begin{cases} 0 & \text{if } B_{j \bmod n} = C_{j-1} = 0, \\ B_{j \bmod n} + C_{j-1} - 1, & \text{otherwise.} \end{cases} \tag{9.10.8}$$

Our earlier example with $n = 13$ and $m = 8$, and

$$(B_0, B_1, B_2, \ldots, B_{12}) = (0, 2, 1, 0, 0, 1, 0, 1, 3, 0, 0, 0, 0)$$

gives

$$(C_0, C_1, C_2, \ldots, C_{12}) = (0, 1, 1, 0, 0, 0, 0, 0, 2, 1, 0, 0, 0).$$

Here, one key needs to be 'carried over' from location 1 to location 2, one from location 2 to location 3, none from locations 3, 4, 5, 6, 7, two from location 8 to location 9, one from location 9 to location 10, none from locations 10, 11, 12 and 0.

The number of probes for entering m keys can now be given in terms of the 'carry sequence':

$$C_s(\alpha) = \frac{C_0 + C_1 + C_2 + \ldots + C_{n-1}}{m} + 1, \tag{9.10.9}$$

where 1 stands for the first probe.

Note that both B_j and C_j do not depend on the particular location j to which i keys were hashed but only on the value i itself. This condition ensures a unique solution of equations (9.10.8) for C_j whenever $m < n$. In (9.10.9), to obtain the average number of probes, we need to consider the complete range of values that each C_j can take, the range $0 \leqslant C_j \leqslant n-1$. Denoting by q_k the probability that a particular $C_j = k$, we rewrite (9.10.9) in the form

$$1 + \frac{1}{m} \underbrace{\left[\sum_{k=0}^{n-1} kq_k + \sum_{k=0}^{n-1} kq_k + \ldots + \sum_{k=0}^{n-1} kq_k \right]}_{n \text{ terms}} = 1 + \frac{n}{m} \sum_{k=0}^{n-1} kq_k. \tag{9.10.10}$$

To complete the analysis we need to compute the probability q_k. To do this, we first introduce the probability p_k that $B_j = k$ and then use equations (9.10.8) to determine the q's in terms of the p's. The probability

$$p_k = \Pr(B_j = k, \text{ i.e. exactly } k \text{ of the } A_i \text{ are equal to } j, \text{ for fixed } j)$$

is determined by the assumption that each hash sequence $A_1, A_2, \ldots, A_m$ is equiprobable. This gives

$$P_k = \binom{m}{k}\left(\frac{1}{n}\right)^k\left(1 - \frac{1}{n}\right)^{m-k}, \tag{9.10.11}$$

where the binomial coefficient $\binom{m}{k}$ stands for the number of all possible combinations of k elements of total m, the second term expresses the probability that a particular location A_i will assume a fixed value j and the number of $A_i = j$ is k, and the third term expresses the probability that a particular location A_s from $(m - k)$ remaining will assume any fixed value between 0 and $n - 1$, except j, and the number of such A's is $m - k$.

From equations (9.10.8) the relations for the probabilities $q_k = \text{PR}(C_j = k)$ and $p_k = \Pr(B_j = k)$ follow immediately:

$$\begin{aligned}
\Pr(C_{j \bmod n} = 0) &= \Pr(C_{j-1} = 0)[\Pr(B_{j \bmod n} = 0) + \Pr(B_{j \bmod n} = 1)] \\
&\quad + \Pr(B_{j \bmod n} = 0)\Pr(C_{j-1} = 1), \\
\Pr(C_{j \bmod n} = 1) &= \Pr(C_{j-1} = 0)\Pr(B_{j \bmod n} = 2) + \Pr(C_{j-1} = 1)\Pr(B_{j \bmod n} = 1) \\
&\quad + \Pr(C_{j-1} = 2)\Pr(B_{j \bmod n} = 0),
\end{aligned} \tag{9.10.12}$$

etc.
giving

$$\begin{aligned}
q_0 &= q_0p_0 + q_0p_1 + q_1p_0, \\
q_1 &= q_0q_2 + q_1p_1 + q_2p_0, \\
q_2 &= q_0p_3 + q_1p_2 + q_2q_1 + q_3p_0, \\
&\text{etc.}
\end{aligned} \tag{9.10.13}$$

Introducing the generating functions for these probability distributions,

$$B(z) = \Sigma p_k z^k \quad \text{and} \quad C(z) = \Sigma q_k z^k, \tag{9.10.14}$$

the set of equations (9.10.13) can be expressed in an equivalent form as:

$$B(z)C(z) = p_0q_0 + (q_0 - p_0q_0)z + q_1z^2 + \ldots = p_0q_0(1 - z) + zC(z). \tag{9.10.15}$$

(Equations (9.10.13) are obtained by equating in (9.10.15) the coefficients of like powers of z.)

In terms of (9.10.14) the average number of probes required, (9.10.9), is then given by

$$C_s(\alpha) = 1 + \frac{n}{m}C'(1). \tag{9.10.16}$$

It remains to determine the value $C'(1)$. At this point it is convenient to introduce function $D(z)$ by setting

$$B(z) = 1 + (z-1)D(z), \tag{9.10.17}$$

where $B(1) = 1$ is satisfied. The generating functions of (9.10.15) can then be rewritten as

$$[1 + (z-1)D(z)]\,C(z) = p_0q_0(1-z) + zC(z), \tag{9.10.18}$$

giving

$$C(z) = \frac{p_0q_0}{1 - D(z)}\,. \tag{9.10.19}$$

In (9.10.19) the quantity p_0q_0 is determined from the initial condition $C(1) = 1$, i.e.

$$p_0q_0 = [1 - D(z)]\,C(z)\,|_{z=1} = 1 - D(1). \tag{9.10.20}$$

Hence $C(z) = \dfrac{1 - D(1)}{1 - D(z)}$, $\qquad C'(z) = \dfrac{D'(z)[1 - D(1)]}{[1 - D(z)]^2}$

and $$C'(1) = \frac{D'(1)}{1 - D(1)}\,. \tag{9.10.21}$$

Differentiating twice relation (9.10.17) and setting $z = 1$ we obtain

$$D(1) = B'(1) \quad \text{and} \quad 2D'(1) = B''(1). \tag{9.10.22}$$

Now, using (9.9.14) and (9.10.11) we get

$$B(z) = \sum p_k z^k = \sum \binom{m}{k}\left(\frac{z}{n}\right)^k\left(1 - \frac{1}{n}\right)^{m-k} = \left(\frac{z}{n} + 1 - \frac{1}{n}\right)^m. \tag{9.10.23}$$

It follows that

$$B'(1) = B'(z)\Bigg|_{z=1} = \frac{m}{n}\left(\frac{z-1}{n} + 1\right)^{m-1}\Bigg|_{z=1} = \frac{m}{n}\,, \tag{9.10.24}$$

and

$$B''(1) = B''(z)\Bigg|_{z=1} = \frac{m(m-1)}{n^2}\left(\frac{z-1}{n} + 1\right)^{m-2}\Bigg|_{z=1} = \frac{m(m-1)}{n^2}\,. \tag{9.10.25}$$

Finally, according to (9.10.16), (9.10.21), (9.10.22), (9.10.24) and (9.10.25) the average number of probes required for a successful search using linear probing is given as

$$C_S(\alpha) = \frac{1}{2}\left(1 + \frac{n-1}{n-m}\right) \approx \frac{1}{2}\left(1 + \frac{1}{1-\alpha}\right), \quad \text{where} \quad \alpha = \frac{m}{n}\,. \tag{9.10.26}$$

To obtain the average number of probes required for an unsuccessful search we recall that in open addressing the $C_S(\alpha)$ and $C_U(\alpha)$ are related by

$$C_S\left(\frac{m}{n}\right) = \frac{1}{m} \sum_{k=0}^{m-1} C_U\left(\frac{k}{n}\right) . \tag{9.10.27}$$

To solve this equation for $C_U(\alpha)$ we eliminate the summation terms in the usual way by subtracting (9.10.27) from a similar relation with m replaced by $m+1$. This gives

$$C_U\left(\frac{m}{n}\right) = (m+1)C_S\left(\frac{m+1}{n}\right) - mC_S\left(\frac{m}{n}\right) . \tag{9.10.28}$$

Substituting the expression for C_S as obtained in (9.10.26), we obtain

$$C_U\left(\frac{m}{n}\right) = \frac{1}{2}(m+1)\left(1 + \frac{n-1}{n-m-1}\right) - \frac{1}{2} m\left(1 + \frac{n-1}{n-m}\right) \approx \frac{1}{2}\left[1 + \frac{1}{(1-\alpha)^2}\right] . \tag{9.10.29}$$

Analysis of Double Hashing

The hashing process which completely ignores clustering phenomena is known as uniform hashing. In uniform hashing it is assumed that

(i) the keys go into random locations of the table so that each of the $n!/(m!(n-m)!)$ possible configurations of m occupied locations and $n-m$ empty locations is equally likely,

(ii) occupancy of each location in the table is essentially independent of the others,

(iii) no pile-up or secondary clustering occurs.

An unsuccessful search is completed as soon as an empty location is encountered, the key is then inserted into the table. The probability that exactly r probes are needed to insert $(m+1)$st key is given as

$$p_r = \frac{\text{the number of configurations where } r-1 \text{ given locations are occupied and another is empty}}{\text{the total number of possible configurations of } m \text{ occupied locations and } n-m \text{ empty ones}} = \frac{\binom{n-r}{m-(r-1)}}{\binom{n}{m}}, \tag{9.10.30}$$

and the average number of probes for an unsuccessful search is given by

$$C_U(\alpha) = \sum_{r=1}^{n} r p_r . \tag{9.10.31}$$

In order to obtain the $C_U(\alpha)$ in terms of the load factor α, the following sequence of transformations of the right-hand side of (9.10.31) is carried out.

(a) Write

$$\sum_{r=1}^{n} rp_r = n + 1 - (n+1) \sum_{r=1}^{n} p_r + \sum_{r=1}^{n} rp_r, \tag{9.10.32}$$

where

$$\sum_{r=1}^{n} p_r = 1 \text{ (i.e. the sum of all probabilities equals one).}$$

Then

$$C_U(\alpha) = n + 1 - \sum_{r=1}^{n} (n+1-r)p_r = n + 1 - \sum_{r=1}^{n} (n+1-r) \frac{\binom{n-r}{m-(r-1)}}{\binom{n}{m}}. \tag{9.10.33}$$

(b) Expression (9.10.33) can be written as

$$C_U(\alpha) = n + 1 - \sum_{r=1}^{n} (n+1-r) \frac{\binom{n-r}{n-m-1}}{\binom{n}{m}}, \tag{9.10.34}$$

since

$$\binom{n-r}{m-r+1} = \binom{n-r}{n-m-1}. \tag{9.10.35}$$

(c) Expression (9.10.34) can further be written as

$$C_U(\alpha) = n + 1 - (n-m) \sum_{r=1}^{n} \frac{\binom{n-r+1}{n-m}}{\binom{n}{m}}, \tag{9.10.36}$$

since

$$(n+1-r)\binom{n-r}{n-m-1} = (n-m)\binom{n-r+1}{n-m}. \tag{9.10.37}$$

(d) Expression (9.10.36) can finally be written as

$$C_U(\alpha) = n + 1 - (n-m) \frac{\binom{n+1}{n-m+1}}{\binom{n}{m}} \quad \text{for } 1 \leqslant m < n, \tag{9.10.38}$$

since it can be shown by induction that

$$\sum_{r=1}^{n} \binom{n-r+1}{n-m} = \binom{n+1}{n-m+1} \quad \text{for } 1 \leqslant m \leqslant n. \tag{9.10.39}$$

Expression (9.10.38) gives

$$C_U(\alpha) = \frac{n+1}{n-m+1} \approx \frac{1}{1-\alpha} \quad \text{for reasonably large } n. \tag{9.10.40}$$

The average number of probes for a successful search is then obtained in the usual way:

$$\begin{aligned} C_S\left(\frac{m}{n}\right) &= \frac{1}{m} \sum_{k=0}^{m-1} C_U\left(\frac{k}{n}\right) = \frac{n+1}{m} \left(\frac{1}{n+1} + \frac{1}{n} + \ldots + \frac{1}{n-m+2}\right) \\ &= \frac{n+1}{m} (H_{n+1} - H_{n-m+1}) \\ &\approx \frac{n+1}{m} (\ln(n+1) - \ln(n-m+1) +) \\ &\approx \frac{1}{\alpha} \ln \frac{1}{1-\alpha} . \end{aligned} \tag{9.10.41}$$

The completes the analysis of uniform hashing.

The effect of secondary clustering (double hashing with secondary clustering) increases formulae (9.10.40) and (9.10.41) to

$$C_U(\alpha) \approx \frac{1}{1-\alpha} - \ln(1-\alpha) - \alpha$$

and

$$C_S(\alpha) \approx 1 - \ln(1-\alpha) - \tfrac{1}{2}\alpha,$$

respectively.

9.11 Virtual Hashing

For use with large data systems which due to the system's size are stored on a peripheral memory, such as disc, a method of dynamic hashing is used, which has been named virtual hashing. Virtual hashing allows expansion of the hash table when needed and involves rehashing of only a limited number of keys (records). Development of this technique was prompted by the fact that when a large data system is stored on a peripheral memory it would be useful to be able to retrieve the information in as few disc accesses as possible. Let us recall that a hashing scheme is characterized by the four main attributes: it has a key space containing all possible keys K; it has an address space A, the area available for store; access is performed using a hash function, which is a way of storing a record in the address space A by using a part of the key

K as an identifier; it has a collision resolution mechanism. In this general form we can visualize the table itself as $N+1$ memory cells labelled $\{0, \ldots, N\}$ and each cell has a bucket attached to it. Each bucket may contain up to b records where $b \geqslant 1$, and b is called the capacity of a bucket (Fig. 9.11.1). In order to examine the contents of a specific bucket one (disc) access is required.

When the record is searched for, the scan is begun at the primary address $h(K)$, and the bucket $h(K)$ is the primary bucket. We look at the bucket $h(K)$; if there has been an overflow then we will require a collision resolution mechanism with which to continue the search. (An overflow is said to occur when more than b records are assigned by $h(K)$ to a bucket.) Using hashing as the access method together with a collision resolution mechanism, one (disc) access only will be required to access a specific bucket.

Practice shows that for a low load factor (when the number of stored records is much smaller than the table capacity, $(N+1)b$, the buckets are poorly loaded and in this case the hashing method is very inefficient. When the load factor is in the range $0.7 \geqslant \alpha \geqslant 0.9$ one finds that only very few overflows exist and thus almost all records are retrieved within one access. However if the number of stored records is high or, worse still, is close to the table's capacity, then the number of overflows is high and is rapidly increasing with α. In such situations several (disc) accesses may be required per search and as a result performance of the hashing method is again very poor.

Hashing is found to be efficient if the number of stored records, R, remains close to the mean value of R, for which $(N+1)b$ has been selected to give a high α and a good performance. In the applications where insertions are frequent collisions are eventually created, followed soon after by an overflow, which, in turn, is followed by a deterioration of the hash performance. To remedy the situation a collision resolution method must be able to modify the hash function h.

Let us say that if a search, a deletion or an insertion cannot modify the hash function $h(K)$ then we have what is known as a traditional hashing system. In an expanding data system one can use a hashing method known as extendable hashing (or dynamic or virtual hashing). Virtual hashing may change $h(K)$ during one of the above processes.

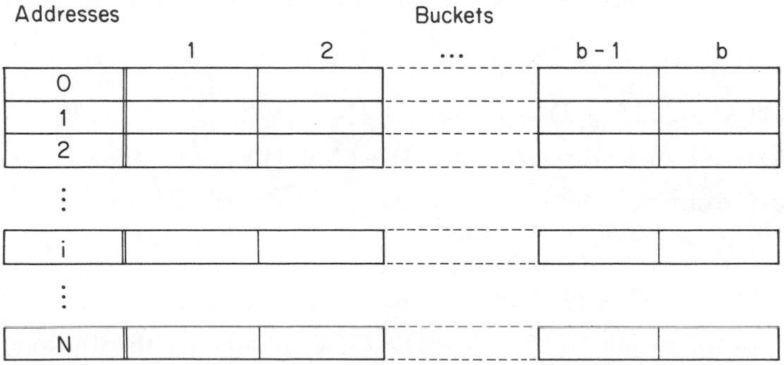

Addresses	Buckets 1	2	...	b - 1	b
0					
1					
2					
⋮					
i					
⋮					
N					

Figure 9.11.1 The setup of a hash table which contains $N+1$ addresses to each of which a bucket of size b is attached

Suppose that the hashing function $h_0(K)$ is used to retrieve the record. If we have a collision in order to resolve it another address, h_1, is computed, and the record can now be stored at $h_1(K)$. The crucial point now is this. If in creating the new address $h_1(K)$ we have created a new hash function then not only the records at $h_1(K)$ have to be reinserted but several of the records stored at other locations may have to be reinserted using the new hash function. In this case, perhaps, it would be more desirable to rehash the whole table. If, however, we resolve to move some selected records only then will $h_1(K)$ become a very complicated function. So, both of the suggested methods are very complex and would result in all basic advantages of hashing being lost.

Instead a compromise method can be used. The method's requirements are

(i) that reinsertions must use as few disc accesses as possible, so that repeated probes should concern the records present in the main memory only—that is, only those in the bucket for which a collision has been detected;

(ii) a simple algorithm to compute which records need to be shifted and to which address they must be moved. This procedure is called a split. Suppose a bucket, $h(K)$, contains b records. The $(b+1)$st record is added to the bucket. This creates an overflow, and we wish to rehash the contents of the bucket.

Mechanics of a Split

For concreteness, we assume that the division hashing function is used. Now, suppose that using the hash function $h_0(K)$ we attempt to store a set of records.

$$h_0 : K \rightarrow G(K, N+1),$$

where $G(K, N+1) = K \bmod (N+1)$,

for example, $K = 15$, $N = 9$, and

$$h_0 : 15 \rightarrow 15 \bmod 10 = 5.$$

In order to resolve a collision we will need a modified hashing function, h_1,

$$h_1 : K \rightarrow G(K, 2(N+1)).$$

We can now calculate a new primary address for K, due to the following property of G.

For all K in the bucket $h(K)$

$$\text{either } G(K, 2(N+1)) = G(K, N+1)$$
$$\text{or } G(K, 2(N+1)) = G(K, N+1) + (N+1). \qquad (9.11.1)$$

So, in our example:

$$25 \bmod 20 = 25 \bmod 10 = 5.$$
$$15 \bmod 20 = 5 \bmod 10 + 10 = 15.$$

Those keys for which $G(K, 2(N+1) = G(K, N+1)$ are then hashed into the original bucket while the keys for which $G(K, 2(N+1)) = G(K, N+1) + (N+1)$ are hashed into the new bucket with h_1.

Example 9.11.1

Let the table size be $N = 9$ and the bucket size be $b = 5$.

0
1
2   2   12   22   102   242
⋮
9

The required operation is to insert key 32. When attempting to do this we encounter a collision at bucket(2). So we proceed as follows.

Remove all the records from bucket(2). Rehash using the new function $h_1(K)$.

h_0: 0
h_0: 1
h_1: 2   2   22   102   242
h_0: 3
⋮
h_0: 9
h_1: 10
h_1 12   12   32
⋮
19

Thus by comparing the key values mod 10 and mode 20 a simple split has been achieved and we have resolved a collision at bucket(2). More important is the fact that $h_1(K)$ is local to bucket(2) and, for example, bucket(0) is still accessed via h_0.

We can also see that the new addresses created are always greater than N due to the condition (9.11.1). This means that record will always be moved into empty buckets. A very important assumption for the virtual hashing to work efficiently is that if hashing by $h_1(K)$ is random and $b \gg 1$ then a change of address of zero or more than $b + 1$ records is very unlikely (this is the Bernoulli scheme).

If the split rehashes all records to the same bucket then the bucket must be split again. This idea implies that a collision will usually cost no more than two (disc) accesses: one for writing each of the new buckets. The method allows up to $2b$ records to have the same primary address $h_0(K)$, and each one will be found in a single disc access. The simple split can be continued to any split in the table.

Full Definition of a Repeated Split

Let $h_j(K): K \rightarrow G(K,\ 2^j(N+1))$ for $j = 0, 1, 2, \ldots$

{for example, $h_2(K): K \rightarrow G(K,\ 4(N+1))$}.

Suppose that an overflow occurred in bucket($h_j(K)$) and its content needs to be split. To split bucket($h_j(K)$) we use $h_{j+1}(K)$ but only for those records which are stored at bucket($h_j(K)$) under the previous function $h_j(K)$, and such that:

$$h_{j+1}(K): K \rightarrow G(K,\ 2^{j+1}(N+1)) = G(K,\ 2^j(N+1)) + 2^j(N+1). \tag{9.11.2}$$

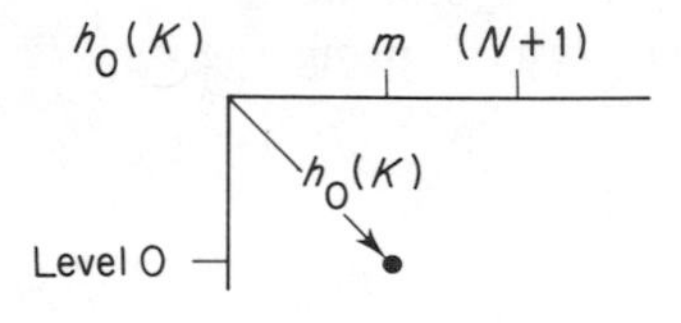

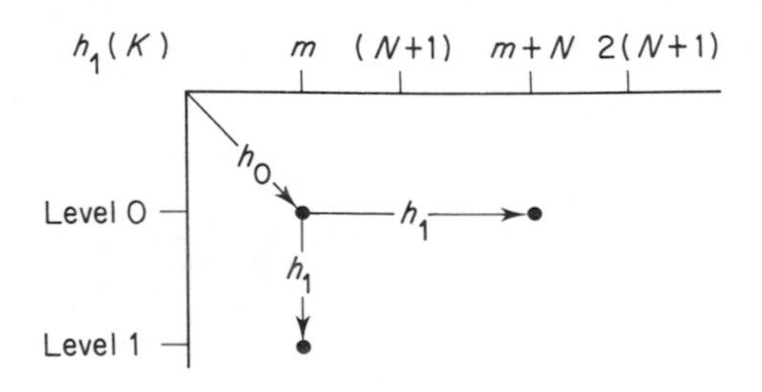

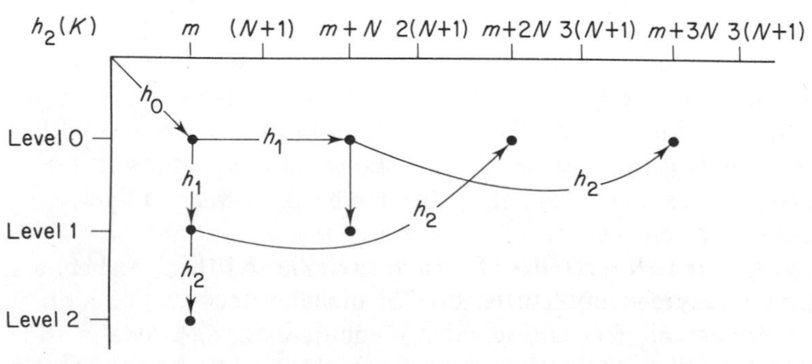

Figure 9.11.2 A step-by-step development of splits to remedy an overflow in bucket (m)

In Fig. 9.11.2 is shown an example of a structure graph of the process of split development. It shows how splits have been applied to the buckets and how relocation addresses are located. The graph corresponds to the split given by the general rule (9.11.2). For each split there are in effect two storage areas, however since one of them is the original address, we need only to create one new relocation 'edge'. The graph shows step-by-step expansion of splits.

Exercises

1. Modify algorithm *seqlistsearch* so that for every unsuccessful search a new key is inserted in the linked list.
2. In sequential list searching substantial savings can be made by ordering the keys intelligently according to the information about the relative frequency of access for various keys. But, even if such information is not available, an approximation to an efficient arrangement of the keys can be achieved using the following mechanism: each time a key is accessed, move it to the beginning of the list.

   Write a modified sequential list search algorithm which incorporates the above mechanism.

   Develop the algorithm into a complete program and experimentally test the average running time for this method of search. Compare the results with the average running time of the basic *seqlistsearch*.
3. Show the tree structure which will be imposed on the keys

   sun mon wed thur fri sat

   if the algorithm *treesearchinsert* is applied to these keys so that the tree is constructed when the keys are inserted one by one into an initially empty tree.

   What is the sequence of nodes visited in the tree if we wish to search for tue?
4. An improvement for binary search is gained by trying to guess more precisely where the search key falls within the current interval of interest. This approach is called interpolation search.

   (a) Assuming numerical key values, modify the algorithm *binsearch* to produce an algorithm *interpolsearch* by replacing the midpoint of the interval as the new position to search with the new guessed position, which is weighted with respect to the search key.

   (b) Interpolation search has been shown to decrease the number of keys examined to about log log n. Verify this estimate experimentally.

   [*Hint*. In the algorithm *binsearch* replace '$index := (i + j)$ *div* 2' by '$index := i + (K - K(i)) \times (j - i)$ *div* $(K(j) - K(k))$'.]
5. Give a recursive implementation of binary search.
6. Implement Fibonaccian search to include insert operation.
7. Draw the binary and Fibonaccian search trees that result from inserting the keys $D\ I\ S\ T\ R\ I\ B\ U\ T\ I\ O\ N$ into an initially empty tree.

8. Rewrite algorithm *eightdelete* so that when searching for a suitable node which would replace the node to be deleted, it would follow the route via the right side branches of the tree searching for a node with the null right subtree.
9. Draw an optimal binary search tree for the following sequence of keys with unequal frequencies of occurrence.

   D, 0.1; *I*, 0.05; *S*, 0.1; *T*, 0.3; *R*, 0.05; *I*, 0.02; *B*, 0.05; *U*, 0.02; *T*, 0.02; *I*, 0.25; *O*, 0.02; *N*, 0.02.
10. Draw the top-down *two–three–four tree* that is built when the keys *IN NS SE ER RT TA AN ND DE EL LE ET TE* are inserted into an initially empty tree.
11. Modify the algorithm *hashinsert* to include the check for the table being full.
12. Modify the algorithm *hashinsert* to *hashsearch* the table. [*Hint*. Add condition '$h(index).key < > K'$' to the **while loop** and delete the following statement which stores *K*. The algorithm next has to check if the search was successful (the table position returned actually contains the search key) or unsuccessful (the table position contains specvalue).]
13. Modify the algorithm *hashinsert* to implement double hashing using Knuth's scheme for computing the second hash function.
    [*Solution*.

Algorith knuthhash

```
//initialize//
for i := 1 to n do
      table[i].key := specvalue
enddo
//double hashing//
g(K) := 1 + K mod (n − 2)
index := h(K)
while table[index].key < > specvalue do
      index := (index + g(K)) mod n
enddo
table[index].key := K
return(index)      ]
```

In a similar way modify the algorithm *hashinsert* to accommodate (a) double hashing using the Bell–Kaman algorithm, (b) Morris's scheme and (c) the Luccio formula (9.9.12).

14. Give the contents of the hash table that results when the keys *D I S T R I B U T I O N* are inserted in that order into an initially empty table of size 17 using the quadratic quotient search (9.9.22), where constant $p = 3$.

[Use $h_0(k) = k \bmod 17$ and $q(k) = k$ mode 17 to compute the hash functions for the kth letter of the alphabet.]

15. A hash table is to be built consisting of n equal keys. Approximately how many probes would be required when (a) linear probing, and (b) double probing is used?

Appendix A

Some Basic Results on the Error Analysis of the Floating-Point Matrix Multiplication and the Solution of Sets of Linear Equations

A.1 Vector and Matrix Norms

In order to analyse computational processes involving matrices, it is convenient to associate with any vector or matrix a non-negative scalar that in some sense provides a measure of its magnitude. Such a scalar is normally called the norm.

Definition. The norm of a vector $\mathbf{x}$ may be defined as a real non-negative number which is denoted by $\| \mathbf{x} \|$ and which satisfies the following relations:

$$\| \mathbf{x} \| > 0, \text{ unless } \mathbf{x} = 0,$$

$$\| k\mathbf{x} \| = | k | \, \| \mathbf{x} \|, \text{ where } k \text{ is a complex number},$$

$$\| \mathbf{x} + \mathbf{y} \| \leqslant \| \mathbf{x} \| + \| \mathbf{y} \| .$$

The three most commonly known vector norms are given by

$$\| \mathbf{x} \|_p = (| x_1 | x_2 |^p + \ldots + | x_n |^p)^{1/p}, \tag{A.1}$$

where

$$\text{for } p = 1, \quad \| \mathbf{x} \|_1 = \sum_{i=1}^{n} | x_i |, \tag{A.2}$$

$$\text{for } p = 2, \quad \| \mathbf{x} \|_2 = \left[\sum_{i=1}^{n} | x_i |^2 \right]^{1/2} \tag{A.3}$$

and

$$\text{for } p = \infty, \quad \| \mathbf{x} \|_\infty = \max_i \{ | x_i | \}. \tag{A.4}$$

Similarly, the norm of a matrix $\mathbf{A}$ may be defined as a real non-negative number which satisfies the relations:

$$\| \mathbf{A} \| > 0, \text{ unless } \mathbf{A} = \mathbf{0}$$

$$\| k\mathbf{A} \| = | k | \, \| \mathbf{A} \|, \text{ where } k \text{ is a complex number,}$$

$$\| \mathbf{A} + \mathbf{B} \| \leqslant \| \mathbf{A} \| + \| \mathbf{B} \|, \tag{A.5}$$

$$\| \mathbf{AB} \| \leqslant \| \mathbf{A} \| \, \| \mathbf{B} \|. \tag{A.6}$$

We say that the matrix norm is subordinate to the vector norm if the following relation holds

$$\| \mathbf{A} \| = \max \frac{\| \mathbf{Ax} \|}{\| \mathbf{x} \|} \tag{A.7}$$

This is equivalent to

$$\| \mathbf{A} \| = \max \| \mathbf{Ax} \|, \ \| \mathbf{x} \| = 1 \tag{A.8}$$

Corresponding to the three vector norms given above we have the matrix norms $\| \mathbf{A} \|_p$, where

$$\| \mathbf{A} \|_1 = \max_j \left\{ \sum_i | a_{ij} | \right\}, \tag{A.9}$$

$$\| \mathbf{A} \|_2 = [\text{maximum eigenvalue of } (\mathbf{A}^{\mathrm{H}}\mathbf{A})]^{1/2}, \tag{A.10}$$

where $\mathbf{A}^{\mathrm{H}}$ is the complex conjugate transpose of $\mathbf{A}$,

$$\| \mathbf{A} \|_\infty = \max_i \left\{ \sum_j | a_{ij} | \right\} \tag{A.11}$$

If the vector and matrix norms satisfy

$$\| \mathbf{Ax} \| \leqslant \| \mathbf{A} \| \, \| \mathbf{x} \| \tag{A.12}$$

then they are consisten.

It is properties (A.6) and (A.12) which make the vector and matrix norms so very useful in analysing the errors involved in solving linear equations. Another matrix norm, the so-called Euclidean norm,

$$\| \mathbf{A} \|_E = \left[\sum_i \sum_j | a_{ij} |^2 \right]^{1/2} \tag{A.13}$$

is also useful in practical error analysis. It has certain advantages over $\| \mathbf{A} \|_2$. These are

(i) the Euclidean norm of $| \mathbf{A} |$ is the same as that of $\mathbf{A}$. ($| \mathbf{A} |$ is the matrix with elements $| a_{ij} |$.)

(ii) $\| \mathbf{A} \|_E$ is easier to calculate than $\| \mathbf{A} \|_2$.

The Euclidean norm is not subordinate to any vector norm but it is consistent with $\| \mathbf{x} \|_2$.

The matrix norms provide the upper bound for the moduli of the eigenvalues of **A**. Consider

$$\mathbf{Ax} = \lambda\mathbf{x}, \text{ where } \lambda \text{ is an eigenvalue of } \mathbf{A}$$

then

$$\| \mathbf{A} \| \, \| \mathbf{x} \| \geqslant \| \mathbf{Ax} \| = \| \lambda\mathbf{x} \| = | \lambda | \, \| \mathbf{x} \|$$

thus

$$\| \mathbf{A} \| \geqslant | \lambda | \tag{A.14}$$

We already know that

$$\begin{aligned} \| \mathbf{A} \|_2^2 &= \text{maximum eigenvalue of } \mathbf{A}^{\mathrm{H}}\mathbf{A} \\ &\leqslant \| \mathbf{A}^{\mathrm{H}}\mathbf{A} \|_\infty \\ &\leqslant \| \mathbf{A}^{\mathrm{H}} \|_\infty \| \mathbf{A} \|_\infty \end{aligned}$$

Thus

$$\| \mathbf{A} \|_2^2 \leqslant \| \mathbf{A} \|_1 \| \mathbf{A} \|_\infty \tag{A.15}$$

For a more detailed review of the properties of norms see Wilkinson (1961), Wilkinson (1963), Householder (1964), Forsythe and Moler (1967).

A.2 Error Analysis of Basic Matrix Operations

(1) Multiplication by a scalar.
Let $\mathbf{B} = k\mathbf{A}$, where **B** and **A** are matrices and k is a scalar.
Then

$$b_{ij} \equiv \mathrm{fl}(ka_{ij}) \equiv ka_{ij}(1 + e_{ij})$$

where

$$| e_{ij} | \leqslant 2^{-t}.$$

Thus

$$b_{ij} - ka_{ij} = ka_{ij}e_{ij}$$

and so

$$\| \mathbf{B} - k\mathbf{A} \|_E \leqslant 2^{-t} | k | \, \| \mathbf{A} \|_E, \tag{A.16}$$

giving

$$\frac{\| \mathbf{B} - k\mathbf{A} \|_E}{| k | - \| \mathbf{A} \|_E} \leqslant 2^{-t}.$$

(2) Matrix multiplication.
Consider $\mathbf{y} = \mathbf{Ax}$, where **A** is an $n \times n$ matrix and **y** and **x** are vectors of dimension n.

Then

$$y_i = \mathrm{fl}\left(\sum_{j=1}^{n} a_{ij}x_j\right)$$

$$= \left[\sum_{j=1}^{n} a_{ij}x_j\right] + e_i$$

where

$$|e_i| \leqslant 2^{-t}[n|a_{i1}||x_1| + n|a_{i2}||x_2| + (n-1) \times |a_{i3}||x_3| + \ldots + 2|a_{in}||x_n|] \tag{A.17}$$

Thus

$$\mathbf{y} = \mathbf{Ax} + \mathbf{e}, \tag{A.18}$$

where $|\mathbf{e}| \leqslant 2^{-t_1}|A|D|\mathbf{x}|$ and D is a diagonal matrix of the form

$$\mathbf{D} = \begin{bmatrix} n & & & & 0 \\ & n & & & \\ & & n-1 & & \\ & & & \ddots & \\ 0 & & & & 2 \end{bmatrix} \tag{A.19}$$

In terms of the norms the error $\mathbf{e}$ is bounded by

$$\|\mathbf{e}\|_2 \leqslant 2^{-t_1}n\|\mathbf{A}\|_E\|\mathbf{x}\|_2 \tag{A.20}$$

Similarly, for the multiplication of two matrices we have

$$\mathbf{C} = \mathrm{fl}(\mathbf{AB}) \equiv \mathbf{AB} + \mathbf{E}$$

where

$$\|\mathbf{E}\|_E \leqslant 2^{-t_1}n\|\mathbf{A}\|_E\|\mathbf{B}\|_E \tag{A.21}$$

Since the product $\|\mathbf{A}\|_E\|\mathbf{B}\|_E$ may be very much greater than $\|\mathbf{AB}\|_E$, the computed $\mathbf{C}$ may well have a very high relative error. If we can accumulate inner products in floating-point and write

$$\mathbf{C} = \mathrm{fl}_2(\mathbf{AB}) \equiv \mathbf{AB} + \mathbf{E}$$

then

$$|\mathbf{C} - \mathbf{AB}| \leqslant 2^{-t}|\mathbf{AB}| + \tfrac{3}{2}2^{-t_2}|\mathbf{A}||\mathbf{D}||\mathbf{B}| \tag{A.22}$$

where $\mathbf{D}$ is as in (A.19).
Hence

$$\|\mathbf{C} - \mathbf{AB}\|_E = \||\mathbf{C} - \mathbf{AB}|\|_E \leqslant 2^{-t}\|\mathbf{AB}\|_E + \tfrac{3}{2}2^{-t_2}n\|\mathbf{A}\|_E\|\mathbf{B}\|_E \tag{A.23}$$

The second term on the right hand side is negligible in comparison with the first and so we have

$$\frac{\| \mathbf{C} - \mathbf{AB} \|_E}{\| \mathbf{AB} \|_E} \leqslant 2^{-t} \tag{A.24}$$

which shows that the computed product has a very low error.

A.3 Error Analysis of the Solution of a Set of Linear Equations and of Matrix Inversion

We will consider the effect of perturbations **A** in **b** on the solution of the set $\mathbf{Ax} = \mathbf{b}$.

(1) Let **k** be a small perturbation in **b** and **h** be the corresponding perturbation in **x**.
We have

$$\mathbf{A}(\mathbf{x} + \mathbf{h}) = \mathbf{b} + \mathbf{k}, \tag{A.25}$$

which gives

$$\mathbf{Ah} = \mathbf{k} \qquad \text{or} \qquad \mathbf{h} = \mathbf{A}^{-1}\mathbf{k}. \tag{A.26}$$

Assuming the matrix subordinate norms, we obtain:

$$\| \mathbf{h} \| = \| \mathbf{A}^{-1}\mathbf{k} \| \leqslant \| \mathbf{A}^{-1} \| \, \| \mathbf{k} \| \tag{A.27}$$

Also, for the matrix subordinate norms,

$$\| \mathbf{b} \| = \| \mathbf{Ax} \| \leqslant \| \mathbf{A} \| \, \| \mathbf{x} \|,$$

and so

$$\| \mathbf{x} \| \geqslant \| \mathbf{b} \| \, \| \mathbf{A} \|^{-1}. \tag{A.28}$$

Therefore the relative change $\| \mathbf{h} \| / \| \mathbf{x} \|$ is given by

$$\frac{\| \mathbf{h} \|}{\| \mathbf{x} \|} \leqslant \frac{\| \mathbf{A}^{-1} \| \, \| \mathbf{k} \|}{\| \mathbf{A} \|^{-1} \| \mathbf{b} \|} = \frac{\| \mathbf{A} \| \, \| \mathbf{A}^{-1} \| \, \| \mathbf{k} \|}{\| \mathbf{b} \|}$$

$\| \mathbf{A} \| \, \| \mathbf{A}^{-1} \|$ is known as the condition number and usually denoted by cond (A). We, thus, have

$$\frac{\| \mathbf{h} \|}{\| \mathbf{x} \|} \leqslant \frac{\| \mathbf{k} \|}{\| \mathbf{b} \|} \text{cond } (A) \tag{A.29}$$

where $\text{cond}(A) = \| \mathbf{A} \| \, \| \mathbf{A}^{-1} \|$.
If cond (A) is very large then the set of equations, $\mathbf{Ax} = \mathbf{b}$, is said to be ill-conditioned and the relative error, $\| \mathbf{h} \| / \| \mathbf{x} \|$, introduced by the small perturbation **k** will be large.

(2) Let **E** be a small perturbation in **A** and **h** be again the corresponding perturbation in **x**.

We have

$$(\mathbf{A}+\mathbf{E})(\mathbf{x}+\mathbf{h})=\mathbf{b}, \tag{A.30}$$

which gives

$$(\mathbf{A}+\mathbf{E})\mathbf{h}=-\mathbf{E}\mathbf{x}. \tag{A.31}$$

Assuming $\mathbf{A}$ is non-singular (i.e. its inverse exists), then $(\mathbf{A}+\mathbf{E})$ may still be singular if $\mathbf{E}$ is not restricted. We shall, thus, assert the conditions on $\mathbf{E}$, under which matrix $(\mathbf{A}+\mathbf{E})$ will be non-singular if $\mathbf{A}$ is non-singular. Note that a matrix is called non-singular if its determinant is not equal to zero.

$(\mathbf{A}+\mathbf{E})$ can be expanded as

$$\mathbf{A}+\mathbf{E}=\mathbf{A}(\mathbf{I}+\mathbf{A}^{-1}\mathbf{E}), \tag{A.32}$$

giving

$$(\mathbf{A}+\mathbf{E})^{-1}=(\mathbf{I}+\mathbf{A}^{-1}\mathbf{E})^{-1}\mathbf{A}^{-1}. \tag{A.33}$$

From (A.33) it follows that, provided $\mathbf{A}$ is non-singular, for the $(\mathbf{A}+\mathbf{E})$ to be non-singular, the inverse $(\mathbf{I}+\mathbf{A}^{-1}\mathbf{E})^{-1}$ must exist, i.e. the determinant $|\mathbf{I}+\mathbf{A}^{-1}\mathbf{E}|$ must be non-zero.
Consider the matrix $(\mathbf{I}+\mathbf{A}^{-1}\mathbf{E})$. Its eigenvalues have the form $1+\lambda_i$, where λ_i, $i=1,\ldots,n$, are the eigenvalues of the matrix $\mathbf{A}^{-1}\mathbf{E}$. Now, if we assume the condition

$$\|\mathbf{A}^{-1}\mathbf{E}\|<1, \tag{A.34}$$

then it follows that all $|\lambda_i|$ are less than unity, since by virtue of the matrix norms properties,

$$\|\mathbf{A}^{-1}\mathbf{E}\|\geqslant|\lambda_i| \tag{A.35}$$

holds for any matrix norm.

But if all $|\lambda_i|$ are less than unity, then $1+\lambda_i$ are all non-zero, and thus the determinant $|\mathbf{I}+\mathbf{A}^{-1}\mathbf{E}|\neq 0$, since the determinant of a matrix is equal to the product of its eigenvalues. We conclude that provided A is non-singular, $(\mathbf{A}+\mathbf{E})$ will be non-singular if condition (10) holds.

Now, assuming that condition (A.34) is satisfied, consider (A.31) again. We have

$$\mathbf{h}=-(\mathbf{A}+\mathbf{E})^{-1}\mathbf{E}\mathbf{x} \tag{A.36}$$

giving

$$\mathbf{h}=-(\mathbf{I}+\mathbf{A}^{-1}\mathbf{E})^{-1}\mathbf{A}^{-1}\mathbf{E}\mathbf{x}, \tag{A.37}$$

where we used relation (A.33).

From (A.37) we, further, have

$$\|\mathbf{h}\|\leqslant\frac{\|\mathbf{A}^{-1}\mathbf{E}\|\,\|\mathbf{x}\|}{\|\mathbf{I}+\mathbf{A}^{-1}\mathbf{E}\|}\leqslant\frac{\|\mathbf{A}^{-1}\|\,\|\mathbf{E}\|\,\|\mathbf{x}\|}{1-\|\mathbf{A}^{-1}\|\,\|\mathbf{E}\|}, \tag{A.38}$$

since

$$\| \mathbf{A}^{-1}\mathbf{E} \| \leqslant \| \mathbf{A}^{-1} \| \, \| \mathbf{E} \| ,$$

$$\| \mathbf{I} + \mathbf{A}^{-1}\mathbf{E} \| \geqslant \| \mathbf{I} \| - \| \mathbf{A}^{-1}\mathbf{E} \| \geqslant \| \mathbf{I} \| - \| \mathbf{A}^{-1} \| \, \| \mathbf{E} \|$$

and

$$\| \mathbf{I} \| = 1.$$

The relative error is given by

$$\frac{\| \mathbf{h} \|}{\| \mathbf{x} \|} \leqslant \frac{\| \mathbf{A} \| \, \| \mathbf{A}^{-1} \| \, \| \mathbf{E} \| / \| \mathbf{A} \|}{1 - \| \mathbf{A} \| \, \| \mathbf{A}^{-1} \| \, \| \mathbf{E} \| / \| \mathbf{A} \|}$$

$$\leqslant \frac{\| \mathbf{E} \|}{\| \mathbf{A} \|} \frac{\text{cond } (A)}{1 - \dfrac{\| \mathbf{E} \|}{\| \mathbf{A} \|} \text{cont } (A)} \tag{A.39}$$

Note that for the Euclidean norms we have $\| \mathbf{I} \|_E = \sqrt{n}$, which means that unity on the right-hand side of (A.38) and (A.39) has to be replaced by n.

From (A.39) we can see that the relative error, $\| \mathbf{h} \| / \| \mathbf{x} \|$ will be large if cond (A) is large.

Appendix B

Some Basic Preliminaries on Laws of Probability and Statistical Analysis

B.1 Probability

The most basic notion encountered in the theory of probability is that of an *event*, by which one means that occurrence of a specified outcome of an experiment, and the notion of probability has to do with the frequency of occurrence of an event X in repeated independent trials of an experiment.

If an experiment is repeated n times and if the event X occurs $n(X)$ times, then one expects the ratio $n(X)/n$ to cluster about a unique number, the probability of X. More precisely, the probability is defined as a real-valued function, $P(X)$, on the events of an experiment, satisfying $0 \leqslant P(X) \leqslant 1$.

B.2 Random Variable and the Distribution Function

The numerical equivalent of an event is called the random variable. A set of probabilities for the different possible values of the random variable is called a probability distribution.

If ξ is a random variable then associated with ξ is the (cumulative) distribution function which is defined as the real-valued function of a real variable and given as

$$F(x) = P(\xi \leqslant x),$$

i.e. it is equal to the probability that the variable assumes a value not greater than x.

The function $f(x) = F'(x)$, if it exists for all x, is called the distribution density of the random variable ξ.

For ξ continuous on an interval $[a, b]$, which may be finite or infinite, its (continuous) distribution function is given by

$$F(x) = \int_a^x f(x)\mathrm{d}x, \quad a \leqslant x \leqslant b.$$

For ξ discrete, i.e. defined as a set of discrete values, $\{x_i, i = 1, 2, \ldots\}$, the probability is given as

$$P(\xi = x_i) = p_i, \quad \sum_i p_i = 1,$$

and the distribution function has the form

$$F(x) = \sum_{x_i \leqslant x} p_i.$$

B.3 Mean Value and Standard Deviation

Associated with each random variable is its mean value defined as

$$\mu[\xi] = \begin{cases} \int_a^b xf(x)\mathrm{d}x = \int_a^b x\,\mathrm{d}F(x), & \text{if } \xi \text{ is continuous in } [a,b], \\ \sum_i x_i p_i, & \text{if } \xi \text{ is defined as a set } \{x_i\}. \end{cases}$$

Another important characteristic is the variance

$$\mathrm{var}[\xi] = \sigma^2[\xi] = \begin{cases} \int_a^b (x-\mu)^2 f(x)\,\mathrm{d}x = \int_a^b (x-\mu)^2\,\mathrm{d}F(x) & \text{f } \xi \text{ is continuous in } [a,b] \\ \sum_i (x_i-\mu)^2 p_i & \text{if } \xi \text{ is given as a set } \{x_i\} \end{cases}$$

The standard deviation is then defined as

$$\sigma = (\mathrm{var}[\xi])^{1/2}.$$

B.4 Uniform Distribution

The standard uniform distribution is defined for $0 \leqslant x \leqslant 1$ as follows: the distribution density is given by

$$f(x) = \begin{cases} 0, & x < 0, \\ 1, & 0 \leqslant x \leqslant 1, \\ 0. & x > 1, \end{cases}$$

and the (cumulative) distribution function is obtainea as

$$F(x) = \int_{-\infty}^{x} f(x)\,\mathrm{d}x = \begin{cases} 0, & x < 0 \\ x, & 0 \leqslant x \leqslant 1. \\ 1 & x > 1 \end{cases}$$

Figure B.1 shows graphical illustrations of both functions. The mean value

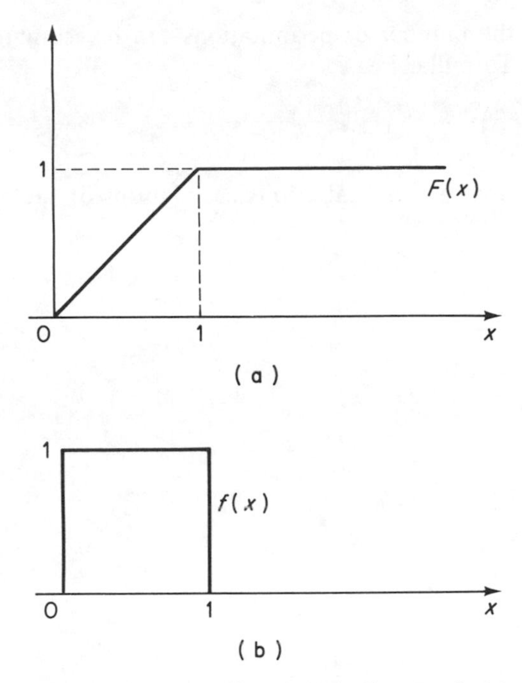

Figure B.1 (a) The standard cumulative distribution function; (b) the standard density function

and the variance of the ξ are, then, given as

$$\mu[\xi] = \int_{-\infty}^{\infty} xf(x)\,\mathrm{d}x = \frac{1}{2},$$

and

$$\mathrm{var}[\xi] = \sigma^2[\xi] = \int_{-\infty}^{\infty} \left(x - \frac{1}{2}\right)^2 f(x)\,\mathrm{d}x = \frac{1}{12},$$

respectively.

In the general case of the uniform random variable y given in $a \leqslant y \leqslant b$, y is generated from the standard distribution using the relation

$$y = a - \xi(a - b).$$

B.5 Permutation

A permutation of n objects is an arrangement or ordering of the objects; the number of permutations is the maximum number of ways the n objects can be arranged or ordered. This number is given as

$$p_n = n(n-1)(n-2)\ldots 1 = n!$$

For large n, the number of permutations can be estimated approximately using Stirling's formula, i.e.

$$n! \approx \sqrt{2\pi n}\left(\frac{n}{e}\right)^n.$$

This formula helps to avoid laborious calculations of factorials for large n.

Bibliography and References

Ackerman, A. F., 1974, Quadratic search for hash tables of size *p*, *CACM*, **17**, No. 3, 164.

Adelson-Velski, G. H. and Landis, E. II., 1962, An algorithm for the organization of information, *Dok. Akad. Nauk SSSR*, **146**, 263–266 (in Russian). English translation in *Soviet Math. Dokl.*, **3**, 1259–1262 (1962).

Aho, A. V., Hopcroft, J. E., and Ullman, J. D., 1974, *The Design and Analysis of Computer Algorithms*, Addison-Wesley.

Aho, A. V., Hopcroft, J. E., and Ullman, J. D., 1983, *Data Structures and Algorithms*, Addison-Wesley, Reading, Mass.

Baer, J.-L., and Schwab, B., 1977, A comparison of tree-balancing algorithms, *CACM*, **20**, No. 5, 322–330.

Bakhvalov, N. S., 1971, On the stable evaluation of polynomials (in Russian), *J. Comp. Math. and Math. Phys.*, **11**, No. 6, 1568–1574.

de Balbine, G., 1968, *PhD Thesis*, Californian Institute of Technology, 1968, see Knuth (1973), p. 521.

Barnes, J., 1965, An algorithm for solving nonlinear equations based on the secant method, *Comp. J.*, **8**, 66–72.

Bayer, R., 1972, Symmetric Binary B-trees: Data structure and maintenance algorithms, *Acta-Informatica*, **1**, No. 4, 290–306.

Belaga, E. C., 1958, Some problems in the computation of polynomials, *Dokl. Akad. Nauk SSSR*, **123**, 775–777 (Russian).

Belaga, E. C., 1961, On computing polynomials in one variable with initial conditioning of the coefficients, *Probl. Kibernetiki*, **5**, 7–15 (Russian). English translation in *Probl. Cybernetics*, **5**, 1–13.

Bell, J. R., 1970, The quadratic quotient method: A hash code eliminating secondary clustering, *CACM*, **13**, No. 2.

Bell, J. R., and Kaman, C. H., 1970, The linear quotient hash code, *CACM*, **13**, No. 11.

Bellman, R. E., and Dreyfus, S. E., 1961, *Applied Dynamic Programming*, Princeton University Press, Princeton, NJ.

Bennett, B. T., and Frazer, W. D., 1971, *Approximating Optimal Direct-Access Merge Performance*, IFIP.

Bergland, G. D., 1968, A fast Fourier transform algorithm using base eight iterations, *Math.* Comput., **22**, 275–279.

Betz, B. K., 1956, *Unpublished Memorandum*, Minneapolis–Honeywell Regulator Co. (Reference taken from Knuth, 1973.)

Betz, B. K., and Carter, W. C., 1959, New merge sorting techniques, in 14th Nat. Mt. ACM. Conference, Cambridge, MA.

Blum, M., Floyd, R. W., Pratt, V. R., Rivest, R. L., and Tarjan, R. E., 1973, Time bounds for selection, *J. Comput. System Sci.*, **1**, No. 4, 448–461.

Borodin, A., 1971, Horner's rule is uniquely optimal, in *Theory of Machines and Computations*, Kohavi, Z., and Paz, A. (eds), Academic Press, New York, 45–58.

Borodin, A., 1973, Computational complexity: theory and practice, in *Currents in the Theory of Computing*, Aho, A. V. ed, Prentice-Hall, Englewood Cliffs, NJ.

Borodin, A., and Munro, I., 1975, *The Computational Complexity of Algebraic and Numerical Problems*, Elsevier, New York.

Brent, R. P., 1970a, Algorithms for matrix multiplication, *Tech. Report CS157*, March 1970, Computer Sci. Dept., Stanford University.

Brent, R. P., 1970b, Error analysis of algorithms for matrix multiplication and triangular decomposition using Winograd's Identity, *Num. Math.*, **16**, 145–156.

Brent, R. P., 1971a, *Algorithms for Finding Zeros and Extrema of Functions Without Calculating Derivatives*, STAN-CS-71–198.

Brent, R. P., 1971b, An algorithm with guaranteed convergence for finding a zero of a function. *Comput. J.*, **14**, 422–425.

Brent, R. P., 1973, Some efficient algorithms for solving systems of nonlinear equations, *SIAM J. Num. Anal.*, **10**, 327–344.

Brent, R., Winograd, S., and Wolfe, P., 1973, Optimal iterative processes for root-finding, *Num.* Math., **20**, 327–341.

Buchholz, W., 1963, FIle organization and addressing, *IBM Systems J.*, **2**, 86–111.

Bus, J. C. P., and Dekker, T. J., 1974, Two efficient algorithms with guaranteed convergence for finding a zero of a function, *Afdeling Numericke Wiskunde*, NW 13/74, Stichting Mathematisch Centrum, Amsterdam.

Cheney, E. W., 1962, *Algorithms for the Evaluation of Polynomials Using a Minimum Number of Multiplications*, Technical Note 2, Computation and Data Processing Center, Aerospace Corporation, El Segundo, California.

Chernous'ko, F. L., 1968, An optimal algorithm of the root search for the function which is evaluated approximately, *J. Comput. Math. and Math. Phys.*, **8**, No. 4. 705–724 (Russian).

Clenshaw, C. W., 1955, A note on the summation of Chebyshev Series, *MTAC*, **9**, 118–120.

Conte, S. D., and de Boor, C., 1972, *Elementary Numerical Analysis–An Algorithmic Approach*, McGraw-Hill, Kogakusha, Second Edition.

Cook, S. A., 1966, On the minimum computation time of functions, *Doctoral Thesis*, Harvard University Cambridge, Massachusetts.

Cooley, J. W., Lewis, P. A. W., and Welch, P. D., 1977, The fast Fourier transform and its application to time series analysis, in *Statistical Methods for Digital Computers*, Enslien, K., Ralston, A., and Wilf, H. S., eds, J. Wiley, London.

Cooley, J. W., and Tukey, J. W., 1965, An algorithm for the machine calculation of complex Fourier series, *Math. Comput.*, **19**, 297–301.

Danielson, G. C., and Lanczos, C., 1942, Some improvements in practical Fourier analysis and their application to X-ray scattering from liquids, *J. Franklin. Inst.*, **233**, 365–380; 435–452.

Dekker, T. J., 1969, Finding a zero by means of successive linear interopolation, in Dejon, B. and Henrici, P. Eds, *Constructive Aspects of the Fundamental Theorem of Algebra*, Wiley Interscience, London.

Ecker, A., 1973, The period of search for quadratic and related hash methods, *Comput. J.*, **17**, No. 4, 340–343.

Ershov, A. P., 1958a, On programming arithmetic operations, *CACM*, **1**, No. 8, 3–6.

Ershov, A. P., 1958b, *Dokl. Akad. Nauk SSSR*, **118**, 427–430.

Feldstein, A. and Firestone, R. M., 1969, A study of Ostrowski efficiency for composite iteration algorithms, *Proceedings of ACM 1969*.

Fiduccia, C. M., 1971, Fast matrix multiplication, in *Proceedings 3rd Annual ACM Symposium on Theory of Computing*, pp. 45–49.

Floyd, R. W., 1964, Algorithm 245: Treesort 3, *CACM*, **7**, No. 12, 701.

Floyd, R. W., 1970, Reference to personal communication in Knott (1975).

Ford, L. R., and Johnson, S. M., 1959, A tournament problem, *Amer. Math. Monthly*, **66**, 387–389.

Forsythe, G. E., and Moler, C. B., 1967, *Computer Solution of Linear Systems*, Prentice-Hall, Englewood Cliffs, NJ.

Foster, C. C., 1973, A generalization of AVL trees, *CACM*, **16**, 8, 513–517.

Fox, L., 1954, Practical solution of linear equations and inversion of matrices, *Appl. Math. Ser. Nat. Bur. Stand.*, **39**, 1–54.

Fox, L., and Mayers, D. F., 1968, *Computing Methods for Scientists and Engineers*, Oxford University Press, Oxford.

Frazer, W. D. and Bennett, B. T., 1972, Bounds on optimal merge performance, and a strategy for optimality, *JACM*, **19**, 4, 641–648.

Frazer, W. D. and McKellar, A. C., 1970, Samplesort: a sampling approach to minimal storage tree sorting, *JACM*, **17**, No. 3, 466–507.

Gentleman, W. M., 1969, Off-the-shelf black boxes for programming, *IEEE Trans. Educ.*, E-12, 43–50.

Gentleman, W. M., 1973, On the relevance of various cost models of complexity, in *Complexity of Sequential and Parallel Numerical Algorithms*, Traub, J. F. ed., Academic Press, New York.

Gentleman, W. M., and Sande, G., 1966, Fast Fourier transforms—for fun and profit, in *Proc. Fall Joint Computer Conf.*, pp. 563–578.

Gilstad, R. L., 1960, Polyphase merge sorting—an advanced technique, in *Proceed. AFIPS Eastern Jt. Computer Conf.*, 18, 143–148.

Givens, W., 1954, Numerical computation of the characteristic values of a real symmetric matrix, *Rep. ORNL-1954*, Oak Ridge Associated Universities, Oak Ridge, Tennessee.

Givens, W., 1958, Computation of plane unitary rotations transforming general matrix to triangular form. *J. Soc. Indust. Appl. Math.*, **6**, 26–50.

Gonnet, G. H., 1983, Balancing binary trees by internal path reduction, *CACM*, **26**, 12, 1074–1081.

Good, I. J., 1958, 1960, The interaction algorithm and practical Fourier series, *J. Roy, Statist. Soc., Series B*, **20**, 361–372 (1958). *Addendum*, **22**, 372–375 (1960).

Gross, O., and Johnson, S. M., 1959, Sequential minimax search for a zero of a convex function, *MTAC* (now *Math. Comp.*), **13**, 44–51.

Hadian, A. and Sobel M., 1969, Selecting the tth largest using binary errorless comparisons, *Tech. Rept. 121*, Dept. of Statistics, University of Minnesota, Minneapolis.

Halton, J. H., 1965, The distribution of the sequence $\{n, n = 0, 1, 2, \dots\}$, *Cambridge Phil. Soc. Proc.*, **61**, 665–670.

Hamming, R. W., 1950, Error detecting and error correcting codes, *BSTJ*, **26**, No. 2, 147–160.

Hart, J. F., *et al.*, 1968, *Computer Approximations*, John Wiley, London.

Harter, R., The optimality of Winograd's formula, *CACM*, **15**, No. 5, 352.

Helms, H. D., 1967, Fast Fourier transform method of computing difference equations and simulating filters, *IEEE Trans.*, **AU-15**, 85–90.

Hoare, C. A. R., 1962, Quicksort, *Comput. J.*, **5**, No. 1, 10–15.

Hopcroft, J. E. and Kerr, L.R., 1969, Some techniques for proving certain simple programs optimal, in *Proc. 1969 IEEE Tenth Annual Symposium on Switching and Automata Theory*, pp 36–45.

Hopcroft, J. E. and Kerr, L. R., 1971, On minimizing the number of multiplications necessary for matrix multiplication, *SIAM J. Appl. Math.*, **20:1**, 30–36.

Hopcroft, I. and Musinski, J., 1973, Duality in determining the complexity of noncommutative matrix multiplication, *Proc. 5th Symposium on Theory of Computing*, Austin 1973.

Hopgood, F. R. A. and Davenport, J., 1972, The qaudratic hash method when the table size is a power of 2, *Comput. J.*, **15**, No. 4, 314–315.

Householder, A. S., 1958a, A class of methods for inverting matrices, *J.* SIAM, **6**, 189–195.

Householder, A. S., 1958b, Unitary triangularization of a non-symmetric matrix, *J. ACM*, **5**, 339–342.

Householder, A. S., 1964, *The theory of Matrices in Numerical Analysis*, Blaisdell Publishing Co., New York.

Hu, T. C., and Tucker, A. C., 1971, Optimal computer search trees and variable-length alphabetic codes, *SIAM J. Appl. Math.*, **21**, 4, 514–532.

Huffman, D. A., 1952, A method for the construction of minimum redundancy codes, *Proc. IRE*, **Sept 1952**, 1098–1101.

Hwang, F. K. and Lin, S., 1972, A simple algorithm for merging two disjoint linearly ordered sets, *SIAM J. Comp.*, **1**, 31–39.

Jarratt, P., 1966a, Multipoint iterative methods for solving certain equations, *Comput. J.*, **8**, 398–400.

Jarratt, P., 1966b, A rational iteration function for solving equations, *Comput. J.*, **9**, 304–307.

Jarratt, P., 1966c, Some fourth order multi-point iterative methods for solving equations, *Math. Comput.*, **20**, 434.

Jarratt, P., 1969, Some efficient fourth order multipoint methods for solving equations, *BIT*, **9**, 2, 119–124.

Jarratt, P., 1973, A review of methods for solving non-linear algebraic equations in one variable, in *Numerical Methods for Nonlinear Algebraic Equations*, Rabinowitz, P. ed., Gordon and Breach, New York.

Jarratt, P. and Nudds, D., 1965, The use of rational functions in the iterative solution of equations on a digital computer, *Comput. J.*, **8**, 62–65.

Kaneko, T. and Liu, B., 1970, Accumulation of round-off error in fast Fourier transforms, *JACM*, **17**, No. 4, 637–654.

Karatsuba, A. and Ofman, Yu., 1962, Multiplication of multiple numbers by means of Automata, *Dokl. Akad. Nauk USSR*, **145**, No. 2, 293–294 (Russian).

Kirkpatrick, D., 1972, On the additions necessary to compute certain functions, *Proc. 4th Annual ACM Symposium on Theory of Computing*, 94–101.

Kislitsyn, S. S., 1964, On the selection of the *k*th element of an ordered set by pairwise comparison, *Sibirsk. Mat. Zh.*, **5**, 557–564.

Klinkhamer, J. F., 1968, *On Key-to-Address Transformation for Mass Storage*, University of Utah, Computer Science Department Report.

Kluyev, V. V., and Kokovkin-Shcherbak, H. I., 1965, Minimization of the number of arithmetic operations in the solution of linear algebraic systems of equations, *Zh. Vychisl. Mat. i Mat. Fiz.*, **5**, 1, pp. 21–33. English translation in *USSR Computational Mathematics and Mathematical Physics*, **5**, 25–43.

Knott, G. D., 1975, Hashing functions, *Comput. J.*, **18**, No. 3.

Knuth, D. E., 1963, Length of strings for a merge sort, *ACM*, **6**, No. 11, 685–688.

Knuth, D. E., 1965, Problem No. 47, *BIE*, **5**, 142.

Knuth, D. E., 1968, *The Art of Computer Programming*, Vol. 1, Addison-Wesley, Reading, MA.

Knuth, D. E., 1969, *The Art of Computer Programming*, Vol. 2, Addison-Wesley, Reading, MA.

Knuth, D. E., 1971, Optimum binary search trees, *Acta Informatica*, **1**, 14–25.

Knuth, D. E., 1973, *The Art of Computer Programming*, Vol. 3, Addison-Wesley, Reading, MA.

Kogan, T. I., 1967, Construction of the high order iterative processes, *J. Comp. Math. and Math. Phys.*, **7**, No. 2, 423–424 (in Russian).

Krautstengel, R., 1968, On one iterative method of calculating a single root of equation $f(x) = 0$, *J. Comp. Math. and Math. Phys.*, **8**, No. 6, 1327–1329 (Russian).

Kreczmar, A., 1976, On memory requirements of Strassen algorithms, in *Algorithms and Complexity: New Dimensions and Recent Results*, Traub, J. F. ed., Academic Press, New York.

Kung, H. T. and Traub, J. F., 1973, Computational complexity of one-point and multi-point iteration, in *Complexity of Real Computation*, Karp, R., ed, American Mathematical Society, Providence, RI.

Kung, H. T. and Traub, J. F., 1974, Optimal order of one-point and multi-point iteration, *J. ACM*, **21**, No. 4.

Laderman, J. D., 1976, A non-commutative algorithm for multiplying 3×3 matrices using 23 multiplications, *Bull. Am. Math. Soc.*, **82**, 126–128.

Luccio. F., 1972, Weighted increment linear search for scatter tables, *CACM*, **15**, No. 12, 1045–1047.

MacCallum, I. R., 1972, A simple analysis of the nth order polyphase sort, *Comput., J.*, **16**, No.1, 16–18.

Makarov, O. M., 1975a, On relation of two multiplication algorithms (Russian), *J. Comp. Math. and Math. Phys.*, **15**, No. 1, 227–231.

Makarov, O. M., 1975b, On relation of the algorithms of the fast Fourier transform and Hadamard with the algorithms of Karatsuba, Strassen, and Winograd (Russian), *J. Comp. Math. and Math. Phys.*, **15**, No. 5, 1095–1105.

Manacher, G. K., 1975, The Ford–Johnson algorithm is not optimal, *Dept. of Information Engineering, University of Illinois, Chicago*, I11.60680.

Martin, R. S., Peters, G., and Wilkinson, J. H., 1965, Symmetric decomposition of a positive definite matrix, *Num. Math.*, **7**, 362–383.

Maurer, W. D., 1968, An improved, hash code for scatter storage, *CACM*, **11**, No. 1.

Mendelson, H., 1982, Analysis of extendable hashing, *IEEE Trans SE*, **SE-8**, **6**, 611–619.

Mesztenyi, C. K. and Witzgall, C., 1967, Stable evaluation of polynomials, *J. Res. Nat. Bureau of Standards*, **71B**, 11–17.

Miller, W., 1973, Computational complexity and numerical stability, *Thomas J. Watson Res. Rep. RC-4480*, Yorktown Heights.

Moler, C. B., 1967, Iterative refinement in floating point, *J. ACM*, **14**, 316–321.

Morris, R., 1968, Scatter storage techniques, *CACM*, **11**, No. 1.

Motzkin, T. S., 1955, Evaluation of polynomials, *Bull. Amer. Math. Soc.*, **61**, 163.

Mukai, H., 1979, Parallel algorithms for solving systems of nonlinear equations, in *Proc. 17th Annual Allerton Conf. on Communication, Control and Computing, 10–12 Oct.*, pp. 37–46.

Muller, D. E., 1956, A method of solving algebraic equations using an automatic computer, *MTAC*, **10**, 208–215.

Muroga, S., 1961, Unpublished report, IBM.

Newbery, A. C. R., 1974, Error analysis for polynomial evaluation, *Math. of Comp.*, **28**, No. 127, 789–793.

von Neumann, J., and Goldstine, H. H., 1947, Numerical inverting of matrices of high order, *Bull. Amer. Math. Soc.*, **53**, 1021–1099.

Nievergelt, J., 1974, Binary search trees and file organization, *Comp. Surv.*, **6**, No. 3, 195–207.

Ofman, Yu., 1962, On the algorithmic complexity of discrete functions, *Dokl. Akad. Nauk USSR*, **145**, No. 1, 48–51 (in Russian).

Ortega, J. M. and Rheinboldt, W. C., 1970, *Iterative Solution of Nonlinear Equations in Several Variables*, Academic Press, New York.

Ostrowski, A. M., 1954, On two problems in abstract algebra connected with Horner's Rule, *Studies in Mathematics and Mechanics*, Presented to R. von Mises, Academic Press, New York, pp. 40–48.

Overholt, K. J., 1973a, Optimal binary search methods, *BIT*, **13**, 84–91.

Overholt, K. J., 1973b, Efficiency of the Fibonacci search method, *BIT*, **13**, 92–96.

Pan, V. Ya., 1959, Schemes for computing polynomials with real coefficients, *Dokl. Akad. Nauk SSSR*, **127**, 266–269 (in Russian). English translation in *Math. Rev.* **23**, 1962, B560.

Pan, V. Ya., 1966, Methods of computing values of polynomials, *Uspekhi Math. Nauk*, **21**, 103–134 (Russian). English translation in *Russian Math. Surv.*, **21**, 105–136.

Pan, V., 1978, Strassen algorithm is not optimal. Trilinear technique of aggregating, uniting and cancelling for constructing fast algorithms for matrix multiplication, *Proc. 19th Annual Symposium on the Foundations of Computer Science*, Ann Arbor, MI, pp. 166–176

Pan, V., 1984, How can we speed up matrix multiplication?, *SIAM Rev.*, **26**, 3, 393–415.

Paterson, M. S., 1972, Efficient iterations for algebraic numbers, in *Complexity of Computer Computations*, Miller, R. E., and Thatcher, J. W., eds, Plenum Press, New York.

Peterson, W. W., 1957, Addressing for random access storage, *IBM J. Res. Develop.*, **1**, No. 2, 130–146.

Polak, E., 1974, A globally converging secant method with applications to boundary value problems, *SIAM J. Num. Analy.*, **11**, 529–537.

Rabin, M. O., 1974, Theoretical impediments to artificial intelligence, in *Information Processing 74* (Invited Paper). North-Holland, Amsterdam.

Rabin, M. and Winograd, S., 1971, *Fast Evaluation of Polynomials by Rational Preparation*, IBM Technical Report RC 3645, Dec. 1971.

Radke, C. E., 1970, The use of quadratic residue research, *CACM*, **13**, No. 2.

Ramamohanara, K. and Lloyd, J. W., 1982, Dynamic hashing schemes, *Comp. J.*, **25**, 478–485.

Ramos, G. U., 1971, Round-off error analysis of the fast Fourier transform, *Math. Comp.*, **25**, No. 16, 757–768.

Reingold, E. M. and Stocks, A. I., 1972, Simple proofs of lower bounds for polynomial evaluation, *Complexity of Computer Computations*, Miller, R. E., and Thatcher, J. W., eds., Plenum Press, New York, pp. 21–30.

Rice, J. R., 1965, On the conditioning of polynomial and rational forms, *Num. Math.*, **7**, 426–435.

Rice J., 1972, *Seminar given at Purdue University*, Spring 1972.

Runge, C. and König, R., 1924, Die Grundlehren der Mathematischen Wissenschaften, in *Vorlesung über Numerisches Rechnen*, 11, Springer, Berlin.

Schachtel, G., 1978, A non-commutative algorithm for multiplying 5×5 matrices using 103 multiplications, *Information Processing Lett.*, **4**, 180–182.

Schay, G., Jr and Raver, N., (1963), A method for key-to-address transformation, *IBM J. Res. Develop.*, **7**, No. 2, 121–126.

Schay, G. and Spruth, W. G., 1962, Analysis of a file addressing method, *CACM*, **5**, 459–462.

Schönhage, A., 1966, Multiplikation Grosser Zahlen, *Computing* (*Arch. Elektron. Rechnen*), **1**, 182–196.

Schönhage, A. and Strassen, V., 1971, Schnelle Multiplikation Grosser Zahlen, *Computing*, **7**, 281–292.

Schreier, J., 1932, On tournament eliminations systems, *Math. Polska*, **7**, 154–160 (Polish).

Sedgewick, R., 1983, *Algorithms*, Addison-Wesley, Reading, MA.
Shamanskii, V. E., 1967, On a modification of Newton's method, *Ukrain. Mat. Z.*, **19**, 133–138.
Shaw, M. and Traub, J. F., 1974, On the number of multiplications for the evaluation of a polynomial and some of its derivatives, *J. ACM*, **21**, No. 1, 161–167.
Singleton, R. C., 1969a, An algorithm for computing the mixed radix fast Fourier transform, *IEEE Trans. Audio and Electroacoustics*, **AU-17**, **2**, 93–103.
Singleton, R. C., 1969b, Algorithm 347: an algorithm for sorting with minimal storage, *CACM*, **19**, No. 3, 185–187.
Sleator, D. D. and Tarjan, R. A., 1985, self-adjusting binary search trees, *J. ACM*, **32**, 3, 652–686.
Slupecki, J., 1951, On the system S of tournaments, *Colloq. Math.*, **2**, 286–290.
Strassen, V., 1969, Gaussian elimination is not optimal, *Num. Math.*, **13**, 354–356.
Sukharev, A. G., 1976, Optimal root search for the function which satisfies the Lipschitz condition, *J. Comp. Math. and Math. Phys.*, **16**, No. 1, 20–29 (in Russian).
Todd, J., 1955, Motivations for working in numerical analysis, *Comm. Pure Appl. Math.*, **8**.
Toom, A. L., 1963, The complexity of a scheme of functional elements realizing the multiplications of integers, *Dokl. Akad. Nauk SSSR*, **150**, 496–498.
Traub, J. F., 1964, *Iterative Methods for the Solution of Equations*, Chapter 5. Prentice-Hall, Englewood Cliffs, NJ.
Weinstein, C. J., 1969, Round-off noise in floating point fast Fourier transform computation, *IEEE Trans. Audio and Electroacoustics*, **AU-17**, 209–215.
Welch, P. D., 1969, A fixed point fast Fourier transform error analysis. *IEEE Trans.* **AU-17**, 151–157.
Wilkinson, J. H., 1961, Error analysis of direct methods of matrix inversion, *J. ACM*, 281–330.
Wilkinson, J. H., 1963, *Rounding Errors in Algebraic Processes*. Prentice-Hall, Englewood Cliffs, NJ.
Wilkinson, J. H., 1965, *The Algebraic Eigenvalue Problem*, Oxford University Press, Oxford.
Williams, J. W. J., 1964, Algorithm 232; Heapsort, *CACM*, **7**, No. 6, 347–348.
Winograd, S., 1967, On the number of multiplications required to compute certain functions, *Proc. Natl. Acad. Sci. USA*, **58**, 1840–1842.
Winograd, S., 1968, A new algorithm for the inner product, *IEEE Trans. Comput.*, C17, 693–694.
Winograd, S., 1970a, On the multiplication of 2 × 2 matrices, *IBM Res. Rep.*, RC 267, Jan. 1970.
Winograd, S., 1970b, On the number of multiplications necessary to compute certain functions, *Comm. Pure Appl. Math.*, **23**, 165–179.
Winograd, S., 1973. Some remarks on fast multiplication of polynomials, in *Complexity of Sequential and Parallel Numerical Algorithms*, Traub, J. F., ed, Academic Press, New York.
Wirth, N., 1976, *Algorithms + Data Structures = Programs*, Prentice-Hall series in Automatic Computation, Prentice-Hall, Englewood Cliffs, NJ.
Wolfe, P., 1959, The secant method for simultaneous nonlinear equations, *CACM*, **2**, 12–14.
Wood, D., 1973, A note on table look-up, *BIT*, **13**, 245–246.
Wozniakowski, H., 1974a Rounding error analysis for the evaluation of a polynomial and some of its derivatives, *SIAM J. Num. Anal.*, **11**, No. 4.
Wozniakowski, H., 1975a, *Maximal Order of Multipoint Iterations Using n Evaluations*, Department of Computer Science, Carnegie-Mellon University.

Wozniakowski, H., 1975b, *Numerical Stability for Solving Nonlinear Equations*, Department of Computer Science, Carnegie-Mellon University.

Wozniakowski, H., 1975c, *Numerical Stability of the Chebyshev Method for the Solution of Large Linear Systems*, Department of Computer Science, Carnegie-Mellon University.

Yeh, R. T. (ed.), 1976, *Applied Computation Theory: Analysis, Design, Modeling*, Prentice-Hall, Englewood Cliffs, NJ.

Zaremba, S. K., 1966, Good lattice points, discrepancy, and numerical integration, *Ann. Mat. Pura. Appl*, Series 4, **73**, 293–317.

Index